NOISE IN ELECTRONIC DEVICES AND SYSTEMS

NOISE IN ELECTRONIC DEVICES AND SYSTEMS

M. J. BUCKINGHAM, B.Sc., Ph.D.
Senior Principal Scientific Officer, Ministry of Defence
Royal Aircraft Establishment
Farnborough, Hampshire

ELLIS HORWOOD LIMITED
Publishers · Chichester

Halsted Press: a division of
JOHN WILEY & SONS
New York · Brisbane · Chichester · Ontario

First published in 1983 by
ELLIS HORWOOD LIMITED
Market Cross House, Cooper Street, Chichester, West Sussex, PO19 1EB, England

The publisher's colophon is reproduced from James Gillison's drawing of the ancient Market Cross, Chichester.

Distributors:

Australia, New Zealand, South-east Asia:
Jacaranda-Wiley Ltd., Jacaranda Press,
JOHN WILEY & SONS INC.,
G.P.O. Box 859, Brisbane, Queensland 40001, Australia

Canada:
JOHN WILEY & SONS CANADA LIMITED
22 Worcester Road, Rexdale, Ontario, Canada.

Europe, Africa:
JOHN WILEY & SONS LIMITED
Baffins Lane, Chichester, West Sussex, England.

North and South America and the rest of the world:
Halsted Press: a division of
JOHN WILEY & SONS
605 Third Avenue, New York, N.Y. 10016, U.S.A.

© 1983 M. J. Buckingham/Ellis Horwood Ltd.

Reprinted 1985

British Library Cataloguing in Publication Data
Buckingham, M. J.
Noise in electronic devices and systems —
(Ellis Horwood series in electrical and electronic engineering)
1. Electronic circuits — Noise
I. Title
621.3815'3   TK7867.5
Library of Congress Card No. 83–8605
ISBN 0-85312-218-0 (Ellis Horwood Limited-Library Edn)
ISBN 0-85312-876-6 (Ellis Horwood Limited-Student Edn)
ISBN 0-470-27467-0 (Halsted Press-Library Edn)
ISBN 0-470-20164-9 (Halsted Press-Student Edn)

Typeset in Press Roman by Ellis Horwood Ltd.
Printed in Great Britain by Unwin Brothers of Woking

Table of Contents

Author's Preface

Solid-state technology has advanced at an almost unbelievable pace since the advent of the germanium transistor. Nowadays, solid-state devices are employed widely in all sorts of applications. The range of devices available is extensive, and, as well as the familiar silicon bipolar transistor and the various types of field-effect transistor, it includes microwave devices such as the Impatt, the Baritt and the Gunn diode, tunnelling devices such as the Esaki diode, the Josephson junction and the SQUID, and opto-electronic devices. All these devices show noise of one form or another, which in small signal applications is often the factor limiting their performance.

The primary objective of this book is to examine the physical characteristics of noise in a number of solid-state devices. Discussions of noise in certain non-linear circuits, such as the parametric amplifier and the van der Pol oscillator, are also included; and the recovery of pulse signals from noise is treated in connection with the detection of gravitational radiation.

The exposition is based on a general mathematical treatment of noise processes, in which the properties of random pulse trains and impulse processes are established. Throughout the book thermal noise and shot noise are recurring themes, but also included are generation–recombination (g–r) noise, the ubiquitous $1/f$ noise, burst noise, non-equilibrium Johnson noise associated with hot electron populations, and avalanche noise due to impact ionization. An extensive bibliography, allowing the interested reader to delve deeper into the subject, is included with each chapter.

In conclusion, I should like to acknowledge with thanks a number of publishers, as specified in the text, for permission to reproduce diagrams from articles which have appeared in their publications. Throughout the course of writing the book Mr Michael Horwood gave constant support and encouragement, for which I am grateful. And finally, may thanks go to my wife for her patience and sustained cheerfulness which contributed greatly to the successful conclusion of the book.

Washington D.C.   M.J.B
December 1982

1

Introduction

Electronic devices and systems exhibit random fluctuations in the voltage (or current) at their terminals, and these fluctuations are usually referred to as *noise*. The noise is not due to faulty contacts, for example, or any other spurious effect which could be eliminated, but is inherent in the system itself. It originates in the random, microscopic behaviour of the charge carriers within the electronic components constituting the system. This is the type of noise with which we shall be primarily concerned in this book.

A noisy electronic circuit with a pair of input and a pair of output terminals (i.e. a 2-port) is shown in Fig. 1.1(a). The noise may be due to one or more sources within the system. A convenient way of representing the noisy system is illustrated in Fig. 1.1(b), where the network is shown as noise-free and the noise is represented by noise current generators $i_{n1}(t)$ and $i_{n2}(t)$ at the input and output ports, respectively. These two current generators may show some degree of correlation because the mechanisms giving rise to the noise at the two ports could be to some extent common. An alternative representation of the noisy system is shown in Fig. 1.1(c), where the network is again shown as noise-free but now the noise is represented by noise voltage generators $v_{n1}(t)$ and $v_{n2}(t)$ (which may be correlated) at the input and output ports, respectively.

In order to specify the noise generators at the ports, the details of the network and the characteristics of the internal noise sources must be known. Now, the internal sources are associated with the electronic devices within the system, and in general these sources are strongly device-dependent, even though the physical mechanisms responsible for the noise may be common to a range of devices. Many of the discussions throughout the book concentrate on the noise-producing mechanisms and the associated noise generators in particular devices. A strong emphasis is placed on the solid state, since noise in vacuum tubes has been fully discussed elsewhere (e.g. Bell, 1960).

Noise in electronic circuits is usually regarded as detrimental, and indeed it often imposes a practical limit to the performance of the circuit. This is true, for example, in a low-noise audio amplifier, where the minimum detectable power

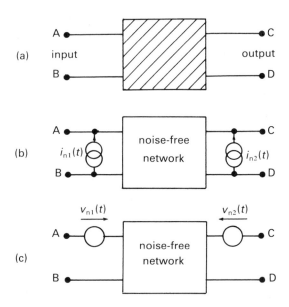

Fig. 1.1 – (a) A noisy 2-port, and two equivalent circuits with the noise represented by (b) current generators at the ports and (c) voltage generators at the ports.

of the input signal is set mainly by the noise level of the input stage. However, noise is not always a nuisance. There are occasions when the noise inherent in a circuit can be exploited as a means for investigating the electrical characteristics of the circuit itself. Such could be the case if, for example, a measurement of the conductivity of an ionic solution were required. Normally, such a measurement would be performed by applying an electric field across the cell containing the solution. The difficulty with this technique is that, under certain circumstances, molecular dissociation may occur due to the presence of the field, which in turn affects the conductivity. An alternative approach is to measure the noise at the terminals of the cell with the specimen in equilibrium (i.e. with zero electric field), from which the conductivity of the solution can be inferred.

Perhaps the two most commonly encountered types of noise are thermal noise and shot noise. Thermal noise arises from the random velocity fluctuations of the charge carriers (electrons and/or holes) in a resistive material. The mechanism is sometimes said to be Brownian motion of the charge carriers due to the thermal energy in the material. Thermal noise is present when the resistive element is in thermal equilibrium with its surroundings, and is often referred to as Johnson noise in recognition of the first observations of the phenomenon†

---

† A non-equilibrium version of thermal noise, associated with hot electron populations, is observed in certain microwave devices which we shall be discussing later. Unless specified otherwise, the terms 'thermal noise' and 'Johnson noise' are used here in connection with the equilibrium state.

(Johnson, 1927a, 1927b, 1928). The thermal noise fluctuations may be regarded as the mechanism by which the state of thermal equilibrium is maintained: a random (microscopic) departure from that state is followed, on average, by a relaxation back towards it, and the very large number of such microscopic 'events' gives rise to the rapidly varying current or voltage fluctuations constituting the noise at the terminals. According to this view, a thermal noise current or voltage waveform must consist of a very large number of individual pulses associated with the discrete 'events' occurring within the resistive material.

Shot noise is associated with the passage of current across a barrier, and in this sense is a non-equilibrium form of noise. It was first discussed by Schottky (1918), who used the analogy of small shot pattering into a container. Shot noise, or at least a form of noise having all the attributes of shot noise, is frequently encountered in solid-state devices, whenever a net current flows across a potential barrier such as the depletion layer of a p–n junction. The details of the physics underlying shot noise in this case will be discussed later. In order to understand the nature of shot noise, it is perhaps easier to consider a thermionic diode in which electrons are randomly emitted from the cathode and then drift across to the anode under the influence of the electric field. The current associated with the stream of electrons fluctuates randomly about a mean level, the fluctuations (i.e. the shot noise) being due to the random, discrete nature of the emission process.

Clearly, the *physical origins* of thermal noise and shot noise are distinct, but the *structures* of the two types of noise waveform are similar: both can be represented as a *random pulse train* consisting of a sequence of similarly shaped pulses randomly distributed in time. An example of a random pulse train in which the pulse shape is a decaying exponential, is illustrated in Fig. 1.2. In the present context, 'random' means that the discrete events giving rise to the pulses are independent, and that the statistical law governing the distribution of the events in time is the Poisson probability density function. The Poisson distribution and the conditions under which it applies are discussed in Appendix 1.

If the noise waveform is represented by the function $x(t)$ and the pulse shape function is $f(t)$ ($f(t) = 0$ for $t < 0$, assuming that the event giving rise to the pulse occurs at $t = 0$, and that the system is causal), then the random pulse train is the linear superposition

$$x(t) = \sum_k a_k f(t - t_k) , \qquad (1.1)$$

where a_k is the amplitude of the kth pulse in the sequence and t_k is the time at which the kth event occurs. The distribution of the t_k is governed by the Poisson law. A waveform whose structure can be represented by equation (1.1) shows some interesting properties, as discussed in Chapter 2. In particular, its power spectral density can be simply expressed as

$$\overline{S_x(\omega)} = 2\nu \overline{a^2} \, |F(j\omega)|^2 , \qquad (1.2)$$

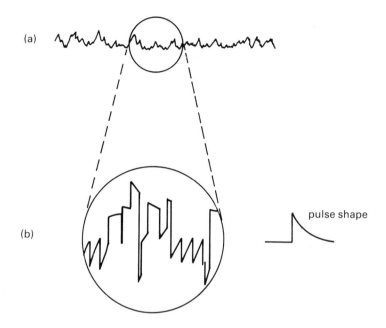

Fig. 1.2 – (a) Schematic illustration of a random pulse train and (b) a much magnified section of the waveform showing the combined effect of the individual pulses.

which is known as Carson's theorem (Rice, 1944, 1945). In this expression, ω is the angular frequency, $F(j\omega)$ is the Fourier transform of the pulse shape function $f(t)$, ν is the mean rate of events, $\overline{a^2}$ is the mean-square value of the pulse amplitudes, and the overbar on the left-hand side indicates an average taken over a large number of trials (i.e. an ensemble average).

In the special case when the constituents of the pulse train are extremely narrow, the shape function is an impulse with infinitesimal width. The random pulse train is then known as an *impulse process* (Beutler and Leneman, 1968). Now, the Fourier transform of an impulse is unity, and hence from equation (1.2) the power spectral density of an impulse process is

$$\overline{S_x(\omega)} = 2\nu\,\overline{a^2} \ . \tag{1.3}$$

This is an important result, showing that the spectrum of an impulse process is flat up to an indefinitely high frequency. Such a spectrum is sometimes said to be 'white'.

The pulses arising from the discrete events responsible for thermal noise and shot noise show flat power spectral densities up to very high frequencies (where 'very high' means comparable with the reciprocal of the actual pulse widths).

The level of the power spectral densities in the two cases, that is the value of the term on the right-hand side of equation (1.3), is determined from consideration of the physics of the noise mechanisms.

In the case of thermal noise, the argument involves statistical mechanics and the law of equipartition of energy, which states that every system at absolute temperature θ which is in thermal equilibrium with its surroundings contains thermal energy amounting, on average, to $k\theta$ per degree of freedom, where k is Boltzmann's constant. The result for the power spectral density of the open-circuit noise voltage across the terminals of a resistor, R, is

$$\overline{S_v(\omega)} = 4k\theta R \ . \tag{1.4a}$$

Thus, the resistor can be represented as in Fig. 1.3(a), where the series noise voltage generator has a spectral density given by equation (1.4a). By a simple circuit transformation, the noisy resistor can be represented as in Fig. 1.3(b), where the parallel noise current generator has a spectral density

$$\overline{S_i(\omega)} = 4k\theta G \ . \tag{1.4b}$$

In this expression the conductance $G = 1/R$.

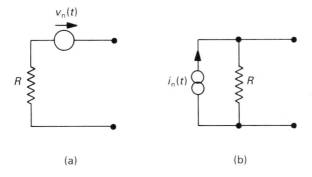

(a)   (b)

Fig. 1.3 – Thermal noise in a resistor R, represented by (a) a series noise voltage generator and (b) a parallel noise current generator.

Equations (1.4) were first derived by Nyquist (1927, 1928) from an argument based on thermodynamics and the exchange of energy between resistive elements in equilibrium. Nyquist's macroscopic approach to the problem of thermal noise – which is quite different from the microscopic treatment outlined above – is described in Appendix 2.

Nyquist's theorem, which is embodied in equations (1.4a) and (1.4b), involves the resistance R. However, the maximum available noise power from the resistance in a frequency interval df is independent of R. This can be proved from Fig. 1.4, which shows the resistance R in parallel with a noise current generator feeding into a noiseless matched load. From simple circuit analysis the power dissipated in the load in a frequency range df is

$$dP = \frac{\overline{S_i(\omega)}\, R}{4}\, df = k\theta\, df \,, \tag{1.4c}$$

where the expression on the right of (1.4b) has been substituted for $\overline{S_i(\omega)}$. The value of Boltzmann's constant is $k = 1.38 \times 10^{-23}$ joule/degK, which when multiplied by $\theta = 293$, corresponding to room temperature, gives the maximum available noise power from a resistance in a frequency band 1 Hz wide as 4×10^{-21} W.

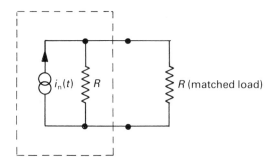

Fig. 1.4 – Noisy resistance, enclosed in broken lines, connected to a noiseless matched load.

The power spectral density of shot noise when the mean current is I, is

$$\overline{S_i(\omega)} = 2qI \,, \tag{1.5}$$

where q is the magnitude of the electronic charge. This result follows almost immediately from equation (1.3), since the mean rate of pulses is I/q and the pulse amplitudes are all equal to q, giving $\overline{a^2} = q^2$.

The microscopic treatment of thermal noise and shot noise, based on the concept of a random pulse train, is discussed further in some detail in Chapter 2.

A type of noise that occurs in a wide variety of systems (electronic, biological, music, etc.), but particularly in solid-state devices, has come to achieve a certain notoriety. This is a consequence of its ubiquity combined with its resilience against theoretical investigation. We refer to $1/f$ noise or, as it is sometimes known for historical reasons, current noise, flicker noise, contact noise or excess noise.

$1/f$ noise is so-called because it shows a power spectral density which varies with frequency as $|f|^{-\alpha}$, where α usually falls between 0.8 and 1.2. This dependence has been observed down to frequencies as low as 10^{-6} Hz. The upper limit of its existence is difficult to establish because it is usually masked by thermal noise or some other type of noise. There are various theoretical difficulties associated with the treatment of $1/f$ noise, mostly concerning convergence of integrals. At present, no entirely satisfactory explanation for the phenomenon exists, although in certain instances (e.g. electron trapping in the oxide layer overlaying a semi-conductor, such as occurs in MOSFETs) specific models have been developed. But it appears that such models have only limited application and do not adequately account for many of the observed $1/f$ noise waveforms.

It is interesting to note that $1/f$ noise can be represented as a random pulse train — or, more accurately, a random pulse train with a certain type of pulse shape function shows a power spectral density which varies as $|f|^{-1}$ over a wide range of frequency. The pulse shape to achieve this result is

$$f(t) = t^{-\frac{1}{2}}u(t) ,\tag{1.6}$$

as introduced by Schönfeld (1955) and discussed more recently by van der Ziel (1979). In this expression $u(t)$ is the unit step function, having a value of unity for $t > 0$ and zero when $t < 0$. The Fourier transform of $f(t)$ exists and is

$$F(j\omega) = \frac{\sqrt{\pi}}{|\omega|^{\frac{1}{2}}} \exp \mp (j\pi/4), \quad \begin{array}{l} - \text{ when } \omega > 0 \\ + \text{ when } \omega < 0 \end{array} .\tag{1.7}$$

On substituting this expression into Carson's theorem (equation (1.2)), the power spectral density of the waveform is found to be

$$\overline{S_x(\omega)} \propto 1/|\omega| ,\tag{1.8}$$

as required. There is a minor problem here, since $f(t)$ is not absolutely integrable, but this is eliminated by very slight changes to $f(t)$ which do not significantly alter the final result.

It would appear from this argument that a theoretical model of $1/f$ noise based on a random pulse train looks promising. The difficulty with this approach lies in finding a physical mechanism which generates pulses having the shape given by equation (1.6). No such mechanism is known at present. The properties, problems and existing theoretical models of $1/f$ noise are discussed at length in Chapter 6.

In addition to thermal noise, shot noise and $1/f$ noise, introduced above, we shall be encountering in the following chapters several other types of noise, including generation—recombination (g–r) noise, due to the random trapping of charge carriers in semiconductor materials, burst noise, avalanche noise due to impact ionization, and non-equilibrium Johnson noise exhibited by hot electrons in high electric fields. Some of these phenomena are related to the noise processes that we have already met — for example, avalanche noise can be regarded as an enhanced form of shot noise, and Johnson noise of hot electrons is obviously a variant of the thermal noise produced by an equilibrium ensemble of electrons. Before discussing these new types of noise and the devices with which they are associated, we proceed by laying the mathematical foundations necessary for developing a satisfactory theoretical account of the propoerties of noise and stochastic processes.

REFERENCES

D. A. Bell (1960), *Electrical Noise,* Van Nostrand.

F. J. Beutler and O. A. Z. Leneman (1968), The spectral analysis of impulse processes, *Information and Control,* **12**, 236–258.

J. B. Johnson (1927a), Thermal agitation of electricity in conductors, *Nature,* **119**, 50–51; (1927b), Thermal agitation of electricity in conductors, *Phys. Rev.,* **29**, 367–368; (1928), Thermal agitation of electricity in conductors, *Phys. Rev.,* **32**, 97–109.

H. Nyquist (1927), Thermal agitation in conductors, *Phys. Rev.,* **29**, 614; (1928), Thermal agitation of electric charge in conductors, *Phys. Rev.,* **32**, 110–113.

S. O. Rice (1944), Mathematical analysis of random noise, *Bell Syst. Tech. J.,* **23**, 282–332; (1945), **24**, 46–156.

H. Schönfeld (1955), Beitrag zum $1/f$-Gesetz beim Rauschen von Halbleitern, *Z. Naturforsch.,* **A 10**, 291–300.

W. Schottky (1918), Über spontane stromschwankungen in verschiedenen elektrizitatsleitern, *Ann.d.Phys. (Leipzig),* **57**, 541–567.

A. van der Ziel (1979), Flicker noise in electronic devices, *Advances in Electronics and Electron Physics,* **49**, 225–297.

2

Mathematical techniques

2.1 INTRODUCTION

Noise in electronic devices is usually observed as a randomly varying function of time. Such a function is known as a *stochastic process* and, since its instantaneous values cannot be predicted, is characterized in terms of the average or statistical properties which may reveal themselves when very many observations of the process are made.

The noise is commonly observed as random fluctuations in either the voltage across the terminals of a device or in the current flowing through the device. Generally it can be attributed to the microscopic behaviour of the charge carriers within the device, which suggests that the observed fluctuations will be very small in relation, say, to the typical signal levels from a waveform generator. Thus, the noise in an active device will usually produce only extremely small excursions around the operating point, in which case 'small-signal' theory may be applied to the noise fluctuations. This means that amplifier noise can often be treated using the well-established techniques of linear system theory (an exception is noise in parametric amplifiers).

Most of the stochastic processes we shall be discussing are *statistically stationary*, which means that their statistical properties are independent of the epoch in which they are measured. There are two degrees of statistical stationarity, *strict sense* and *wide sense*, the distinction between them concerning only third- and higher-order probability measures. A process which is stationary in the strict sense is also stationary in the wide sense, though the converse is not necessarily true. An exception, where it is true, is found in those processes whose amplitude probability density function is the *normal* or *Gaussian* distribution: their higher-order probability measures are determined entirely by their first- and second-order statistics. Gaussian processes are extremely important in connection with noise in electronic devices, where they are encountered frequently. Examples are thermal noise and shot noise; and there is overwhelming evidence, beginning with the measurements of Bell (1955), that $1/f$ noise is also normally distributed.

The mathematics of stochastic processes is concerned with probability measures in the time and frequency domains. (The spectral content of the noise in electronic components is of concern to the circuit designer, since his aim is often to minimize the noise in a particular band of interest.) Apart from the mean value (first-order), the principal statistical quantities used to specify a noise process are the *power spectral density*, giving the average spectral content of the fluctuating waveform, and the *autocorrelation function*, which provides a measure of the correlation time or 'memory' of the process. Both of these quantities are second-order and, in the case of a statistically stationary process†, are uniquely related through the *Wiener–Khintchine theorem*. A generalized form of this theorem can be derived for non-stationary processes; and an extended version of the Wiener–Khintchine theorem expresses the unique relationship connecting the *cross-correlation function* and the *cross-spectral density* between two statistically stationary processes.

The development of these theorems has a firm foundation in Fourier analysis techniques. In establishing the essential tools of Fourier analysis, the Dirac delta function – one of the class of generalized functions – plays an important role. It is also very useful in the mathematical representation of certain noise waveforms. As a prelude to the discussion of Fourier techniques and their application to noise processes, some of the properties of the delta function are now established.

2.2 SINGULARITY FUNCTIONS

The Dirac delta function, denoted by the symbol $\delta(t)$, is defined by the relationship

$$\int_{-\infty}^{\infty} \delta(t) x(t) \, dt = x(0) , \tag{2.1}$$

where $x(t)$ is an ordinary function of time which is continuous at $t = 0$.‡ According to this definition, $\delta(t)$ is a functional, having the property of assigning the value $x(0)$ to the function $x(t)$. Otherwise, if $\delta(t)$ were an ordinary function, the integral in equation (2.1) would have no meaning. Although it is a functional, the delta function may be manipulated like an ordinary mathematical function provided the operations performed are consistent with the definition given above.

On setting $x(t) = 1$ for all t, it follows from equation (2.1) that

$$\int_{-\infty}^{\infty} \delta(t) \, dt = 1 . \tag{2.2}$$

† Since we are concerned here with second-order statistics, it is unnecessary to distinguish between wide-sense and strict-sense stationarity. This will also be the case in many of the discussions to follow.
‡ The variable here has been chosen as time. Of course, the properties of the delta function do not depend on the variable: delta functions in frequency, space, etc. behave as above, with the appropriate variable substituted for t.

From this result, in conjunction with the defining relationship, the delta function may be interpreted as a pulse of unit area, centred on $t = 0$, having infinite height and infinitesimal width. This is illustrated schematically in Fig. 2.1. Such a curve is of course a mathematical abstraction. It is also the basic constituent of an *impulse process,* consisting of a random train of delta functions, and such a waveform is an extremely good approximation to many actual noise processes.

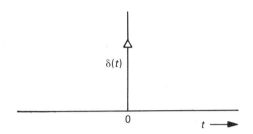

Fig. 2.1 – Schematic of the Dirac delta function.

A useful result that derives directly from equation (2.2) is

$$\delta(at - \tau) = \frac{1}{|a|} \delta\left(t - \frac{\tau}{a}\right) , \qquad (2.3)$$

where a is a constant. It follows from equation (2.3) that

$$\delta(t) = \delta(-t) . \qquad (2.4)$$

A further result, from equation (2.1), concerns the case when the delta function is off-set from zero and located at $t = t_0$. Then we have

$$\int_{-\infty}^{\infty} \delta(t - t_0) x(t) dt = \int_{-\infty}^{\infty} \delta(t) x(t + t_0) dt = x(t_0) , \qquad (2.5)$$

which demonstrates the sampling property of the delta function.

The delta function may be identified as the limiting form of an ordinary, even function of time, with a peak at $t = 0$, whose area remains constant and equal to unity, but which becomes progressively taller and thinner as a particular parameter, N say, increases in value. Eventually, in the limit of high N, the function becomes infinite at $t = 0$ and takes the value zero for all $t \neq 0$. If $g_N(t)$ is such a function, and

$$\lim_{N \to \infty} \int_{-\infty}^{\infty} g_N(t) x(t) dt = x(0) , \qquad (2.6)$$

where $x(t)$ is continuous at $t = 0$, then on comparing this expression with equation (2.1), we have

$$\delta(t) = \lim_{N \to \infty} g_N(t) \ . \tag{2.7}$$

This is to be interpreted as meaning that the number $x(0)$ assigned to the function $x(t)$ by the delta function is equal to the limit of the integral in equation (2.6).

An example of a function satisfying the condition in equation (2.6) is

$$g_N(t) = \frac{\sin (Nt)}{\pi t} \ , \tag{2.8}$$

and hence, according to equation (2.7),

$$\delta(t) = \lim_{N \to \infty} \frac{\sin (Nt)}{\pi t} \ . \tag{2.9}$$

The function in equation (2.8) is plotted in Fig. 2.2, which illustrates the approach to the limiting form as N increases. Equation (2.9) is an important result for establishing the relationship between a Fourier transform and its inverse.

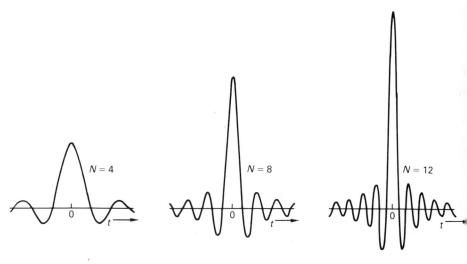

Fig. 2.2 – The function $\dfrac{\sin (Nt)}{\pi t}$

The above discussion outlines some of the properties of the Dirac delta function. A comprehensive account of singularity functions is given in the monograph by Lighthill (1958) on generalised functions.

2.3 THE FOURIER TRANSFORM

The (bilateral) Fourier transform of a function $x(t)$ is

$$X(j\omega) = \int_{-\infty}^{\infty} x(t) \exp{-(j\omega t)} \, \mathrm{d}t \; , \qquad (2.10)$$

where $x(t)$ may be either real or complex. A necessary and sufficient condition for the existence of $X(j\omega)$ is that $x(t)$ should be absolutely integrable; that is,

$$\int_{-\infty}^{\infty} |x(t)| \mathrm{d}t < \infty \; , \qquad (2.11)$$

which is equivalent to the requirement that the total energy in $x(t)$ should be finite.

This condition is not satisfied when $x(t)$ is a stationary stochastic process, since such a process does not decay to zero as $t \to \pm \infty$. It would appear that this presents a serious obstacle to the Fourier analysis of noise, which in turn implies a conceptual difficulty in defining the power spectral density of a random process; and indeed the question of whether noise can be analysed using Fourier techniques was once keenly debated in the literature. The difficulty was eventually resolved on the basis of the following argument.

Although the *total* energy in a stationary process is infinite, this is so because the waveform exists for an infinite time. But the *power* (i.e. the energy per unit time) in the process is finite, and of course in practice the observation time is never infinite. Thus the energy in a real noise waveform is always finite. If the observation time is T, then the observed fluctuation is $x_T(t)$, say, which equals $x(t)$ within the observation interval and is zero beyond it. Therefore, the gated function $x_T(t)$ is absolutely integrable and its Fourier transform, $X_T(j\omega)$, does exist. The power spectral density of $x_T(t)$, constructed from $X_T(j\omega)$, converges in most cases of practical importance to a unique limiting form as $T \to \infty$. This limit function is interpreted as the power spectral density of the original stationary process $x(t)$. The details of the limiting procedure are described in §2.5, though for the moment it is enough to note that stochastic processes are amenable to Fourier analysis, and that there is no fundamental reason prohibiting the definition of the power spectral density of such a process.

The inverse of the integral transform in equation (2.10) is

$$x(t) = (1/2\pi) \int_{-\infty}^{\infty} X(j\omega) \exp{(j\omega t)} \, \mathrm{d}\omega \; , \qquad (2.12)$$

which is valid for all t provided $x(t)$ is continuous. If $x(t)$ has a discontinuity at $t = t_0$, then at that point the inversion integral equals the mean of $x(t_0+)$ and

$x(t_0-)$. In general, therefore, equation (2.12) is valid for all t if it is assumed that

$$x(t) = \{x(t+) + x(t-)\}/2 \ . \tag{2.13}$$

The inversion formula is proved by taking the integral in equation (2.12) over the limits $\pm N$, substituting for $X(j\omega)$ from equation (2.10) and interchanging the order of integration to obtain

$$\int_{-N}^{N} X(j\omega) \exp (j\omega t)d\omega = 2 \int_{-\infty}^{\infty} x(t') \frac{\sin \{N(t'-t)\}}{(t'-t)} dt' \ . \tag{2.14}$$

In the limit as $N \to \infty$, this becomes

$$\int_{-\infty}^{\infty} X(j\omega) \exp (j\omega t)d\omega = 2 \lim_{N \to \infty} \int_{-\infty}^{\infty} x(t') \frac{\sin \{N(t'-t)\}}{(t'-t)} dt'$$

$$= 2\pi \int_{-\infty}^{\infty} x(t') \, \delta(t'-t) \, dt'$$

$$= 2\pi \, x(t) \ , \tag{2.15}$$

where the delta function has been introduced from the expression in equation (2.9), and the sampling property in equation (2.5) has been employed. Equation (2.15) is the required result for the inversion integral.

The kernel of the inversion integral is the function $\exp(j\omega t)$, used to represent harmonic waves, and so a natural interpretation of $X(j\omega)$ is that it is the amplitude spectrum of $x(t)$, giving a measure of the contribution to $x(t)$ from waves with angular frequencies between ω and $\omega + d\omega$. In general, $X(j\omega)$ is a complex quantity, the ratio of its real and imaginary parts being a measure of the phase of the harmonic component at angular frequency ω.

When $x(t)$ is a real function of time, as it always is in the case of an observable waveform, its Fourier transform shows conjugate symmetry:

$$X(j\omega) = X^*(-j\omega) \ , \tag{2.16}$$

where the asterisk denotes complex conjugate. In this case, it follows that $|X(j\omega)|^2$ and the real part of $X(j\omega)$ are even functions of ω, and that the imaginary part of $X(j\omega)$ is an odd function of ω. When $x(t)$ is an even function of t ($x(t) = x(-t)$), equation (2.10) becomes

$$X(j\omega) = 2 \int_{0}^{\infty} x(t) \cos(\omega t) \, dt \ , \tag{2.17a}$$

and when $x(t)$ is an odd function of t $(x(t) = -x(-t))$, the transform can be expressed as

$$X(j\omega) = -2j \int_0^\infty x(t) \sin{(\omega t)}\, dt \ . \tag{2.17b}$$

The integrals in equations (2.17a) and (2.17b) are the cosine and sine Fourier transforms, respectively.

The functions $x(t)$ and $X(j\omega)$ are Fourier mates and together they constitute a Fourier transform pair. Their relationship is conveniently denoted using the symbolism

$$x(t) \leftrightarrow X(j\omega) \ ,$$

which is an abbreviated means of expressing the integral transforms in equations (2.10) and (2.12). In terms of this notation, it is apparent from the sampling property of the delta function that

$$\delta(t) \leftrightarrow 1 \ , \tag{2.18a}$$

or more generally that

$$\delta(t - t') \leftrightarrow \exp{-(j\omega t')} \ , \tag{2.18b}$$

where t' is the fixed point in time at which the impulse occurs. An analogous result to that in equation (2.18b) is derived from the inversion integral when $X(j\omega) = 2\pi\, \delta(\omega - \omega')$, where ω' is a fixed angular frequency, for then we find that

$$\exp{(j\omega' t)} \leftrightarrow 2\pi\delta(\omega - \omega') \ . \tag{2.19a}$$

It is readily shown from this result, by considering ω' to be positive and then negative, that

$$\cos(\omega' t) \leftrightarrow \pi\,[\delta(\omega - \omega') + \delta(\omega + \omega')] \tag{2.19b}$$

and

$$\sin(\omega' t) \leftrightarrow -j\pi\,[\delta(\omega - \omega') - \delta(\omega + \omega')] \ . \tag{2.19c}$$

Thus the transforms of the cosine and sine functions each consist of two delta functions symmetrically placed about the origin at frequencies whose numerical value is equal to the frequency of the harmonic functions themselves.

The transform of the derivative $dx(t)/dt$ is often required, and can be found from the defining integral in equation (2.10) on integrating by parts. Provided $x(t) \to 0$ as $t \to \pm\infty$, this gives

$$\frac{dx(t)}{dt} \leftrightarrow j\omega\, X\,(j\omega) \tag{2.20a}$$

which, by using a similar procedure, may be generalized to the nth derivative. The result is

$$\frac{d^n x(t)}{dt^n} \leftrightarrow (j\omega)^n X(j\omega) \ , \tag{2.20b}$$

which is valid provided all the integrated parts vanish as $t \to \pm \infty$.

The properties of the Fourier integral are extensively discussed in the treatise by Titchmarsh (1937), and Papoulis (1965, 1968) demonstrates the power of the Fourier transform method in connection with a wide range of problems in physics.

2.4 STOCHASTIC PROCESSES

The statistical properties of a stochastic process are those regular features that (may) become apparent over a large number of trials or observations. This suggests that a mathematical treatment of such a process can be developed on the basis of a notional *ensemble* of statistically similar processes, observed simultaneously over the same epoch. The member functions of the ensemble, $x^{(1)}(t)$, $x^{(2)}(t)$, . . . , $x^{(N)}(t)$, have associated with them various probability measures from which predictions can be made about the statistical properties of a single process in the ensemble. Such predictions enable comparisions to be made with observable quantities in the real world. Thus, the quality of the predictions, and hence the validity of the theoretical models giving rise to them, can be assessed in a procedure which should lead to a better understanding of the physical phenomena underlying the observed fluctuations.

The probability measures referred to above are the hierarchy of probability density functions associated with the member functions of the ensemble. In general, these measures depend on the times at which they are computed. However, stationary processes are exceptions to the general rule. Strict sense stationarity means that *all* the probability density functions associated with the process are invariant under a shift in the origin in time. A similar definition applies to wide sense stationarity, except that it refers only to the first- and second-order probability density functions. The important point to note in connection with actual noise fluctuations is that statistically stationary processes, whether wide sense or strict sense, show second-order probability measures which depend only on the *difference* between the observation times, while the first-order measures are entirely independent of time. An example of such a first-order measure is the normal or Gaussian function

$$(1/\sqrt{2\pi\sigma^2}) \exp\{-(x - \bar{x})^2/2\sigma^2\} \ ,$$

where $\bar{x}$ is the mean and $\sigma^2 \equiv \overline{x^2(t)} - \bar{x}^2$ is the variance of the process $x(t)$.

All the statistical measures of a Gaussian process are determined by the first- and second-order probability density functions, so that knowledge of the mean value and the covariance function is sufficient to completely specify all the statistical properties of the process.

There are two different averaging procedures that can be applied to the member functions of an ensemble: one is *time averaging,* performed on a single member function of the ensemble, and the second is *ensemble averaging,* where the averages are expected values computed at fixed times throughout the observation interval from the probability measures discussed above.

The first- and second-order time averages of the ith member of the ensemble are the mean value,

$$\langle x^{(i)}(t)\rangle = \lim_{T\to\infty} \frac{1}{T}\int_{-T/2}^{T/2} x^{(i)}(t)\,dt \; , \tag{2.21}$$

and the autocorrelation function,

$$\phi_n^{(i)}(\tau) \equiv \langle x^{(i)}(t)\,x^{(i)}(t+\tau)\rangle = \lim_{T\to\infty} \frac{1}{T}\int_{-T/2}^{T/2} x^{(i)}(t)\,x^{(i)}(t+\tau)\,dt \; , \tag{2.22}$$

the latter providing a measure of the 'memory' of the process. The symbol $\langle\,\rangle$ in these expressions denotes a time average, and T is the duration of the observation interval. Note that the autocorrelation function is an even function of the time delay, τ, and that its value when $\tau = 0$ is the mean-square value of the process.

The ensemble averages corresponding to the time averages in equations (2.21) and (2.22) are the mean value taken at a time $t = t_1$,

$$\overline{x(t_1)} \equiv E\{x^{(i)}(t_1)\} = \lim_{N\to\infty} \frac{1}{N}\sum_{i=1}^{N} x^{(i)}(t_1) = \int_{-\infty}^{\infty} x_1 p_1(x_1,t_1)\,dx_1 \; , \tag{2.23}$$

and the second-order joint moment, or covariance, taken at times $t = t_1$ and $t = t_2 \neq t_1$,

$$\overline{x(t_1)\,x(t_2)} \equiv E\{x^{(i)}(t_1)\,x^{(i)}(t_2)\}$$

$$= \lim_{N\to\infty} \frac{1}{N}\sum_{i=1}^{N} \{x^{(i)}(t_1)\,x^{(i)}(t_2)\}$$

$$= \int_{-\infty}^{\infty} x_1 x_2 p_2(x_1,t_1;x_2,t_2)\,dx_1 dx_2 \; . \tag{2.24}$$

The mean-square value at $t = t_1$ is

$$\overline{x^2(t_1)} \equiv E\{[x^{(i)}(t_1)]^2\} = \lim_{N \to \infty} \frac{1}{N} \sum_{i=1}^{N} [x^{(i)}(t_1)]^2 = \int_{-\infty}^{\infty} x_1^2 p_1(x_1, t_1) \, dx_1 \ .$$

$$(2.25)$$

The overbar in these expressions denotes an ensemble average, the symbol $E\{\square\}$ means expected value, N is the number of functions in the ensemble, x_1 and x_2 are abbreviations for $x(t_1)$ and $x(t_2)$, $p_1(x_1, t_1)$ is the first-order joint probability density function, and $p_2(x_1, t_1; x_2, t_2)$ is the second-order joint probability density function of the process, giving the probability of finding a particular member of the ensemble with a value lying between x_1 and $x_1 + dx_1$ at time t_1 *and* a value between x_2 and $x_2 + dx_2$ at time t_2.

For stationary processes, the ensemble averages in equations (2.23) and (2.25) are independent of the time at which they are computed, and the average in equation (2.24) depends, not on the absolute times t_1 and t_2, but only on the time difference $|t_1 - t_2|$. By way of contrast, the ensemble averages of a non-stationary process depend on the absolute time scale, which implies that, in general, the concept of time averaging fails when applied to such processes.

In view of the fact that an observation of an actual random process usually involves a single time function, it may seem that time averaging is more closely related to physical reality than ensemble averaging. On the other hand, an ensemble average is a convenient theoretical concept, since it is directly related to the probability density functions, which themselves can often be determined from theoretical considerations. In order to be able to compare theoretical predictions based on ensemble averages with experimental measurements of time averages, it is necessary to know under what conditions time and ensemble averages are equivalent. Usually the observed process is assumed to be a member of an *ergodic* ensemble, and hence that the ergodic theorem applies. This is a reasonable assumption for most of the stochastic processes associated with electronic devices, which are at least wide sense stationary, and means that corresponding time and ensemble averages can be regarded as equivalent. Middleton (1960) gives a much more rigorous account of the ergodic theorem, which includes a bibliography of important references, and Born (1949) presents an interesting discussion of ergodicity.

2.5 ENERGY THEOREMS

The power in a noise process is a second-order statistical property of the process. It is conveniently discussed on the basis of Parseval's theorem, which states that if $x_1(t)$ and $x_2(t)$ are two time functions whose Fourier transforms are $X_1(j\omega)$ and $X_2(j\omega)$, respectively, then

$$\int_{-\infty}^{\infty} x_1(t)\, x_2^*(t)\, dt = (1/2\pi) \int_{-\infty}^{\infty} X_1(j\omega)\, X_2^*(j\omega)\, d\omega \ , \qquad (2.26)$$

where the transforms are assumed to exist. The asterisk denotes complex conjugate. The theorem is proved by substituting the inversion formula (equation (2.12)) for $x_2(t)$ in the integrand on the left, interchanging the order of integration and replacing the resultant integral over time with $X_1(j\omega)$, to obtain the expression on the right of equation (2.26). Note that Parseval's theorem is quite general, involving no conditions on the time functions $x_1(t)$ and $x_2(t)$ except that they should be absolutely integrable.

Suppose now that a noise process is observed over the interval $[-T/2, T/2]$, so that outside this time 'window' its ordinate values can be regarded as zero. We denote the gated process as $x_T(t)$,† and set

$$\begin{aligned} x_1(t) &= x_T(t+\tau) \\ x_2(t) &= x_T(t) \end{aligned} \qquad (2.27)$$

where τ is a delay time. As $x_T(t)$ is zero when $t \to \pm \infty$, its Fourier transform, $X_T(j\omega)$, exists and hence from equation (2.26),

$$\int_{-\infty}^{\infty} x_T(t+\tau)x_T(t)\, dt = (1/2\pi) \int_{-\infty}^{\infty} |X_T(j\omega)|^2 \exp(j\omega\tau)d\omega \ , \qquad (2.28)$$

where the asterisk on the time function has been omitted since $x_T(t)$ is a real process. When $\tau = 0$, equation (2.28) reduces to

$$\int_{-\infty}^{\infty} [x_T(t)]^2\, dt = (1/2\pi) \int_{-\infty}^{\infty} |X_T(j\omega)|^2\, d\omega \ , \qquad (2.29)$$

which is Plancherel's theorem or, as it is sometimes known, the energy theorem.

Each side of equation (2.29) equals the *total energy* in $x_T(t)$. This suggests that $|X_T(j\omega)|^2$ can be interpreted as the *energy density* of the process, with units of energy per Hertz, and this is finite provided $T < \infty$. The average *power* in the gated noise process is the total energy divided by T, which as $T \to \infty$ becomes

$$\lim_{T\to\infty} \frac{1}{T} \int_{-\infty}^{\infty} [x_T(t)]^2\, dt = \lim_{T\to\infty} (1/2\pi) \int_{0}^{\infty} \frac{2|X_T(j\omega)|^2}{T}\, d\omega \ , \qquad (2.30)$$

† Although the process is to be treated as a member function of an ensemble, the superscript indicating this status has now been omitted.

where the limits are postulated to exist. The unilateral form for the integral on the right is possible here because, as $x_T(t)$ is real, the integrand is an even function of frequency. The $\lim_{T \to \infty}$ and the integral on the right of equation (2.30) may be interchanged, provided an ensemble average is performed first (Middleton, 1960). The (unilateral) *power spectral density* of the stationary process $x_T(t)$ is then defined as the ensemble average

$$\overline{S_x(\omega)} = \lim_{T \to \infty} \frac{2|X_T(j\omega)|^2}{T} \, , \tag{2.31}$$

which converges to a specific value. Thus the power spectral density of a stationary process is expressed as a property of the ensemble as a whole, not as a property of an individual member function of the ensemble.

An important property of the function defined in equation (2.31) is that it is an *even* function of frequency. Thus, the power spectral density of any (real) process, irrespective of its detailed nature, is an even function. The functional form in equation (2.31) is often referred to as the *unilateral* power spectral density of $x_T(t)$, as distinct from the *bilateral* form in which the factor of 2 on the right-hand side is absent. In the latter case, by way of compensation, the lower limit of zero on integrals over frequency such as that in equation (2.30) must be replaced by $-\infty$. Incidentally, the inclusion of the negative range of frequencies in this context should not lead to conceptual difficulties concerning the meaning of a negative frequency: since $\overline{S_x(\omega)}$ is an even function of frequency, no new information is gained by looking at frequencies below zero, it is merely a means of ensuring that the correct scaling factor appears in the appropriate places.

The power spectral density of a stationary process is uniquely related to the autocorrelation function of the process. The form of the relationship is found from equation (2.28) by dividing both sides by T, ensemble averaging, and taking the limit as $T \to \infty$, to obtain

$$\lim_{T \to \infty} \frac{1}{T} \int_{-\infty}^{\infty} \overline{x_T(t + \tau) x_T(t)} \, dt = \lim_{T \to \infty} (1/2\pi) \int_0^{\infty} \frac{2|X_T(j\omega)|^2}{T} \cos \omega\tau \, d\omega \, , \tag{2.32}$$

where the unilateral form for the integral on the right is possible because $|X_T(j\omega|^2$ is an even function of frequency. The left-hand side of this expression is the autocorrelation function, $\overline{\phi_x(\tau)}$, of the process.† On interchanging the order of

---

† The ensemble-averaged autocorrelation function of a stationary process is equal to the autocorrelation function of any one of the member functions of the ensemble.

the limit and integration on the right, and bearing in mind the definition in
equation (2.31), it follows that

$$\overline{\phi_x(\tau)} = (1/2\pi) \int_0^\infty \overline{S_x(\omega)} \cos\omega\tau \, d\omega \ . \tag{2.33a}$$

This is a Fourier transform relationship, whose inverse can be written directly,
by analogy with the inversion integral in equation (2.10), as

$$\overline{S_x(\omega)} = 4 \int_0^\infty \overline{\phi_x(\tau)} \cos\omega\tau \, d\tau \ . \tag{2.33b}$$

Equations (2.33) constitute the Wiener–Khintchine theorem, so-called after the
work of Wiener (1930) and Khintchine (1934).

The Wiener–Khintchine theorem is an important analytical tool. As an
example of its utility, consider a *relaxation process*. Such processes are often
encountered in solid state devices, and are characterized by an exponentially
decaying autocorrelation function:

$$\overline{\phi_x(\tau)} = \overline{\phi_x(0)} \exp - (|\tau|/\tau_1) \ , \tag{2.34}$$

where $\overline{\phi_x(0)}$ is the variance of the process and τ_1 is the decay constant. The
inversion formula in equation (2.33b) immediately gives the power spectral
density of the process as

$$\overline{S_x(\omega)} = \frac{4 \, \overline{\phi_x(0)} \, \tau_1}{(1 + \omega^2\tau_1^2)} \ . \tag{2.35}$$

Thus, the spectrum of a relaxation process is flat when $\omega \ll 1/\tau_1$, it rolls off as
$1/\omega^2$ when $\omega \gg 1/\tau_1$, and the half-power point falls at $\omega = 1/\tau_1$.

Of course, not all noise processes are of the relaxation type. A notable
exception is $1/f$ noise, which occurs in most electronic devices as well as a
number of other systems. Its power spectral density varies as $|f|^{-\alpha}$, where α
usually lies between 0.8 and 1.2, and this dependence has been observed over a
wide range of frequencies, limited at the high end by Johnson noise and at the low
end by the observation time employed in the experiment. Despite its ubiquitous
nature and the amount of attention it has received over the last fifty years or so
since the phenomenon has been known, no entirely satisfactory explanation of
$1/f$ noise exists.

It is perhaps worth commenting that the spectral shape of $1/f$ noise is often
loosely described as $f^{-\alpha}$, where $\alpha \simeq 1$. This (strictly erroneous) representation is
almost universally understood to mean $|f|^{-\alpha}$, which of course is an *even* function
of frequency, irrespective of the value of α, as any power spectral density

function must be. By way of contrast, the function $f^{-\alpha}$ is an *odd* function of frequency when $\alpha = 1$, and when α is non-integer it is not a real function at all but is an infinite set of complex functions. Clearly, such a function has no meaning as a power spectral density. In order to avoid any possible confusion, we repeat that the power spectral density of $1/f$ noise over the observed range of frequencies is of the form $|f|^{-\alpha}$, which is an even function of frequency and as such is compatible with the definition of the power spectral density function in equation (2.31).

2.6 RANDOM PULSE TRAINS

Random noise often originates in a very large number of independent, discrete 'events'. Each event produces a pulse of given shape, and the random superposition of all such pulses constitutes the noise waveform. Such a waveform is known as a random pulse train. Shot noise and thermal noise can both be treated as random pulse trains, and many other processes in electronic devices and elsewhere, including generation–recombination noise, precipitation, and wind-generated acoustic noise in the ocean, can be represented in a similar fashion.

If $f(t)$ is the pulse-shape function, then the noise waveform is the superposition†

$$x(t) = \sum_{k=1}^{K} a_k f(t - t_k) \; , \qquad (2.36)$$

where a_k is the amplitude of the kth pulse, t_k is the time at which the kth event occurs, K is the number of pulses in the train of duration T, and causality requires that $f(t) = 0$ when $t < 0$. The statistical properties of the process in equation (2.36) are derived on the assumptions that the summation converges and that the shape function takes appreciable values for a time which is very much less than the observation time, T.

Since the events are independent, the t_k are Poisson distributed, with a probability density function equal to $1/T$. (The Poisson distribution is discussed in Appendix 1.) Thus the expected value of the process is

$$\overline{x(t)} = v\overline{a} \int_{-\infty}^{\infty} f(t) \, dt \; , \qquad (2.37)$$

where $v = \lim_{T \to \infty} (k/T)$ is the mean number of events per second and $\overline{a}$ is the mean

---

† The formal procedure of putting the subscript T on noise processes, indicating that the observation time is finite, will now be dropped for brevity.

value of a_k‡. The result in equation (2.37) is sometimes known as Campbell's theorem of the mean. It is apparent from the theorem that, if the amplitudes are symmetrically distributed about zero, the mean value of the process is zero.

The Fourier transform of $x(t)$ is

$$X(j\omega) = \sum_{k=1}^{K} \{a_k \exp - (j\omega t_k) F(j\omega)\}$$

$$= F(j\omega) \sum_{k=1}^{K} a_k \exp - (j\omega t_k) . \qquad (2.38)$$

Now, according to the definition in equation (2.31), the power spectral density of $x(t)$ is

$$\overline{S_x(\omega)} = \lim_{T \to \infty} \frac{2|F(j\omega)|^2}{T} \sum_{k,m=1}^{K} \overline{a_k a_m \exp - j\omega(t_k - t_m)} , \qquad (2.39)$$

where the dummy summation index m allows cross-terms to be included in the double sum. The summation in equation (2.39) can be expressed as a sum over terms with $k = m$ plus a double sum of terms with $k \neq m$, allowing the power spectral density to be written as

$$\overline{S_x(\omega)} = \lim_{T \to \infty} \frac{2}{T} |F(j\omega)|^2 \left\{ \sum_{k=1}^{K} \overline{a_k^2} + \sum_{k,m=1}^{K}{}' \overline{a_k a_m \exp - j\omega(t_k - t_m)} \right\} , \qquad (2.40)$$

where the prime indicates $k \neq m$. Since the a_k and a_m are independent for all $k \neq m$ (i.e. they are pair-wise independent), and the a_k are independent of the t_k, the product $\overline{a_k a_m}$ can be taken outside the primed summation and set equal to $\overline{a}^2$. Thus, for a symmetrical distribution of the a_k about zero, the primed term is zero and the power spectral density is

$$\overline{S_x(\omega)} = 2\nu \overline{a^2} |F(j\omega)|^2 , \qquad (2.41)$$

where $\overline{a^2}$ is the mean-square value of the a_k. Equation (2.41) is Carson's theorem (Rice, 1944, 1945).

For the more general case, when the a_k are not symmetrically distributed about zero, an extra term representing the dc level appears in the expression for the power spectral density. This additional term derives from the primed summation in equation (2.40). The general result is

$$\overline{S_x(\omega)} = 2\nu \overline{a^2} |F(j\omega)|^2 + 4\pi \overline{x(t)}^2 \delta(\omega) . \qquad (2.42)$$

‡ The probability density function governing the distribution of a_k in the ensemble is assumed to be independence of k.

The term containing the delta function is arrived at by evaluating $\overline{\exp - (j\omega t_k)}$ using the Poisson probability density function, $1/T$, then identifying $F(0)$ with $\int_{-\infty}^{\infty} f(t)dt$, and finally replacing $\lim\limits_{T \to \infty} \dfrac{2 \sin^2 (\omega T/2)}{\omega^2 T}$ with $\pi\delta(\omega)$.†

The Wiener–Khintchine theorem can now be applied to $S_x(\omega)$ in equation (2.42), to obtain the autocorrelation function of the random pulse train. From the inversion integral in equation (2.33b) we have

$$\overline{\phi_x(\tau)} = \frac{v\overline{a^2}}{\pi} \int_0^\infty |F(j\omega)|^2 \cos \omega\tau \, d\omega + 2 \overline{x(t)}^2 \int_0^\infty \delta(\omega) \cos \omega\tau \, d\omega$$

$$= \frac{v\overline{a^2}}{2\pi} \int_{-\infty}^\infty |F(j\omega)|^2 \cos \omega\tau \, d\omega + \overline{x(t)}^2 \; . \tag{2.43}$$

Through Parseval's theorem (equation (2.26)), the integral here can be expressed as an integral over time, which leads to the alternative representation

$$\overline{\phi_x(\tau)} = v\,\overline{a^2} \int_{-\infty}^\infty f(t)f(t + \tau) \, dt + \overline{x(t)}^2 \; . \tag{2.44}$$

When $\tau = 0$, the autocorrelation function is equal to the mean-square value, and hence we have from the general result in equation (2.44)

$$[\overline{x^2(t)} -- \overline{x(t)}^2] = v\,\overline{a^2} \int_{-\infty}^\infty f^2(t) dt \; , \tag{2.45}$$

which is Campbell's theorem of the mean-square. The theorems named after him have been discussed at length in the literature by Campbell (1909, 1910a, b, 1939), and also by several other authors, including Rowland (1936), and Campbell and Francis (1946). Rice (1944, 1946) has extended the theorem of the mean-square to include n^{th}-order averages.

When the shape function is a delta function, the transform $F(j\omega)$ is unity and the random pulse train is said to be an impulse process. From equation (2.42), the power spectral density of such a process is

$$\overline{S_x(\omega)} = 2v\,\overline{a^2} + 4\pi \,\overline{x(t)}^2 \delta(\omega) \; , \tag{2.46}$$

---

† The function $\dfrac{\sin^2 (Nt)}{\pi N t^2}$ is another example of a function, $g_N(t)$, which satisfies the condition in equation (2.7): it shows the properties of a delta function in the limit of high N.

and equation (2.43) gives the autocorrelation function as

$$\overline{\phi_x(\tau)} = \nu \overline{a^2} \int_{-\infty}^{\infty} \cos \omega\tau \, d\omega + \overline{x(t)}^2$$

$$= \nu \overline{a^2} \, \delta(\tau) + \overline{x(t)}^2 \quad , \tag{2.47}$$

where the integral has been replaced with the delta function in accord with the Fourier relationship in equation (2.18a). Equations (2.46) and (2.47) are illustrated in Fig. 2.3. Note the appearance of the delta function at the origin in both the power spectral density and the autocorrelation function. In the latter case, the implication is that the mean-square value of an impulse process is undefined,† which is consistent with the fact that the spectral density extends uniformly over an infinitely wide frequency band. Of course, the shape functions encountered in the physical world are never truly impulsive. However narrow they may be, they always have a finite width. As a consequence the power spectral density of the pulse train rolls-off rapidly at frequencies beyond the reciprocal of the pulse width, and the delta function at the origin in the autocorrelation function disappears, thereby removing the difficulty with the mean-square value, which becomes finite.

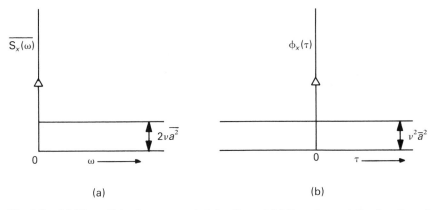

(a)   (b)

Fig. 2.3 – (a) The unilateral power spectral density and (b) the autocorrelation function of an impulse process.

2.7 SIMPLE SHOT NOISE

The random emission of electrons from the cathode of a thermionic diode gives rise to a current in the external circuit which can be represented as a random pulse train. When an electron emitted from the cathode makes the transit to the

† It follows that Campbell's theorem of the mean-square fails for an impulse process.

anode, a displacement-current pulse flows, and the sum of all such pulses is the total current. By analogy with the pattering of small shot into a container, the fluctuation in the diode current is called 'shot noise' (Schottky, 1918). *Simple shot noise* is the current fluctuation produced when the electrons, which are emitted randomly and independently of one another, undergo no mutual interactions during their transit to the anode.

Assuming that the transit time is infinitesimal, each of the current pulses can be represented as an impulse whose area is equal to the electronic charge. The current in the circuit at any instant, $i(t)$, is then the impulse process

$$i(t) = -q \sum_{k=1}^{K} \delta(t - t_k) \; , \tag{2.48}$$

where q is the magnitude of the electronic charge, t_k is the time at which the kth electron is emitted from the cathode, and K is the total number of pulses in the pulse train of duration T. The linear superposition in equation (2.48) is a special case of the random pulse train in equation (2.36), in which the shape function is a delta function.

From Campbell's theorem of the mean, the dc level of the shot noise is

$$-I \equiv \overline{i(t)} = -q\nu \; , \tag{2.49}$$

where ν is the mean rate of emission from the cathode. The sign convention here is such that I is positive, representing the magnitude of the mean current. Since $i(t)$ is an impulse process, Campbell's theorem of the mean-square cannot be applied, but, from equation (2.47), the autocorrelation function of the shot noise is

$$\overline{\phi_i(\tau)} = qI \, \delta(\tau) + I^2 \; , \tag{2.50}$$

and from equation (2.46) the power spectral density is

$$\overline{S_i(\omega)} = 2qI + 4\pi I^2 \, \delta(\omega) \; . \tag{2.51}$$

The expression $2qI$ appearing here, giving the spectral density of shot noise at frequencies above zero, is characteristic of simple shot noise processes in general, and will be encountered again in connection with current fluctuations in p-n junction devices.

In any physical device showing shot noise, a certain amount of pulse broadening occurs. The transit time in a thermionic diode, for example, is short but nevertheless non-zero, leading to current pulses of finite width. Then, as we have seen in §2.6, the power spectral density no longer extends uniformly to indefinitely high frequencies, nor is there a delta function at the origin of the autocorrelation function, which instead shows a mean-square value that is finite.

A departure from simple shot noise occurs when there is some degree of correlation between the current pulses in the pulse train. This could happen in the vacuum diode as a result of mutual interactions between the electrons in transit. This produces space-charge smoothing and a reduction in the noise.

2.8 THERMAL NOISE

A resistor which is in thermal equilibrium with its surroundings shows at its terminals open-circuit voltage (or short-circuit current) fluctuations. The noise was first observed by Johnson (1927a, b, 1928), and is now usually referred to as Johnson noise or thermal noise. It is analogous to Brownian movement, whose statistical properties had been derived by Einstein (1906a, b) twenty years before Johnson's observations. Einstein's analysis is based on a random-walk model, and shows that the mean-square displacement of a Brownian particle is proportional to the observation time.†

The electrons in a resistor are thermally energetic and move randomly through the material, experiencing collisions with the host atoms as they go. The random motions are responsible for the thermal noise. The fluctuations can be conceptualized as the result of a very large number of independent, random 'events'. Each event comprises an initial action, which results in a departure from the equilibrium state, followed by a relaxation back towards that state. The initial action is the transit of an electron between collisions, which produces a non-equilibrium charge distribution within the resistive material, and the relaxation is the subsequent flow of charge which restores the equilibrium state. The occurrence of an event gives rise to a current or voltage pulse at the terminals, and the superposition of all such pulses is the thermal noise fluctuation. According to this model, thermal noise is another example of a random pulse train.

The description of an individual electronic event given above does not accurately represent the microscopic behaviour of an electron in a resistive material; but on averaging over a large number of such events, the correct statistical properties of the terminal noise processes are obtained. Moreover, the idea of an initial action followed by a relaxation helps in understanding the role of thermal fluctuations in a dissipative system: they maintain the system in thermal equilibrium by ensuring that any departure from that state is followed on average by a relaxation back towards it.

The statistical properties of Johnson noise can be derived from a one-dimensional model of a resistor with cross-sectional area A and length L. What is required first is the pulse shape (or its transform) at the terminals due to a single event, comprising an electron transit between collisions of length l_f and the subsequent charge relaxation. Now, the initial action can be visualized as creating

---

† Thus the *displacement* of a Brownian particle, which is analogous to the *charge* fluctuation in a resistor, is a non-stationary process, as discussed in §2.10. The *velocity* fluctuation, analogous to *current* in a resistor, is statistically stationary.

instantaneously two charge sheets with charge density $\pm q/A$, separated by a distance l_f. Assuming the resistor is open-circuited, the charge will then decay away through a flow back over the path between the charge sheets. An equivalent circuit for the event is shown in Fig. 2.4, where R_f and C_f are the resistance and capacitance, respectively, of the region between the charge sheets, $R \equiv R_f L/l_f$ is the bulk resistance of the device, and the current generator $q\delta(t)$ represents the initial action.

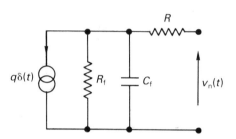

Fig. 2.4 – Equivalent circuit of a single 'event' in a resistor with terminals A–A.

In terms of the voltage pulse, $v_n(t)$, at the terminals due to a single event, the flow equation from Fig. 2.4 is† (Buckingham and Faulkner, 1974)

$$C \frac{dv_n(t)}{dt} = -\frac{v_n(t)}{R} + q\delta(t) \ , \tag{2.52}$$

where $C = C_f l_f/L$. On Fourier transforming both sides of equation (2.52), the transform of the voltage pulse is found to be

$$V_n(j\omega) = \frac{q\,l_f\,(R/L)}{(1 + j\omega\tau_1)} \ , \tag{2.53}$$

where $\tau_1 = RC = \rho\epsilon$ is the *dielectric relaxation time*. ρ is the resistivity and ϵ is the relative permittivity of the material. By electronic standards, τ_1 is extremely short, being of the order of picoseconds.

Equation (2.53) is the transform of a voltage pulse at the terminals due to a single event within the resistor. Its inverse transform, the pulse shape function, is

---

† This equation is analogous to Langevin's (1908) equation of motion for a Brownian particle, which takes into account inertial as well as dissipative (viscous) forces acting on the particle. Langevin's innovation was to assume that the effect of the environment could be split into an average viscous force and a very rapidly varying force associated with the frequent molecular impacts experienced by the particle. The profound reasoning underlying Langevin's argument has been discussed by Chandrasekar (1943). Uhlenbeck and Ornstein (1930) show from a solution of Langevin's equation that the velocity and displacement of a Brownian particle both show Gaussian distribution functions.

a decaying exponential with a decay time equal to τ_1. Since there is an equal probability of finding l_f positive and negative, the mean value of the terminal fluctuation is zero. From Carson's theorem in equation (2.41), with $F(j\omega)$ identified from equation (2.53) as $q(R/L)/(1 + j\omega\tau_1)$, the power spectral density of the terminal voltage fluctuation is

$$\overline{S_{v_n}(\omega)} = \frac{2\nu q^2 (R/L)^2 \overline{l_f^2}}{(1 + \omega^2 \tau_1^2)} , \tag{2.54}$$

where ν is the mean number of events per second within the volume of the resistor. Now, if n is the density of electrons in the material and $\overline{\tau}_f$ is the mean-free time between collisions, then

$$\nu = nAL/\overline{\tau}_f . \tag{2.55}$$

This result follows from the addition theorem for the Poisson distribution (Appendix 3). We also have that

$$R = L/(nq\mu A) , \tag{2.56}$$

where μ is the mobility. A statistical mechanical argument, detailed in Appendix 3, shows that the mobility can be expressed as

$$\mu = q \overline{l_f^2}/(2 \overline{\tau}_f k\theta) , \tag{2.57}$$

where k is Boltzmann's constant and θ is the absolute temperature. On combining equations (2.54) to (2.57), the power spectral density of the open-circuit voltage fluctuations is found to be

$$\overline{S_{v_n}(\omega)} = \frac{4 k\theta R}{(1 + \omega^2 \tau_1^2)} , \tag{2.58}$$

and from a simple circuit transformation it follows that the power spectral density of the short-circuit current fluctuation is

$$\overline{S_{i_n}(\omega)} = \frac{(4k\theta/R)}{(1 + \omega^2\tau_1^2)} . \tag{2.59}$$

For all frequencies of practical interest, the term $\omega^2\tau_1^2$ is negligibly small. The two expressions in equations (2.58) and (2.59) then reduce to $4k\theta R$ and $4k\theta/R$, respectively, which are the classical forms derived originally by Nyquist (1927, 1928) from an argument based on the second law of thermodynamics and the exchange of energy between resistive elements in equilibrium. The two formulae, known as Nyquist's theorem, are discussed further in Appendix 2.

The autocorrelation functions of the thermal noise voltage and current fluctuations follow directly from equations (2.58) and (2.59), through the Wiener–Khintchine theorem. For example, the autocorrelation function of the voltage fluctuation is

$$\overline{\phi_{v_n}(\tau)} = \frac{2\,k\theta R}{\pi} \int_0^\infty \frac{\cos(\omega\tau)}{(1 + \omega^2 \tau_1^2)}\, d\omega = (k\theta R/\tau_1) \exp - (|\tau|/\tau_1)\ . \qquad (2.60)$$

The exponential form of this expression is characteristic of a relaxation process. The mean-square value from equation (2.60) is $k\theta R/\tau_1$, which becomes infinitely large as the relaxation time becomes vanishingly small. In this limit, thermal noise is an impulse process.

Dissipative systems are said to be *irreversible*,† at least for times exceeding the relaxation time. Einstein established that, in maintaining the equilibrium state, such systems always produce random fluctuations in some associated parameter. Thus, thermal noise is an inherent and inevitable feature of resistive materials, and often imposes a fundamental limit on the reduction in noise that can be achieved in an electrical circuit.

2.9 NON-STATIONARY PROCESSES

As we have seen in § 2.5, the autocorrelation function and the power spectral density of a stationary stochastic process are defined in terms of ensemble averages, taken in the limit as the observation time, T, goes to infinity. When the process is stationary these limits converge to specific finite values.

When the process is non-stationary, however, the concept of average power taken in the limit as $T \to \infty$ fails because, in general, the limit does not exist. This does not mean that a non-stationary process cannot be described in terms of an autocorrelation function and a power spectral density; but it does mean that these quantities must be defined in an appropriate way.

In fact, the required formalism is exactly the same as that applied to stationary processes, except that the observation time now remains finite instead of being taken to infinity. The ordinates of the process are set to zero outside the observation interval. The concept of ensemble averaging is employed, as before, on the assumption that the member functions of the ensemble all have the same probability measures. The Wiener–Khintchine theorem in equations

---

† A system such as an ensemble of electrons in a resistor can be specified in terms of its momentum/position coordinates in phase space. Each point represents the momentum and position of *all* the particles in the assembly. If, by the introduction of energy, the system is disturbed from its equilibrium state, it moves to a new region in phase space, but returns as it relaxes back towards the equilibrium condition to the region it originally occupied. The probability that it should then revert spontaneously to its previous non-equilibrium coordinates is negligibly small. This is what is meant by saying that the system is irreversible.

(2.33) still applies, but now the limits on the integral over τ are finite, and both $\overline{\phi_x(\tau)}$ and $\overline{S_x(\omega)}$ are functions of the observation time T (Lampard, 1954).

One property of non-stationary processes is especially interesting. It concerns the autocorrelation function which, when averaged over the ensemble, is defined by analogy with the stationary case as

$$\overline{\phi_x(\tau, T)} = \frac{1}{T} \int_0^{T-|\tau|} \overline{x(t + \tau) x(t)} \, dt \ . \tag{2.61}$$

The finite limits on the integral are a consequence of assuming that $x(t)$ is zero outside the observation interval $[0, T]$. The integrand in equation (2.61) is the covariance function of the process $x(t)$. If $x(t)$ were stationary, the covariance function would be independent of t and hence equal, in the limit as $T \to \infty$, to the autocorrelation function. But when $x(t)$ is non-stationary, the covariance function is not independent of t, nor is it equal to the autocorrelation function. This is a clear demonstration of the failure of the ergodic theorem in connection with non-stationary processes.

2.10 THE WIENER–LÉVY PROCESS

A particular form of non-stationarity is exhibited by the Wiener–Lévy process, which is the limiting form of the random walk, approached as the time between successive steps goes to zero. An example of such a process is the displacement of a particle undergoing Brownian movement. This problem was treated by Einstein in a classic analysis giving the mean-square displacement of the particle. He then went on to show that diffusion and Brownian movement are essentially the same thing, in that both arise from very rapid molecular motion.

A process analogous to the displacement of a Brownian particle is the fluctuation in the charge, due to the thermal energy of the electrons, which is transferred around the external circuit of a resistor. The charge fluctuation, $q(t)$, can be expressed as an integral over the (finite) interval $[0, t]$ of the thermal noise current:

$$q(t) = \int_0^t i(t') \, dt' \ . \tag{2.62}$$

This formulation provides an alternative definition of a Wiener–Lévy process: it is the integral of a stationary process showing a uniform or 'white' power spectral density. Although integrals of stationary processes are non-stationary, they are nevertheless a small class of processes, and are not representative of non-stationary processes in general.

The mean-square value of the charge fluctuation is given by a simple argument. Consider again the one-dimensional model of a resistor described in §2.8.

An initial action consisting of an electron transit over a free path l_f between collisions gives rise to a transfer of charge in the external circuit equal to $q(l_f/L)$. If there are m independent events in time t, then the mean-square value of the charge fluctuation is

$$\overline{q^2(t)} = \overline{m}\, q^2\, \overline{l_f^2}/L^2$$

$$= \nu q^2\, \overline{l_f^2}\, t/L^2 \, , \qquad (2.63)$$

where ν is the mean rate of events and we have set $\overline{m} = \nu t$. From equations (2.55) to (2.57) it follows that

$$\overline{q^2(t)} = 2(k\theta/R)t \, . \qquad (2.64)$$

The appearance of t in this expression confirms that the charge fluctuations are indeed non-stationary. Equation (2.64) is the result that was originally derived by Einstein.

We are now interested in the covariance, the autocorrelation function and the power spectral density of the charge fluctuation $q(t)$. The covariance is calculated first, equation (2.61) is then employed to evaluate the autocorrelation function, and from this the power spectral density is derived, using the Wiener–Khintchine theorem. This procedure, in which the distinction between the covariance function and the autocorrelation function is not only emphasized but also exploited in calculating the power spectral density, is applicable to non-stationary processes in general.

Since $q(t)$ is a cumulative process, we may write

$$q(t + \tau) = q(t) + z(t, \tau) \, , \qquad (2.65)$$

where $z(t, \tau)$ is the contribution made to the charge fluctuation between times t and $(t + \tau)$. As the correlation time of the additive fluctuations is zero in a Wiener–Lévy process, the two processes on the right of equation (2.65) are uncorrelated. It follows that the covariance function is

$$\overline{q(t + \tau)\, q(t)} = \overline{q^2(t)} \, . \qquad (2.66)$$

According to this result, the covariance function of a Wiener–Lévy process is independent of the delay time τ and is equal to the mean-square value. Thus, the covariance function of the charge fluctuation is

$$\overline{q(t + \tau)\, q(t)} = \frac{2k\theta t}{R} \, . \qquad (2.67)$$

The autocorrelation function of $q(t)$ can now be obtained from equation (2.61):

$$\overline{\phi_q(\tau, T)} = \frac{1}{T} \int_0^{T-|\tau|} \frac{2k\theta}{R} t \, dt = \frac{k\theta T}{R} \left(1 - \frac{|\tau|}{T}\right)^2, \; |\tau| \leqslant T \; , \qquad (2.68)$$

where T is the observation time of the process $q(t)$. When the inequality here is not satisfied, the autocorrelation function is zero. Figure 2.5(a) shows $\phi_q(\tau, T)$, normalized to $\phi_q(0, T)$, as a function of τ/T.

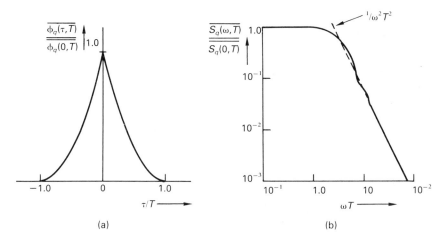

Fig. 2.5 – (a) The normalized autocorrelation function and (b) the normalized (unilateral) power spectral density of the charge fluctuations in a resistor.

Equations (2.67) and (2.68) make an interesting comparison: whereas the covariance is independent of the delay, the autocorrelation depends explicitly on τ in a way which shows that the 'correlation time' of the process is proportional to the observation time T. This seems a strange result at first, but, on recalling the cumulative nature of the process, it can be understood; for, with fixed τ, the proportionate commonality between $q(t + \tau)$ and $q(t)$ increases on average over the interval $[0, T]$ as the observation time gets longer.

The power spectral density of $q(t)$ can now be found from equation (2.68) through the Wiener–Khintchine theorem. The appropriate form for the inversion integral in equation (2.33b) is

$$\overline{S_q(\omega, T)} = 4 \int_0^T \overline{\phi_q(\tau, T)} \cos(\omega\tau) \, d\tau \; , \qquad (2.69)$$

which in conjunction with equation (2.68) gives

$$\overline{S_q(\omega, T)} = \frac{8k\theta}{\omega^2 R} \left\{ 1 - \frac{\sin(\omega T)}{\omega T} \right\}. \tag{2.70}$$

This function, normalized to its zero-frequency value is plotted in Fig. 2.5(b) against ωT.

When $\omega T \gg 1$, the second term in parenthesis in equation (2.70) is negligible and the spectral density varies as ω^{-2}, in agreement with Bell's (1960) spectrum of a random-walk process derived using the method of squared differences. When $\omega T \leqslant 1$, the gating effect of the observation interval governs the shape of the spectrum, which in this region is more or less flat. Thus, the conceptual difficulty with infinite energy at zero frequency (the 'infra-red' catastrophe), which would be encountered if T were allowed to become infinitely long in equation (2.70), is avoided here by keeping the observation time finite. Since practical measurements are always made in a finite time, this treatment of the Wiener–Lévy process would seem to be more realistic than any contrived analysis in which the observation time is allowed, either implicitly or explicitly, to go to infinity.

2.11 MACDONALD'S FUNCTION

The covariance of the non-stationary, cumulative process $q(t)$ defined in equation (2.62) can be expressed in terms of the power spectral density of the stationary process $i(t)$. This is true even when $q(t)$ is not a Wiener–Lévy process. The result is

$$\overline{q(t)q(t + \tau)} = (1/2\pi) \int_0^\infty \frac{S_i(\omega)}{\omega^2} \left\{ 1 + \cos(\omega\tau) - \cos(\omega t) - \cos[\omega(t + |\tau|)] \right\} d\omega, \tag{2.71}$$

where $\overline{S_i(\omega)}$ is the power spectral density of $i(t)$. Equation (2.71) is proved by writing the covariance as

$$\overline{q(t) q(t + \tau)} = \int_0^t \int_0^{t+|\tau|} \overline{i(t') i(t'')} \, dt' \, dt''$$

$$= \int_0^t \int_0^{t+|\tau|} \overline{\phi_i(t' - t'')} \, dt' \, dt'', \tag{2.72}$$

where $\overline{\phi_i(t' - t'')}$ is the autocorrelation function of the current fluctuation. From the Wiener–Khintchine theorem, $\overline{\phi_i(t' - t'')}$ is then written in terms of the inversion integral over frequency, containing $\overline{S_i(\omega)}$ in the integrand. On interchanging the order of integration and integrating over t' and t'', equation (2.71) is obtained.

When $\tau = 0$, equation (2.71) reduces to the expression

$$\overline{q^2(t)} = (1/\pi) \int_0^\infty \frac{\overline{S_i(\omega)}}{\omega^2} \{1 - \cos(\omega t)\}\, d\omega \;, \tag{2.73}$$

which was originally derived by MacDonald (1949). For the case of a Wiener–Lévy process, $\overline{S_i(\omega)}$ is independent of frequency and can be taken outside the integrals in equations (2.71) and (2.73). On performing the integrations,† the equality expressed in equation (2.66) is confirmed. The result is

$$\overline{q(t)q(t+\tau)} = \overline{q^2(t)} = t\,\overline{S_i(\omega)}/2$$

$$= \frac{2k\theta t}{R} \;, \tag{2.74}$$

where we have substituted Nyquist's expression for $\overline{S_i(\omega)}$. This is Einstein's result, as expressed in equations (2.64) and (2.67), though derived from a different argument.

Equation (2.71) provides a systematic means of obtaining the covariance of a random-walk process, irrespective of the spectral shape of the stationary process from which it is derived. It is then possible, as exemplified above for the special case of a Wiener–Lévy process, to obtain the autocorrelation function and the power spectral density of the non-stationary fluctuation. Alternatively, $\overline{S_q(\omega, T)}$ can be expressed as an integral involving $\overline{S_i(\omega)}$ in a general formulation akin to that for the covariance in equation (2.71). However, this involves lengthy algebra, and the result is not very edifying.

2.12 CROSS-CORRELATION, CROSS-SPECTRA AND COHERENCE

The second-order statistics of two stationary processes, $x(t)$ and $y(t)$, are expressed in terms of the cross-correlation function and the cross-spectral density, defined by analogy with the autocorrelation function and the power spectral density of a single stationary process. Thus, the cross-correlation function is

$$\overline{\phi_{xy}(\tau)} = \lim_{T \to \infty} \frac{1}{T} \int_{-T/2}^{T/2} \overline{x(t+\tau)\,y(t)}\, dt \;, \tag{2.75}$$

---

† These integrals can be evaluated by replacing ω^2 in the denominator of the integrand by $(\omega^2 + a^2)$ and then using the identity $\int_0^\infty \dfrac{\cos(\omega t)}{(\omega^2 + a^2)}\, d\omega = (\pi/2|a|)\,\exp{-|at|}$ to obtain each of the terms. The value of the whole integral is then the limit as $a \to 0$.

and the cross-spectral density is

$$\overline{S_{xy}(\omega)} = \lim_{T \to \infty} \left\{ \frac{2\,\overline{X(j\omega)\,Y^*(j\omega)}}{T} \right\} \tag{2.76}$$

where $X(j\omega)$ and $Y(j\omega)$ are the Fourier transforms of $x(t)$ and $y(t)$, respectively. It follows from equation (2.75) that

$$\overline{\phi_{xy}(\tau)} = \overline{\phi_{yx}(-\tau)} \ , \tag{2.77}$$

and from equation (2.76),

$$\overline{S_{xy}(\omega)} = \overline{S_{xy}^*(-\omega)} = \overline{S_{yx}(-\omega)} = \overline{S_{yx}^*(\omega)} \ . \tag{2.78}$$

It is implicit in these equations that $x(t)$ and $y(t)$ are real processes. Note that the cross-spectral density is complex, with real and imaginary parts which can go negative and which, from the symmetry relations in equation (2.78), are seen to be even and odd functions of ω, respectively.

The cross-correlation function and the cross-spectral density are Fourier mates and are related through a pair of inversion integrals similar to those in the Wiener–Khintchine theorem. Beginning with Parseval's theorem and following an analogous derivation to that leading to the Wiener–Khintchine integrals, we find that

$$\overline{\phi_{xy}(\tau)} = (1/4\pi) \int_{-\infty}^{\infty} \overline{S_{xy}(\omega)} \exp(j\omega\tau)\,d\omega \tag{2.79}$$

and

$$\overline{S_{xy}(\omega)} = 2 \int_{-\infty}^{\infty} \overline{\phi_{xy}(\tau)} \exp -(j\omega\tau)\,d\tau \ , \tag{2.80}$$

which is sometimes referred to as the generalized Wiener–Khintchine theorem. The bilateral form for the integrals is used here because of its brevity. It does not represent a conceptual difficulty – or indeed a practical one in connection with a measurement – concerning the meaning of negative frequencies, since the symmetry of the cross-spectral density expressed in equations (2.78) allows the right-hand side of equation (2.79) to be written as the sum of a sine and a cosine Fourier transform where the integrals are taken over only positive frequencies.

The normalized cross-spectral density is defined as

$$\Gamma_{xy}(\omega) = \overline{S_{xy}(\omega)}/[\overline{S_{xx}(\omega)}\,\overline{S_{yy}(\omega)}]^{\frac{1}{2}} \ , \tag{2.81}$$

where $\overline{S_{xx}(\omega)}$ and $\overline{S_{yy}(\omega)}$ are the power spectral densities of $x(t)$ and $y(t)$, respectively. $\Gamma_{xy}(\omega)$ expresses the mutual coherence between the two waveforms, and hence is often known as the coherence function. From the Cauchy–Schwartz inequality it follows that $|\Gamma_{xy}|$ falls in the interval $[0, 1]$, and that the real and imaginary parts fall in the interval $[-1, 1]$. The coherence function obeys the same symmetry relations as $\overline{S_{xy}(\omega)}$ (equations (2.78)), but of particular note is that it is Hermitian, satisfying the condition $\Gamma_{xy}(\omega) = \Gamma^*_{yx}(\omega)$.

2.13 LINEAR SYSTEMS

Two stationary processes, $x(t)$ and $y(t)$, can often be treated as the input and output of a linear system. The system might be, for example, a measuring instrument used in the observation of a noise process. Such a system, in which the principle of superposition is obeyed, is characterized by an *impulse response function*, $h(t)$, which causality requires to be zero for $t < 0$, or by the *system function*

$$H(j\omega) = \int_{-\infty}^{\infty} h(t) \exp - j\omega t \, dt \ , \qquad (2.82)$$

which is the Fourier transform of $h(t)$.

The output of the system, $y(t)$, is the convolution of the input, $x(t)$, with the impulse response function:

$$y(t) = x(t) \otimes h(t) = \int_{-\infty}^{\infty} x(t - \alpha) h(\alpha) \, d\alpha \ , \qquad (2.83)$$

which, on being Fourier transformed, becomes

$$Y(j\omega) = X(j\omega) H(j\omega) \ . \qquad (2.84)$$

Certain relationships between the statistical quantities characterizing the input and output noise processes can now be derived.

The expected value of the output is

$$\overline{y(t)} = \int_{-\infty}^{\infty} \overline{x(t - \alpha)} h(\alpha) \, d\alpha$$

$$= \overline{x(t)} \int_{-\infty}^{\infty} h(\alpha) \, d\alpha$$

$$= \overline{x(t)} H(0) \ . \qquad (2.85)$$

Thus the dc output is just the dc input times the gain of the system at zero frequency.

The expected value of the product $x(t + \tau)y(t)$ is the average

$$\overline{x(t + \tau)y(t)} = \int_{-\infty}^{\infty} \overline{x(t + \tau)x(t - \alpha)}\, h(\alpha)\, d\alpha$$

$$= \int_{-\infty}^{\infty} \overline{\phi_{xx}(\tau + \alpha)}\, h(\alpha)\, d\alpha \ . \tag{2.86}$$

The second integral here can be identified as the convolution of the autocorrelation function of the input, $\overline{\phi_{xx}(\tau)}$, with the impulse response function $h(-\tau)$; and since the right-hand side of equation (2.86) is independent of t, the left-hand side is equal to the cross-correlation function $\overline{\phi_{xy}(\tau)}$. Hence

$$\overline{\phi_{yx}(\tau)} = \overline{\phi_{xx}(\tau)} \otimes h(\tau) \ . \tag{2.87}$$

A similar argument shows that the autocorrelation function of the output is

$$\overline{\phi_{yy}(\tau)} = \overline{\phi_{xy}(\tau)} \otimes h(\tau) \ . \tag{2.88}$$

It can be inferred from these results that if the input to a linear system is a stationary process, then the output is also stationary.

The cross-spectral density between the input and the output is obtained by Fourier transforming both sides of equation (2.87). Bearing in mind the Wiener–Khintchine inversion integrals, we have

$$\overline{S_{xy}(\omega)} = \overline{S_{xx}(\omega)}\, H^*(j\omega) \ ; \tag{2.89}$$

and similarly, the spectral density of the output, obtained by transforming equation (2.88), is

$$\overline{S_{yy}(\omega)} = \overline{S_{xx}(\omega)}\, |H(j\omega)|^2 \ . \tag{2.90}$$

It follows from the definition of the coherence function in equation (2.81), that for the input and output processes $x(t)$ and $y(t)$,

$$\Gamma_{xy}(\omega) = H^*(j\omega)/|H(j\omega)| \ . \tag{2.91}$$

The modulus of this expression is unity, which is consistent with the fact that there is a full causal connection between the input and the output. If the system were itself producing noise, so that the output contained a random component fluctuating independently of the input, then the modulus of the coherence function would drop below unity.

In an observation of a noise waveform, the measuring instrument (detector) always has a non-zero response time or, equivalently, a finite bandwidth. Since certain frequencies in the input fluctuation will in general lie outside the passband of the instrument, the output noise waveform will not be a facsimile of the input. Thus the measuring instrument itself influences what is observed at the output, by acting as a filter.

As an example of the effect, suppose that the system function of a (linear) measuring instrument is

$$H(j\omega) = 1/(1 + j\omega\tau_m) , \tag{2.92}$$

where τ_m is the response time, and that the input fluctuation is a relaxation process whose relaxation time is τ_x, where $\tau_x \ll \tau_m$. Thus, the bandwidth of the input signal is very much greater than that of the detector. The power spectral density of $x(t)$ can be written as

$$\overline{S_{xx}(\omega)} = \frac{S_0}{(1 + \omega^2\tau_x^2)} , \tag{2.93}$$

where S_0 is independent of frequency, and hence from equation (2.90), the power spectral density of the output is

$$\overline{S_{yy}(\omega)} \simeq \frac{S_0}{(1 + \omega^2\tau_m^2)} , \tag{2.94}$$

where the term $(1 + \omega^2\tau_x^2)$ has been approximated as unity. The Wiener–Khintchine theorem gives the autocorrelation functions corresponding to these spectral densities as

$$\overline{\phi_{xx}(\tau)} = \frac{S_0}{4\tau_x} \exp - (|\tau|/\tau_x) \tag{2.95}$$

and

$$\overline{\phi_{yy}(\tau)} = \frac{S_0}{4\tau_m} \exp - (|\tau|/\tau_m) . \tag{2.96}$$

It is apparent from these expressions that, by filtering out the high frequency energy of the input waveform, the detector has *increased* the correlation time by the factor (τ_m/τ_x) and *reduced* the mean-square value by the reciprocal of this factor, (τ_x/τ_m).

2.14 PULSE-TRAIN PAIRS

The theorems for the autocorrelation function and the power spectral density of a random pulse train, discussed in §2.6, can be extended to give the cross-correlation function and the cross-spectral density between two such trains, provided there is a one-to-one correspondence between the pulses in each train. Suppose, for example, that two linear systems had a common input consisting of a random train of impulses, as illustrated in Fig. 2.6. If the impulse response functions of the systems were $f_1(t)$ and $f_2(t)$, then the output pulse trains would be

$$x_1(t) = \sum_{k=1}^{K} a_k f_1(t - t_k) \tag{2.97}$$

and

$$x_2(t) = \sum_{k=1}^{K} a_k f_2(t - t_k) \ , \tag{2.98}$$

Apart from the subscripts 1 and 2, the notation here is the same as that in §2.6. The one-to-one correspondence between the pulses in $x_1(t)$ and $x_2(t)$ is apparent from the formulation in questions (2.97) and (2.98).

From an argument analogous to that leading to Carson's theorem, the cross-spectral density between $x_1(t)$ and $x_2(t)$ is

$$\overline{S_{12}(\omega)} = 2\nu \overline{a^2} F_1(j\omega) F_2^*(j\omega) \ , \tag{2.99}$$

where ν is the mean rate of the pulses. $F_1(j\omega)$ and $F_2(j\omega)$ are the Fourier transforms of the shape functions $f_1(t)$ and $f_2(t)$. It follows from the inversion integral in equation (2.80) that the cross-correlation function is

$$\overline{\phi_{12}(\tau)} = \frac{\nu \overline{a^2}}{2\pi} \int_{-\infty}^{\infty} F_1(j\omega) F_2^*(j\omega) \exp(j\omega\tau) \, d\omega$$

$$= \nu \overline{a^2} \int_{-\infty}^{\infty} f_1(t + \tau) f_2(t) \, dt \ . \tag{2.100}$$

If the mean values of *both* processes are non-zero, then additional terms analogous to those in equations (2.42) and (2.44) must be included in these results.

According to the definition in equation (2.81), the coherence function between the two fluctuations is

$$\Gamma_{12}(\omega) = \frac{F_1(j\omega) F_2^*(j\omega)}{|F_1(j\omega)||F_2(j\omega)|} \ . \tag{2.101}$$

This has a modulus of unity, which could have been anticipated since the two processes are derived from a common source, and hence show a full causal connection.

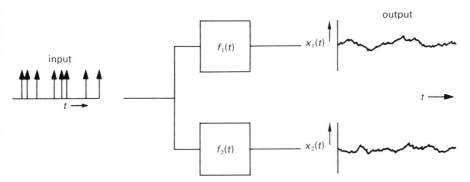

Fig. 2.6 – Two linear systems with impulse response functions $f_1(t)$ and $f_2(t)$ and a common random impulse train input, producing coherent pulse train outputs.

REFERENCES

D. A. Bell (1955), Distribution function of semiconductor noise, *Proc. Phys. Soc. B,* **68**, 690–691.

D. A. Bell (1960), *Electrical Noise,* Van Nostrand.

M. Born (1949), *Natural Philosophy of Cause and Chance,* O.U.P.

M. J. Buckingham and E. A. Faulkner (1974), The theory of inherent noise in p-n junction diodes and bipolar transistors, *The Radio and Elect. Eng.,* **44**, 125–140.

N. R. Campbell (1909), The study of discontinuous phenomena, *Proc. Camb. Phil. Soc.,* **15**, 117–136; (1910a), Discontinuities in light emission, pt 1', *Proc. Camb. Phil. Soc.,* **15**, 310–328; (1910b), Discontinuities in light emission, pt 2, *Proc. Camb. Phil. Soc.,* **15**, 513–525; (1939), the fluctuation theorem. (Shot effect), *Proc. Camb. Phil. Soc.,* **35**, 127–129.

N. R. Campbell and V. J. Francis (1946), A theory of valve and circuit noise, *J. Inst. Elect. Eng.,* **93**, pt III, 45–52.

S. Chandrasekhar (1943), Stochastic problems in physics and astronomy, *Rev. Mod. Phys.,* **15**, 1–89.

A. Einstein (1906a), Eine neue bestimmung moleküldimensionen (A new determination of molecular dimensions), *Ann. d. Phys.,* **19**, 289–305; (1906b), Zur theories der Brownschen bewegung (Theory of Brownian motion), *Ann. d. Phys.,* **19**, 371–379.

J. B. Johnson (1927a), Thermal agitation of electricity in conductors, *Nature,* **119**, 50–51; (1927b), Thermal agitation of electricity in conductors, *Phys. Rev.,* **29**, 367–368; (1928), Thermal agitation of electricity in conductors, *Phys. Rev.,* **32**, 97–109.

A. Khintchine (1934), Korrelationstheorie der stationaren stochastichen prozesses, *Math. Annalen,* **109**, 604–615.

D. G. Lampard (1954), Generalization of the Wiener–Khintchine theorem to non-stationary processes, *J. Appl. Phys.*, **25**, 802–803.

M. P. Langevin (1908), Sur la theorie du mouvement brownien, *Comptes Rend. Acad. Sci. Paris,* **146**, 530–533.

M. J. Lighthill (1958), *An Introduction to Fourier Analysis and Generalized Functions,* Cambridge University Press.

D. K. C. MacDonald (1949), Transit-time deterioration of space-charge reduction of shot effect, *Phil. Mag.*, **40**, 561–568.

D. Middleton (1960), *Introduction to Statistical Communication Theory,* McGraw-Hill.

H. Nyquist (1927), Thermal agitation in conductors, *Phys. Rev.*, **29**, 614; (1928), Thermal agitation of electric charge in conductors, *Phys. Rev.*, **32**, 110–113.

A. Papoulis (1965), *Probability, Random Variables and Stochastic Processes,* McGraw-Hill; (1968), *Systems and Transforms with Applications in Optics,* McGraw-Hill.

S. O. Rice (1944), Mathematical analysis of random noise, *Bell Syst. Tech. J.,* **23**, 282–332; (1945), **24**, 46–156.

E. N. Rowland (1936), The theory of the mean square variation of a function formed by adding known functions with random phase, and applications to the theories of the shot effect and of light, *Proc. Camb. Phil. Soc.,* **32**, 580–597.

W. Schottky (1918), Über spontane stromschwankungen in verschiedenen elektrizitatsleitern, *Ann. d. Phys. (Leipzig),* **57**, 541–567.

E. C. Titchmarsh (1937), *Introduction to the Theory of Fourier Integrals,* O.U.P.

G. E. Uhlenbeck and L. S. Ornstein (1930), On the theory of the Brownian motion, *Phys. Rev.*, **36**, 823–841.

N. Wiener (1930), Generalized harmonic analysis, *Acta Math.,* **55**, 117–258.

3

Noise in linear networks

3.1 INTRODUCTION

An inherently noisy electronic network (active or passive) can be represented as a *noise-free* network with external noise generators. These noise generators may in turn be represented as equivalent thermal noise generators, in which case it is possible to talk of an equivalent noise resistance or an equivalent noise conductance or an equivalent noise temperature, whichever is appropriate to the circuit under consideration.

The noise from a four-terminal network, or two-port, can be represented by two external generators which, in general, are partially correlated. The noise figure (or factor) is a figure of merit for such a network, expressing the intrinsic noisiness of the circuit, taking into account the gain of the system and impedance matching conditions at the input port.

3.2 TWO-TERMINAL NETWORKS

A noisy, two-terminal network with impedance $Z(\omega) = R(\omega) + jX(\omega)$ is illustrated in Fig. 3.1(a), where $v(t)$, the open-circuit voltage fluctuation between the terminals A–B, is due to one or more internal noise sources.† An equivalent circuit for the noisy network, derived from Thévenin's (1883) theorem and consisting of a noise-free network with impedance $Z(\omega)$ in series with a voltage generator $v(t)$, is shown in Fig. 3.1(b). The spectral density of $v(t)$ can be expressed as

$$\overline{S_v(\omega)} = 4k\theta R_n \ , \tag{3.1}$$

† It is difficult to avoid mixing functions of frequency (impedances) and functions of time (generators) in circuit diagrams of noisy networks. The impedances could be replaced by their corresponding impulse response functions, but such a representation is unfamiliar and unattractive because working in the time domain implies the use of convolution integrals; or the noise generators could be represented by their spectral densities, but this leads to a cumbersome notation which is not consonant with circuit analysis (particularly when two or more partially correlated generators are present in the circuit). Thus, a mixture of time-dependent and frequency-dependent quantities is used in the circuit diagrams of this and subsequent chapters, as exemplified in Fig. 3.1.

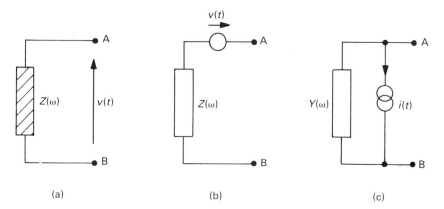

Fig. 3.1 – (a) A noisy two-terminal network with (b) its Thévenin equivalent circuit and (c) the dual of (b).

where θ is the absolute temperature of the circuit and R_n is the *equivalent (thermal) noise resistance* of the network. If the circuit is linear and passive, and contains only thermal noise sources at temperature θ, then $R_n = R(\omega)$; but in non-linear circuits, or circuits containing noise sources other than thermal noise sources, this equality may not hold.

The dual of the Thévenin circuit in Fig. 3.1(b) is shown in Fig. 3.1(c). In this equivalent circuit the noisy network is represented as a noise-free network with admittance $Y(\omega) = 1/Z(\omega) = G(\omega) + jB(\omega)$ in parallel with a current generator $i(t)$. The spectral density of $i(t)$ can be expressed as

$$\overline{S_i(\omega)} = 4k\theta G_n \ , \tag{3.2}$$

where G_n is the *equivalent (thermal) noise conductance* of the noisy network. If the network is linear and passive, and contains only thermal noise sources at temperature θ, then $G_n = G(\omega) = R/(R^2 + X^2)$; but in non-linear circuits, or circuits containing noise sources other than thermal noise sources, this equality may not hold.

In some problems, for example, those involving noise in hot-electron devices, where the equivalent temperature of the carriers differs from the ambient temperature, it may be convenient to express equations (3.1) and (3.2) in the alternative forms

$$\overline{S_v(\omega)} = 4k\theta_n R, \quad \theta_n = R_n\theta/R \tag{3.3}$$

and

$$\overline{S_i(\omega)} = 4k\theta_n G, \quad \theta_n = G_n\theta/G \ , \tag{3.4}$$

where θ_n is the *equivalent noise temperature*. The choice of formulation depends very much on the nature of the problem and the characteristics of the noise.

In circuits containing shot noise sources as the primary sources of noise it is often convenient to express the power spectral density of the current fluctuations at the terminals of the network in the form

$$\overline{S_i(\omega)} = 2q\xi I \tag{3.5}$$

where I is the terminal current and ξ is the *suppression factor*. Obviously, the suppression factor of full shot noise is unity, but if smoothing effects are significant, due for example to space-charge or trapping phenomena, then ξ falls below unity.

An example of a simple two-terminal network is provided by the parallel RC circuit, shown in Fig. 3.2(a). The thermal noise of the resistor, represented by the parallel current generator $i(t)$, transforms in the Thévenin equivalent circuit to a series voltage generator $v(t)$, as shown in Fig. 3.2(b). The spectral density of the voltage generator is

$$\overline{S_v(\omega)} = \overline{S_i(\omega)}\ R^2/(1 + \omega^2 C^2 R^2) \tag{3.6}$$

where $\overline{S_i(\omega)} = 4k\theta/R$ is the spectral density of the thermal noise generator $i(t)$. Notice that, although the circuit contains only a single thermal noise source, the voltage fluctuations at the terminals exhibit a power spectrum which is frequency dependent. This effect is due to the impedance of the capacitor. In general, circuit impedances affect the frequency dependence of the spectral density of the noise at the terminals of the network.

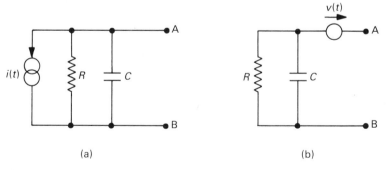

(a)   (b)

Fig. 3.2 – (a) Parallel RC combination with thermal noise current generators and (b) its Thévenin equivalent circuit.

From the Wiener–Khintchine theorem the mean-square value of the voltage generator in Fig. 3.2(b) is

$$\overline{v^2(t)} = \frac{1}{2\pi}\int_0^\infty \overline{S_v(\omega)}\, d\omega = \frac{4k\theta R}{2\pi}\int_0^\infty \frac{d\omega}{(1 + \omega^2 R^2 C^2)} = \frac{k\theta}{C}. \tag{3.7}$$

It is noteworthy that the resistance R does not affect the value of $\overline{v^2(t)}$, which depends only on the capacitance C and the temperature θ; but R is significant in deterninining the magnitude and the bandwidth of the spectral density given in equation (3.6).

3.3 LINEAR TWO-PORTS

A network with two pairs of terminals, constituting input and output ports, is variously known as a fourpole, a four-terminal network, or a two-port. A linear two-port is one in which the principle of superposition is obeyed. For small signals and noise a number of important electronic devices, including the bipolar transistor, the junction field effect transistor, the MOS transistor, the vacuum triode and the vacuum pentode, may be classified as linear two-ports; and multistage amplifiers also fall into the same category.

In the early days of the transistor little was known about the physical mechanisms responsible for producing the noise at the terminals of the device. Thus, it was impossible to predict the noise properties from theoretical argu-ments, and instead the linear two-port representation was used as the basis of an empirical treatment of the problem. Ryder and Kircher (1949) and Montgomery (1952) represented the noisy transistor as a noise-free circuit having the same impedance properties as the transistor itself, with two external voltage generators, one in series with the input and the other in series with the output, representing the noise. Their technique was essentially similar to that of Peterson (1947, 1948) who had previously discussed the noise of a tetrode in terms of parallel noise current generators at the input and output of a noise-free circuit.

It was recognized early on that a noisy two-port could be represented by a variety of equivalent circuits, differing only in the configuration of the noise generators at the ports of the (noise-free) network. In Montgomery's Fig. 2, for example, three different arrangements of the noise generators at the input and output ports are shown.

From the point of view of calculating the noise figure of a network, an equivalent circuit of particular interest is one in which both generators are referred to the input port. Becking et al. (1955) established the equivalence of such a configuration with representations in which a noise generator appears at each port; and they also showed that four quantities, namely the spectral densities of the two noise generators and the real and imaginary parts of the cross-spectral density between them, are sufficient to characterize completely the noise properties of a two-port at a single frequency. These four quantities can all be determined from measurements made at the terminals of the network, and thus it is possible to obtain a complete description of the noise characteristics of a two-port without knowing anything of the structure of the network or of the underlying physical mechanisms responsible for producing the noise. For the purpose of circuit design this is satisfactory, but if one wishes to investigate the relationship between the structure of a device and the noise at the terminals, a different approach must be used.

The currents and voltages at the terminals of a two-port are related to each other through a pair of linear equations. For a noiseless two-port,† that is, one containing no internal noise sources, these equations may be formulated in terms of the impedance matrix, $\mathbf{Z}$, of the network or in terms of the admittance matrix, $\mathbf{Y}$, as follows:

$$V_1 = Z_{11}I_1 + Z_{12}I_2$$
$$V_2 = Z_{21}I_1 + Z_{22}I_2 \qquad (3.8a)$$

$$I_1 = Y_{11}V_1 + Y_{12}V_2$$
$$I_2 = Y_{21}V_1 + Y_{22}V_2 \qquad (3.8b)$$

The upper case letters I and V in these equations indicate Fourier transforms or Fourier amplitudes, according to whether the terminal signals are periodic or aperiodic. The subscripts 1 and 2 refer to the input and output ports, respectively, and the sign convention used in constructing the equations is that currents flowing into the network are positive, as illustrated in Fig. 3.3. It should be noted that all the quantities in equations (3.8) are in general dependent on frequency.

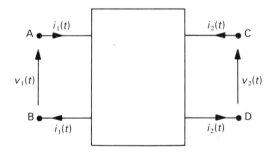

Fig. 3.3 – Schematic of a noiseless two-port showing the terminal currents and voltages.

When the two-port is not noiseless but contains internal noise generators, that is, when the noise level is comparable with the signal level at the ports, equations (3.8) must be modified to take account of the random fluctuations at the ports. This is achieved by the now familiar method of representing the noisy two-port (illustrated in Fig. 3.4(a)) as a noiseless network with externally placed noise generators. By an extension of Thévenin's theorem we can obtain the

† Such a network is fictitious since all circuits in reality produce some noise. However, when the signal levels are sufficiently high the effect of the noise is negligible and equations (3.8) are then valid.

equivalent circuit in Fig. 3.4(b) in which a series noise voltage gnerator appears at each of the two ports. In general, some degree of correlation will exist between these generators since the same physical mechanism may be responsible, at least in part, for the noise fluctuations at the two ports. The dual of Fig. 3.4(b) is the equivalent circuit shown in Fig. 3.4(c) in which the internal noise is represented by parallel current generators connected across the ports. Again, these two generators will in general show some degree of correlation.

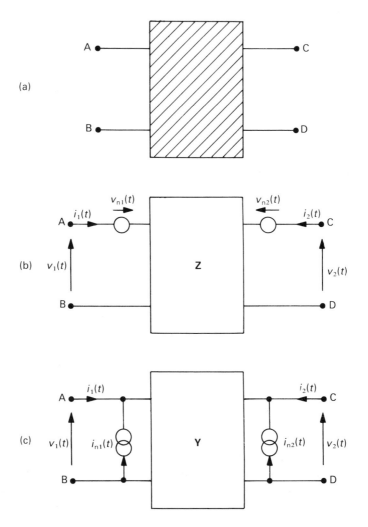

Fig. 3.4 — (a) Linear two-port with internal noise generators and (b) its Thévenin equivalent circuit with external series noise voltage generators. (c) The dual of the equivalent circuit in (b) with external parallel noise current generators.

When the effects of the series noise voltage generators in Fig. 3.4(b) are included we have, instead of equations (3.8a), the current/voltage relationships

$$V_1 = Z_{11}I_1 + Z_{12}I_2 - V_{n1}$$
$$V_2 = Z_{21}I_1 + Z_{22}I_2 - V_{n2} \quad , \tag{3.9a}$$

and similarly, when the effects of the parallel noise current generators in Fig. 3.4(c) are taken into account equations (3.8b) become

$$I_1 = Y_{11}V_1 + Y_{12}V_2 - I_{n1}$$
$$I_2 = Y_{21}V_1 + Y_{22}V_2 - I_{n2} \quad . \tag{3.9b}$$

The terms representing the noise in equations (3.9) are the Fourier transforms of the random time series constituting the current or voltage noise fluctuations at the ports. The question of whether such transforms exist has been discussed in the previous chapter (§2.3 and §2.5). It was concluded there that, although the integral transforms over an infinite interval do not converge, it is nevertheless legitimate to construct the power spectra and cross-spectra of stationary time series taken over a finite interval, provided that the ensemble and time averaging procedures are carried out in the correct order. Since the ultimate purpose of equations (3.9) is to enable expressions to be formulated for the power spectra and cross-spectra of the noise generators associated with the two-port, we may in this context safely refer to the 'Fourier transform' of a stationary random time series on the understanding that the transform is to be manipulated as outlined above. There is then no need for concern about the fact that stationary noise waveforms are not absolutely integrable.

Instead of placing a noise generator at each of the ports, it is often more convenient to refer both generators to the input port. On choosing the configuration shown in Fig. 3.5, where the two-port is characterized by its admittance matrix, $\mathbf{Y}$, and the noise is represented by a series voltage generator and a parallel current generator at the input, the current/voltage relationships for the network are

$$I_1 = Y_{11}(V_1 + V_{na}) + Y_{12}V_2 - I_{na}$$
$$I_2 = Y_{21}(V_1 + V_{na}) + Y_{22}V_2 \quad . \tag{3.10}$$

Analogous expressions can, of course, be written for the dual of the equivalent circuit in Fig. 3.5, in which case the network would have to be represented by its impedance matrix, $\mathbf{Z}$.

When the expressions in equations (3.10) are compared with those in equations (3.9b), the Fourier transform of the voltage generator, $v_{na}(t)$, is seen to be

$$V_{na} = -I_{n2}/Y_{21} \tag{3.11a}$$

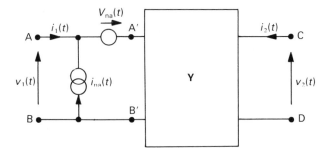

Fig. 3.5 – Representation of a noisy two-port in which the external noise generators are both referred to the input port.

and the transform of the current generator, $i_{na}(t)$, is

$$I_{na} = I_{n1} - (Y_{11}/Y_{21}) I_{n2} \ . \tag{3.11b}$$

Thus, through equations (3.11) the transforms of the noise generators at the input port in Fig. 3.5 can be related to the transforms of the noise current generators at the input and output ports in Fig. 3.4(c). It should be noticed, however, that the equivalent circuit in Fig. 3.5 is valid only for calculating the noise in the *output* circuit. It does not give a correct description of the noise fluctuations at the input port, as may be seen from the fact that $i_{na}(t)$ is different from the true current generator $i_{n1}(t)$. The arrangement of noise generators in Fig. 3.5 is particularly convenient for calculating the noise figure of the network.

Becking *et al.* (1955) have pointed out that equations (3.11) have meaning only if Y_{21} is non-zero, or in other words, only if the input is coupled to the output. This will always be the case with any practical electronic device or amplifier.

3.4 NOISE FIGURE OF A LINEAR TWO-PORT

Before the advent of the transistor the concept of a noise figure had been used in connection with noise in radio receivers (Friiss, 1944). It was not, however, widely used to characterize the noise of thermionic valves, mainly because a single noise generator was often sufficient to describe the noise properties of electron tubes. As we have already mentioned, this is not the case with transistors, whose noise behaviour is characterized by four parameters, and nor is it the case with high-frequency electron tubes. When these devices were developed it was soon appreciated that their noise characteristics could not be adequately represented by one generator and as a result several analyses appeared† in which the noise figure was introduced as a figure of merit quantifying the noise properties of four terminal networks.

† See for example Peterson (1947), Ryder and Kircher (1949), Montgomery (1952), Becking *et al.* (1955) and Rothe and Dalke (1956).

The noise figure of a two-port is defined for a specified frequency as the ratio

total output noise power per unit bandwidth
———
output noise power per unit bandwidth due to the input termination

at the standard temperature θ_0. Thus defined the noise figure, F, depends upon the input termination, that is to say the source admittance, Y_s, (see Fig. 3.6) but not on the output termination of the two-port. F also depends on frequency, and both frequency and Y_s must be specified when the noise figure of a system is quoted.

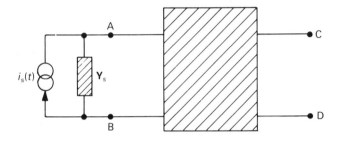

Fig. 3.6 – Signal source, $i_s(t)$, with (noisy) source admittance Y_s, connected to the input of a noisy two-port.

Usually the noise figure is expressed in decibels, with the value 0 dB corresponding to a noise-free system. In reality all electronic networks and devices produce some noise and noise figures in excess of 0dB will always be found in practice. Naturally, if a device is to be used in certain applications, such as in the input stage of a low noise amplifier, it is desirable for its noise figure to be as close to 0 dB as possible. At audio frequencies it is not uncommon to find silicon bipolar transistors with minimum noise figures of 0.5 dB or less (Faulkner and Harding, 1968), whilst junction field effect transistors can do even better with minimum noise figures as low as 0.02 dB (Robinson, 1969). This represents a remarkable improvement over the noise performance of the very early transistors. Ryder and Kircher (1949), for example, cited a representative noise figure of 60 dB at 1 kHz for a type A 'cat's-whisker' germanium transistor.† For comparison they pointed out that, at the same frequency, a good electron tube could have a noise figure of close to 0 dB. In view of these figures it is not surprising that in its infancy the transistor had the reputation of being a 'noisy' device!

† This was a point contact device in which two small phosphor-bronze filaments made contact with a block of germanium.

At a given frequency the noise figure of a linear two-port can be expressed in terms of the source admittance $Y_s = (G_s + jB_s)$ as follows (Haus *et al.* (1960a)):

$$F = F_0 + \frac{1}{G_s G_{nv}} |Y_s - Y_{s0}|^2 \quad , \tag{3.12}$$

where F_0 is the minimum noise figure that can be obtained at the specified frequency by adjustment of Y_s, and $Y_{s0} = (G_{s0} + jB_{s0})$ is the value of Y_s giving the minimum value of F. The remaining two parameters in equation (3.12) are the source conductance, G_s, and an equivalent noise conductance, G_{nv}. When the source is purely resistive it is easy to see that the curve of $(F - F_0)$ versus $(G_s - G_{s0})$ is a parabola, as sketched in Fig. 3.7, and that the slope of the curve depends on G_{nv}. Obviously the condition for the minimum in the noise figure in this case is simply $G_s = G_{s0}$.

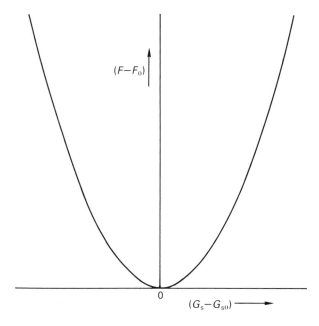

Fig. 3.7 – Sketch depicting the parabolic relationship between $(F - F_0)$ and $(G_s - G_{s0})$.

In general Y_s is complex and the source conductance, G_s, and the source susceptance, B_s, can be adjusted independently. It can be seen from equation (3.12) that $(F - F_0)$ then varies as $(G_s - G_{s0})^2$ or as $(B_s - B_{s0})^2$. In order to achieve the minimum noise figure two matching conditions must now be satisfied, namely $G_s = G_{s0}$ and $B_s = B_{s0}$; and the equivalent noise conductance

G_{nv} appears as a scaling factor in the slopes of the curves of $(F - F_0)$ versus $(G_s - G_{s0})$ and $(B_s - B_{s0})$.

The expression for the noise figure in equation (3.12) is derived from the equivalent circuit shown in Fig. 3.8. In this circuit both of the noise generators associated with the two-port have been referred to the input port, as in Fig. 3.5. As the noise fluctuations in the source admittance and the two-port are independent, the noise figure can immediately be written in the form

$$F = \overline{|I_{ns} + I_{na} + Y_s V_{na}|^2} / \overline{|I_{ns}|^2}$$

$$= 1 + \frac{\overline{S}_{ia}}{\overline{S}_{is}} + |Y_s|^2 \frac{\overline{S}_{va}}{\overline{S}_{is}} + 2\,\mathrm{Re}\,(\Gamma_{iv}\,Y_s^*) \left[\overline{S}_{ia} \cdot \overline{S}_{va}\right]^{\frac{1}{2}} / \overline{S}_{is} \quad , \qquad (3.13)$$

where $\overline{S}_{ia}$, $\overline{S}_{va}$ and $\overline{S}_{is}$ are the power spectra of the generators $i_{na}(t)$, $v_{na}(t)$ and $i_{ns}(t)$, respectively, and Γ_{iv} is the normalized cross-spectral density between $i_{na}(t)$ and $v_{na}(t)$. The term $Y_s V_{na}$ in the first of equations (3.13) arises from the transformation of the series voltage generator $v_{na}(t)$ in Fig. 3.8 into an equivalent parallel current generator, achieved by the application of Norton's theorem.

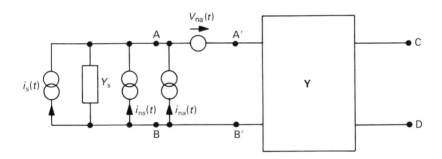

Fig. 3.8 – Circuit for calculating the noise figure of an amplifier with admittance matrix **Y**. The current generator $i_{ns}(t)$ represents the noise in the source admittance, Y_s.

The power spectral densities in equation (3.13) can be expressed in terms of equivalent thermal conductances as follows:

$$\overline{S}_{ia} = 4k\theta\,G_{ni}$$

$$\overline{S}_{va} = 4k\theta/G_{nv} \qquad\qquad (3.14)$$

$$\overline{S}_{is} = 4k\theta\,G_s \qquad .$$

G_{ni} and G_{nv} are not necessarily actual conductances, in the sense of being circuit elements, but are merely quantities which are representative of associated noise generators. In contrast, the conductance G_s specifying the noise of the source is the actual source conductance (cf. equation (3.2) above). Often G_{ni} and/or G_{nv} will depend on the biasing levels of a circuit, in which case the noise figure will be a function of the operating point of the network.

In treating the normalized cross-spectral density in equation (3.13) it is convenient to split the current generator $i_{na}(t)$ into two, one part of which is independent of $v_{na}(t)$ whilst the other part is fully correlated with $v_{na}(t)$. We then have

$$I_{na} = I_{nb} + Y_c V_{na} \quad , \tag{3.15}$$

where Y_c is a complex quantity having the dimensions of admittance, which Rothe and Dalke (1956) designate the *correlation admittance* of the generators $i_{na}(t)$ and $v_{na}(t)$. Since by definition $\overline{I_{nb} V_{na}^*}$ is zero, the normalized cross-spectral density between $i_{na}(t)$ and $v_{na}(t)$ is

$$\Gamma_{iv} \equiv \frac{\overline{I_{na} V_{na}^*}}{[\overline{|I_{na}|^2} \; \overline{|V_{na}|^2}]^{\frac{1}{2}}} = Y_c [\overline{|V_{na}|^2}/\overline{|I_{na}|^2}]^{\frac{1}{2}} = Y_c/(G_{ni} G_{nv})^{\frac{1}{2}} \quad . \tag{3.16}$$

The noise figure in equation (3.13) can now be expressed as

$$F = 1 + \frac{G_{ni}}{G_s} + \frac{\{(G_s + G_c)^2 + (B_s + B_c)^2 - (G_c^2 + B_c^2)\}}{G_{nv} G_s} \quad , \tag{3.17}$$

where G_c and B_c are the real and imaginary parts of the correlation admittance; that is, G_c and B_c are the correlation conductance and correlation susceptance, respectively.

The optimum source admittance, $Y_{s0} = (G_{s0} + jB_{s0})$, for which F is a minimum is found by differentiating equation (3.17) twice, once with respect to B_s and then with respect to G_s. This gives the optimum source susceptance as

$$B_{s0} = -B_c \tag{3.18a}$$

and the optimum source conductance as

$$G_{s0} = \{G_{nv} G_{ni} - B_c^2\}^{\frac{1}{2}} \quad . \tag{3.18b}$$

These conditions are referred to by Rothe and Dalke as *noise tuning* and *noise matching,* respectively. When both conditions are satisfied simultaneously the minimum noise figure is found to be

$$F_0 = 1 + \frac{2}{G_{nv}}(G_{s0} + G_c) \quad , \tag{3.19}$$

which, when combined with the expression for the noise figure in equation (3.17), gives

$$F = F_0 + \frac{\{(G_s - G_{s0})^2 + (B_s - B_{s0})^2\}}{G_{nv}\,G_s} . \tag{3.20}$$

This result is precisely the expression for the noise figure in equation (3.12).

It is apparent from the expressions given above that when Γ_{iv} is real the correlation susceptance is zero and hence the minimum noise figure can be achieved with a purely resistive source. This is obviously not the case when Γ_{iv} is complex, that is when the correlation susceptance is non-zero. Then noise tuning – which can be implemented at a *spot frequency* by shunting an inductance $L = 1/\omega B_c$ across a resistive source – can lead to a significant reduction in the noise figure of the two-port.

The four parameters F_0, G_{s0}, B_{s0} and G_{nv} in the expression for the noise figure in equation (3.20) completely characterize the noise fluctuations at the terminals of the two-port. Once known they may be used to evaluate G_{ni}, G_c and B_c from equations (3.18) and (3.19), and then it is a simple matter to determine the spectral densities of the generators $v_{na}(t)$ and $i_{na}(t)$ from equations (3.14) and the normalized cross-spectral density from equation (3.16). A procedure for measuring the four characteristic parameters has been described by Haus *et al.* (1960b). The method is to adjust the source conductance and susceptance until a minimum in the noise figure is achieved, at which point F_0, G_{s0} and B_{s0} are simply read off, and then G_{nv} is determined by making a measure of the noise figure for some non-optimum source admittance and using this measure in conjunction with equation (3.20) to calculate G_{nv}.

3.5 NOISE FIGURE OF AMPLIFIERS IN CASCADE

When several amplifiers are connected in cascade, as illustrated in Fig. 3.9, the overall noise figure of the system depends on the noise figures and the available power gains of the individual amplifiers in the cascade.

The available power gain of an amplifier is defined in terms of the *available power* of the signal source; and the available power of the source is defined as the power that can be drawn from the source when it is connected to a matched load. For a source consisting of an e.m.f. v_s in series with an internal (positive) resistance R_s the available power is

$$P = \frac{v_s^2}{4R_s}, \quad R_s > 0 . \tag{3.21}$$

The available power gain of an amplifier can now be defined as

$$\eta = P_{out}/P_s \tag{3.22}$$

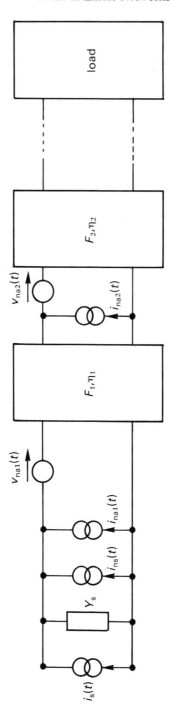

Fig. 3.9 – Noisy linear two-ports connected in cascade.

where P_{out} is the output power fed into a matched load and P_s is the available power of the source.

It is reasonable to suppose that each amplifier in the cascade is connected to a matched load, or in other words that the output and input admittances of adjoining amplifiers are equal. On representing the noise of the ith amplifier by equivalent noise generators $i_{nai}(t)$ and $v_{nai}(t)$ at its input, as in Fig. 3.5, the noise figure of the whole system can be written as

$$F = \frac{\overline{|I_{ns} + I_{na1} + Y_s V_{na1}|^2}}{\overline{|I_{ns}|^2}} + \frac{\overline{|I_{na2} + Y_1 V_{na2}|^2}}{\eta_1 \overline{|I_{ns}|^2}} + \frac{\overline{|I_{na3} + Y_2 V_{na3}|^2}}{\eta_1 \eta_2 \overline{|I_{ns}|^2}} + \dots, (3.23)$$

where Y_i and η_i are the input admittance and the available power gain of the ith amplifier. It follows immediately from equation (3.23) that, if F_i is the noise figure of the ith amplifier, the overall noise figure is

$$F = F_1 + (F_2 - 1)/\eta_1 + (F_3 - 1)/\eta_1 \eta_2 + \dots. \qquad (3.24)$$

Equation (3.24) is known as Friiss's formula. A cursory examination of the expression reveals that the noise figure of the system is essentially given by the noise figure of the first stage, provided that the available power gain of the first stage is sufficiently high. Thus it is important in the design of low-noise amplifiers to ensure that the input stage shows a noise figure close to unity (0 dB) and a power gain considerably greater than unity if a satisfactory performance is to be achieved.

If any of the amplifier stages in the cascade shows a negative output conductance certain difficulties are encountered with the formulation of the overall noise figure in equation (3.24). Haus and Adler (1957) have shown that the expression becomes indeterminate, the reason being that the available power of a source with a negative internal resistance is infinite. This is easily seen to be true by considering a source with negative internal resistance $-R_s$ feeding into a 'matched' load, R_s. In this case the infinite value obtained for the available power does not correspond to a turning point: it is neither an extremum nor a stationary value of the power output as a function of the terminal current. Haus and Adler overcome the difficulty with the noise figure arising from the definition of available power, by introducing the concept of *exchangeable power,* defined as the stationary value, or extremum, of the power output from the source obtained by arbitrary variation of the terminal current or voltage. This definition corresponds exactly to the definition of available power when the internal resistance is positive, but when the internal resistance is negative the exchangeable power (which is a turning point) remains finite but takes a negative value. This negative value indicates that, when the internal resistance is negative, the exchangeable power is the maximum power that can be pushed *into* the 'source' by connecting an active impedance to the terminals.

Haus and Adler have shown that Friiss's formula is valid in general, even when some of the amplifier stages in the cascade have negative output conductances, provided that the power gains η_i are interpreted as exchangeable power gains and the noise figures F_i are defined on the basis of exchangeable power rather than available power. If negative output conductances are present in the cascade it is possible for some of the η_i to be negative, and this could have a significant effect on the numerical value of the overall noise figure of the system.

REFERENCES

A. G. Th. Becking, H. Groendijk and K. S. Knol, (1955), The noise factor of four-terminal networks, *Philips Res. Rep.*, **10**, 349–357.

E. A. Faulkner and D. W. Harding (1968), Some measurements on low-noise transistors for audio-frequency applications, *The Radio and Elect. Eng.*, **36**, 31–33.

H. T. Friiss (1944), Noise figure of radio receivers, *Proc. IRE*, **32**, 419–423.

H. A. Haus and R. B. Adler (1957), An extension of the noise figure definition, *Proc. IRE*, **45**, 690–691.

H. A. Haus, *et al.* (1960a), Representation of noise in linear twoports, *Proc. IRE*, **48**, 69–74.

H. A. Haus, *et al.* (1960b), IRE standards on methods of measuring noise in linear twoports, 1959, *Proc. IRE*, **48**, 60–68.

H. C. Montgomery (1952), Transistor noise in circuit applications, *Proc. IRE*, **40**, 1461–1471.

L. C. Peterson (1947), Space-charge and transit-time effects on signal and noise in microwave tetrodes, *Proc. IRE*, **35**, 1264–1272.

L. C. Peterson (1948), Equivalent circuit analysis of active four terminal networks, *Bell Syst. Tech. J.*, **27**, 593–622.

F. N. H. Robinson (1969), Noise in common source amplifiers at moderately high frequencies, *Elect. Eng.*, **41**, 77–79.

H. Rothe and W. Dalke (1956), Theory of noisy fourpoles, *Proc. IRE*, **44**, 811–818.

R. M. Ryder and R. J. Kircher (1949), Some circuit aspects of the transistor, *Bell Syst. Tech. J.*, **28**, 367–400.

L. Thévenin (1883), Sur un nouveau théorème d'électricité dynamique, *Comptes Rend. Acad. Sci., Paris*, **97**, 159–161.

4

Inherent noise in junction diodes and bipolar transistors

4.1 INTRODUCTION

The publication of Shockley's (1949) paper on 'The theory of p–n junctions in semiconductors and p–n junction transistors' heralded a new era in the history of device physics. The transistor had arrived and the advantages of the new device, along with some of the problems, were quickly recognized.

One of the undesirable features of the early transistors was their high level of noise. This was first reported by Ryder and Kircher (1949) in a discussion which appeared in the same issue of the Bell System Technical Journal as Shockley's pioneering paper. They presented measurements of the noise over the frequency band 20 Hz – 20 kHz in a type A point contact transistor.† Their curves show the spectral density of the noise varying as $1/f^\alpha$, with $\alpha = 1.1$. A similar dependence on frequency was later reported by Wallace and Pietonpol (1951) and Montgomery (1952a) for a number of n–p–n 1752 germanium transistors: Montgomery found that over the frequency range 20 Hz – 50 kHz the spectral density of the noise at the emitter or the collector terminals for a variety of bias conditions obeyed a $1/f^\alpha$ law, with α a little greater than unity and usually about 1.2. He also showed that the level of the noise in the n–p–n transistors was considerably lower than that in the point contact transistor discussed by Ryder and Kircher, and quoted noise figures at 1 kHz of about 18–25 dB, depending on bias conditions, compared with 50–70 dB for the point contact devices.

Power spectra which show a $1/f^\alpha$ type of dependence with α close to unity, are characteristic of so-called $1/f$ noise. When this type of noise was first observed in transistors the underlying physical mechanism responsible for the fluctuation was not understood (Montgomery, 1952b), and even today, when $1/f$ noise is known to exist in a wide variety of devices and systems, there is no entirely satisfactory explanation of the phenomenon (see Chapter 6).

† See the footnote on page 63, Chapter 3.

As transistor technology improved the $1/f$ noise component of the noise diminished to such an extent that it was not long before the first measurements of the 'inherent' noise in a transistor were reported: Montgomery and Clark (1953) found that, at frequencies above about 1 kHz, depending on bias conditions, the $1/f$ noise in a p–n–p alloy transistor was insignificant and that the device showed a noise figure which was in agreement with a calculation based on thermal noise in the bulk regions and shot noise in the junctions. These basic physical mechanisms set a fundamental limit to the noise performance that can be expected of any transistor. In the device examined by Montgomery and Clark this limit corresponded to a noise figure of about 3 dB at frequencies of 1 kHz and above.

The first theoretical treatment of shot noise in junction diodes and transistors was proposed by van der Ziel (1955, 1957) on the basis of a distributed R–C transmission line analogy. He considered an 'ideal' diode – that is, one in which depletion layer recombination–generation and surface effects are negligible –and showed that the noise can be represented by a current generator in parallel with the junction whose spectral density depends on the terminal current and the junction conductance. Under strong forward or reverse bias and at low frequencies, van der Ziel's expression for the spectral density of this parallel noise current generator reduces to the simple shot noise form $2qI_D$, where I_D is the steady terminal current through the diode and q is the magnitude of the electronic charge; and for zero applied bias ($I_D = 0$) it simplifies to the Nyquist formula for thermal noise in a passive component with conductance equal to the junction conductance.

Van der Ziel's analysis was quickly followed by a number of theoretical and experimental investigations of noise in junction diodes and transistors: van der Ziel and Becking (1958) produced a theory based on the so-called 'corpuscular approach'; measurements of shot noise in transistors, showing good agreement with van der Ziel's theory, were reported by Hanson and van der Ziel (1957); Guggenbuehl and Strutt (1957) presented theoretical and experimental results on shot noise in semiconductor junction diodes and transistors; and Schneider and Strutt (1959) showed that silicon diodes and transistors exhibit a noise component due to carrier generation and recombination in the depletion layer. It took almost ten years from the time of these last measurements for a satisfactory explanation of generation–recombination noise to appear. Then Lauritzen (1968) proposed a theory based upon the Hall–Shockley–Read (1952) model of carrier recombination and generation through centres with a single energy level in the forbidden gap of the semiconductor. The spectrum of the generation–recombination noise is usually close to that of the shot noise, and Lauritzen makes a comment which underscores the fact that the original formulations for the noise in germanium transistors can be applied to silicon transistors by introducing minor modifications to account for the variation of gain with bias current.

By the time of Lauritzen's paper, silicon technology was well-established and silicon planar transistors were readily available from normal commercial sources. The noise performance of a random selection of such devices was reported by Faulkner and Harding (1968), who found minimum noise figures at audio frequencies consistently better than 0.5 dB. The battle to produce transistors showing good low-noise characteristics had apparently been won.

However, there was still the problem of excess noise at low and intermediate frequencies to investigate. This excess noise appears in p–n junction diodes, bipolar transistors and integrated circuits, and consists of two components (besides carrier recombination–generation noise), namely $1/f$ noise and burst noise. $1/f$ noise is discussed in Chapter 6 and burst noise in Chapter 7, whilst in the present chapter we shall confine our attention to inherent noise (including generation–recombination noise) in junction diodes and transistors.

A theory of junction noise was proposed by Buckingham and Faulkner (1974) in an attempt to reconcile the noise behaviour of junction devices with the accepted physical principles of device operation. The essential mechanism invoked in the theory is charge-carrier diffusion arising from local, random fluctuations in the carrier population. In several respects the diffusion theory of noise differs from the theory of van der Ziel (1955, 1957) – one in particular being that it does not involve a transmission line analogy – although both theories ultimately lead to the same results. On the other hand, the 'corpuscular theory' of van der Ziel and Becking (1958) appears to be based on a misinterpretation of the noise due to carriers crossing the depletion layer, making it incompatible with the diffusion theory. In response to this objection, van der Ziel (1975a) claimed that it 'must be based on a misunderstanding', and subsequently in a review paper van der Ziel and Chenette (1978) restated the corpuscular theory in much the same form as it appeared originally.

The diffusion theory of noise is described below. Throughout the discussion the fundamental differences between the diffusion theory and previously held theories are indicated, and emphasis is placed on the connection between the noise and the physics of p–n junctions. Robinson (1974) also gives an account of the diffusion theory of noise in a treatment which emphasizes the physical significance of the carrier fluxes crossing the junction.

4.2 THE DIFFUSION THEORY OF NOISE IN A P–N JUNCTION

An 'ideal' junction diode is one in which depletion layer recombination–generation and surface effects are negligible, and whose current/voltage characteristic is accurately described by the Shockley (1949) equation

$$I_D = I_s \left\{ \exp\left(\frac{qV}{k\theta}\right) - 1 \right\} , \tag{4.1}$$

where I_D is the terminal current, I_s is the reverse saturation current, q is the magnitude of the electronic charge, and V is the applied voltage. The noise in such a diode can be represented by a current generator, $i_n(t)$, in parallel with the junction whose spectral density is

$$\overline{S_i(\omega)} = 2q(I_D + 2I_s) + 4k\theta\,(G_j - G_0) \tag{4.2}$$

where G_j is the conductance of the junction and

$$G_0 = \frac{dI_D}{dV} = \frac{q}{k\theta}(I_D + I_s) \tag{4.3}$$

is the low frequency value of G_j.

The formula in equation (4.2), which is van der Ziel's (1955, 1957) expression for the noise in an ideal p–n junction, has certain essential properties: at low frequencies under strong forward and reverse bias it reduces to the familiar shot noise forms $2qI_D$ and $2qI_s$, respectively, and for zero applied bias ($I_D = 0$) it simplifies to $4kG_j$, which is the Nyquist expression for the spectral density of the thermal noise in any passive component of conductance G_j in equilibrium with its environment.

A simple but erroneous explanation of equation (4.2) is as follows. It would appear from equation (4.1) that there are two currents, $(I_D + I_s)$ and I_s, flowing in opposite directions across the junction due to minority carrier recombination and generation, respectively, in the bulk regions. If these currents fluctuate independently and each shows full shot noise, then the total noise is $\{2q(I_D + I_s) + 2qI_s\} = 2q(I_D + 2I_s)$, which is the low frequency term on the right of equation (4.2). It must be emphasized that this argument is incorrect; it does not describe the mechanism that is actually responsible for generating the noise and, as pointed out by van der Ziel (1955), the fact that it leads to the correct result is accidental. (In fact, the result is only partially correct since it does not include the high frequency term which appears in equation (4.2).)

According to Shockley's (1949) theory of junction action the currents that actually flow across the junction, I_F and I_R, illustrated in Fig. 4.1, are very much larger than the terminal current, by a factor approximately equal to the ratio of the diffusion length to the mean free path length. The terminal current, I_D, is the difference between I_F and I_R and it is relatively so small that in calculations of the carrier distributions across the junction it is usually set to zero. Under forward bias, this corresponds to the assumption that the quasi-Fermi levels are constant across the depletion layer and separated by the applied voltage.

The role of these large fluxes of carriers crossing the junction is to maintain the equilibrium between the majority carrier population on one side of the transition region and the minority carriers on the other side. The flux of majority carriers consists of those carriers which are incident on the junction, due to their

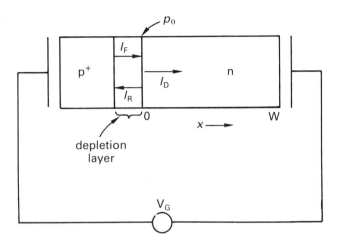

Fig. 4.1 – Current flows I_F and I_R sketched for a p^+–n junction.

random motion in the bulk material, *and* have sufficient energy to cross it; while the flux of minority carriers in the opposite direction consists of *all* those minority carriers which are incident on the junction due to their random motion in the bulk material. A statistical argument, analogous to that used in the kinetic theory of gases for establishing the pressure in an ideal gas, can be used to determine I_F and I_R. For a one-dimensional, p^+-n junction – in which electron flow can be safely neglected – the result is

$$\left.\begin{aligned} I_F &= (qp_nDA/\bar{l}_f)\exp{(qV/k\theta)} \\ I_R &= (qp_0DA/\bar{l}_f) \end{aligned}\right\} \quad , \tag{4.4}$$

where A is the area of the junction, V is the applied voltage, D is the diffusion constant for holes, p_n is the equilibrium concentration of holes on the n-side of the junction and $\bar{l}_f$ is the mean free path length. The hole concentration p_0 is the density of holes at the plane $x = 0$, illustrated in Fig. 4.1, representing the edge of the depletion layer on the n-region side of the junction. In the theory of junction action it is usually assumed that p_0 follows the applied voltage exactly according to the relation

$$p_0 = p_n \exp{(qV/k\theta)} \quad . \tag{4.5}$$

Equation (4.5) is an expression of the equilibrium that is maintained between the hole concentrations on either side of the transition region. However, it is clear from equations (4.4) and (4.5) that, if a net current $I_D = (I_F - I_R)$ is to flow, equation (4.5) cannot be precisely true and that V should be replaced by

$(V - \Delta V)$, where $\Delta V \ll k\theta/q$. By expanding I_R to first order in ΔV we then find from equation (4.4) that

$$\frac{\Delta V}{I_D} = \frac{k\theta}{qI_F} \, . \tag{4.6}$$

The quantity in equation (4.6) can be interpreted as a resistance in series with the junction, whose value, namely $k\theta/qI_F$, is very much less than the low frequency incremental resistance of the junction, which is equal to $k\theta/q(I_D + I_s)$. This effective series resistance represents the relaxation mechanism by which equilibrium is restored after a carrier crosses the depletion layer. It is so small that in impedance calculations it is normally neglected.

It has recently been shown (Buckingham and Faulkner, 1974) that the noise due to carriers crossing the depletion layer is exactly equivalent to the thermal noise in this effective series resistance. Being so small, this resistance produces a proportionately small amount of thermal noise which is negligible compared with the total noise produced within the device. On the basis of this argument it is clear that shot noise in the carrier fluxes crossing the junction is not the mechanism responsible for the noise in p–n junction diodes. This conclusion is borne out by the discussion in Appendix 4 where an analysis of noise due to carriers crossing the depletion layer is given and which, for comparison, includes an interpretation of the corpuscular theory of van der Ziel and Becking (1958).

There are two mechanisms responsible for the noise represented by equation (4.2), namely thermal fluctuations in the minority carrier flow and minority carrier recombination, both in the *bulk regions* close to the edges of the depletion layer. At first it may appear strange that fluctuations in the terminal current – whose constituents are the minority carrier diffusion currents at the two planes bounding the depletion layer – should originate in regions of the device extending away from the junction. The explanation of this apparent anomaly has to do with the minority carrier relaxation mechanism by which equilibrium is restored after the minority carrier distribution has been perturbed. Far from the junction, the minority carrier relaxation does not produce a flow in the external circuit because it is entirely compensated by a majority carrier flow. (The discussion of majority carrier relaxation in § 2.8 shows that this contributes to the thermal noise of the bulk resistance of the device, which we shall assume is negligible compared with the incremental resistance of the junction.) But in the vicinity of the junction, a disturbance in the minority carrier distribution results in a change in the *gradient* of the distribution at the edge of the depletion layer, which means that, in this case, the minority carrier relaxation also involves a flow from across the junction, giving rise to an accompanying flow of charge around the circuit. The contributions to this external flow from all the minority-carrier disturbances in the bulk material manifest themselves as the observed noise in the terminal current.

In the discussion to follow we examine the two components of junction noise in more detail. Since it would introduce unnecessary duplication if holes in the n-region and electrons in the p-region were to be treated separately, we consider a p^+-n junction in which only holes contribute significantly to the current. The extension of the argument to the general situation of a more or less symmetrical junction is obvious and in both cases the results take exactly the same form. A one-dimensional model of the junction is assumed, as illustrated in Fig. 4.2, and potential drops in the bulk regions are ignored. The voltage generator in the figure produces a voltage V across the terminals of the device which maintains an enhanced hole concentration at the $x = 0$ plane given by equation (4.5). At the other boundary of the n-region, at $x = W$, the hole concentration is maintained at p_W by either a metal contact or possibly a second junction. It is important to note that p_0 and p_W are held fixed, and that minority carrier fluctuations occur only *within* the n-region and not on the boundaries at $x = 0$ and $x = W$.

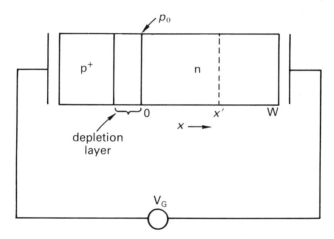

Fig. 4.2 – One-dimensional model of a p^+–n junction (not shown to scale). The plane $x = x'$ is the location at which an event occurs.

4.2.1 Thermal fluctuations in minority carrier flow

The effect of thermal motion among the minority carriers in the n-region is to cause a departure from the unperturbed hole distribution. This leads to relaxation hole currents across the junction and within the bulk material which tend to restore the hole distribution to its unperturbed shape. It is these relaxation currents with which we are now concerned. The technique we shall employ is analogous to Langevin's analysis of thermal noise which was based on an argument involving majority carrier relaxation: in our case the terminal current pulse from a single minority carrier 'event' is calculated on the supposition that the overall noise can be represented as a random superposition of such pulses.

Consider a single 'initial action' consisting of a minority carrier free path of length l_f at $x = x'$. This transit perturbs the minority carrier distribution as illustrated in Fig. 4.3(a), where p' represents the departure from the steady state hole concentration. Since the boundary conditions are fixed, the value of p' at either end of the n-region is zero, whilst elsewhere p' is distributed in a way which is consistent with the excess concentrations p'_1 at $x = x'$ and p'_2 at $x = x' + l_f$.

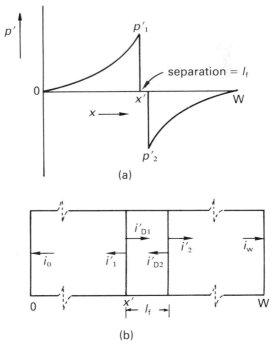

Fig. 4.3 – (a) Departure from the unperturbed minority carrier concentration due to an event consisting of a minority carrier path of length l_f at $x = x'$, and (b) the corresponding relaxation currents.

The initial action is equivalent to a current flow $q\delta(t)$ from right to left (Fig. 4.3(a)) over the path of length l_f. The minority carrier relaxation which follows the initial action occurs mainly by direct return flows i'_{D1} and i'_{D2} and also by flows i'_1 and i'_2 towards $x = 0$ and $x = W$, respectively (Fig. 4.3(b)). These currents, and those at the boundaries, are found by solving the time-dependent diffusion equation (Shockley, 1949) for p' in the three regions shown in Fig. 4.3(b), and substituting these solutions into the equation

$$i = -qDA \frac{\partial p}{\partial x} , \tag{4.7}$$

which is then evaluated at the appropriate value of x.

The time-dependent diffusion equation is

$$\frac{\partial p}{\partial t} = -\frac{(p-p_n)}{\tau_R} + D\frac{\partial^2 p}{\partial x^2} \, , \tag{4.8a}$$

where τ_R is the hole lifetime in the n-region. On Fourier transforming equation (4.8a) and removing the steady-state component of p, we obtain

$$\frac{\partial^2 p'(j\omega)}{\partial^2 x^2} - \frac{p'(j\omega)}{L^2} = 0, \tag{4.8b}$$

where

$$L = \{D\tau_R/(1+j\omega\tau_R)\}^{\frac{1}{2}} = L_0(1+j\omega\tau_R)^{-\frac{1}{2}} \tag{4.9}$$

and $L_0 = \sqrt{D\tau_R}$ is the low frequency diffusion length. For convenience later we express the frequency dependent term in this expression in the form

$$(1+j\omega\tau_R)^{\frac{1}{2}} = (a+jb) \, , \tag{4 10a}$$

where

$$\begin{aligned}
a &= [\tfrac{1}{2}+\tfrac{1}{2}\sqrt{(1+\omega^2\tau_R^2)}]^{\frac{1}{2}} \\
b &= [-\tfrac{1}{2}+\tfrac{1}{2}\sqrt{(1+\omega^2\tau_R^2)}]^{\frac{1}{2}}.
\end{aligned} \tag{4.10b}$$

In transform terminology, the return currents i'_{D1} and i'_{D2} obtained from equations (4.7) and (4.8) are

$$i'_{D1}(j\omega) = \frac{q\,DA}{L}\left[\frac{p_1(j\omega)\cosh\,(l_f/L)-p'_2(j\omega)}{\sinh\,(l_f/L)}\right] \tag{4.11a}$$

and

$$i'_{D2}(j\omega) = -\frac{q\,DA}{L}\left[\frac{p'(j\omega)-p'_2(j\omega)\cosh\,(l_f/L)}{\sinh\,(l_f/L)}\right] . \tag{4.11b}$$

Bearing in mind that for all frequencies of interest $|l_f/L| \ll 1$, it is easily shown that

$$i'_{D1}(j\omega) = -i'_{D2}(j\omega) = i'_D(j\omega) = \frac{q\,DA}{l_f}[p'_1(j\omega)-p'_2(j\omega)]. \tag{4.12}$$

The flows i_1' and i_2' away from the region of the event are similarly found from equations (4.7) and (4.8) to be

$$i_1'(j\omega) = -A k_1(j\omega) p_1'(j\omega) \tag{4.13a}$$

and

$$i_2'(j\omega) = A k_2(j\omega) p_2'(j\omega), \tag{4.13b}$$

where

$$k_1 = \frac{qD}{L} \coth(x'/L) \tag{4.14a}$$

and

$$k_2 = \frac{qD}{L} \coth[(W - x')/L] \ . \tag{4.14b}$$

Since there can be no accumulation of charge at any point in the n-region, we must have current continuity at $x = x'$ and $x = x' + l_f$. In transform terminology this is expressed through the conditions

$$i_1'(j\omega) + i_D'(j\omega) - q = i_2'(j\omega) + i_D'(j\omega) - q = 0, \tag{4.15}$$

and when these expressions are combined with equations (4.12) and (4.13), remembering that $|k_1|$ and $|k_2|$ are both very much less than (qD/l_f), we find that

$$i_1'(j\omega) = i_2'(j\omega) = -\frac{l_f}{D} \frac{k_1 k_2}{(k_1 + k_2)} \ . \tag{4.16}$$

Now we are primarily interested in the flows i_0 and i_W across the boundaries of the n-region. These are given from the time-dependent diffusion equation in conjunction with equations (4.13) and (4.16) as follows:

$$i_0'(j\omega) = -A k_0(j\omega) p_1'(j\omega) = -\frac{l_f}{D} \cdot \frac{k_0 k_2}{(k_1 + k_2)} \tag{4.17a}$$

and

$$i_W'(j\omega) = -A k_W(j\omega) p_2'(j\omega) = +\frac{l_f}{D} \frac{k_1 k_W}{(k_1 + k_2)} , \tag{4.17b}$$

where

$$k_0 = \frac{qD}{L} \operatorname{cosech}(x'/L) \tag{4.18a}$$

and

$$k_W = \frac{qD}{L} \operatorname{cosech}[(W - x')/L] \ . \tag{4.18b}$$

Equations (4.17) are the transforms of the current pulses at the boundaries due to a single minority carrier event in the n-region. (Note that the signs are such that currents flowing into the n-region are positive.) The power spectra of the fluctuations at the boundaries due to a random succession of events in an element of length $\Delta x'$ in the n-region are, from Carson's theorem (see § 2.6, equation (2.41)),

$$\Delta \overline{S'_{i_0}(\omega)} = \frac{4pA}{D} \left| \frac{k_0 \, k_2}{k_1 + k_2} \right|^2 \Delta x' \tag{4.19a}$$

and

$$\Delta \overline{S'_{iw}(\omega)} = \frac{4pA}{D} \left| \frac{k_1 \, k_w}{k_1 + k_2} \right|^2 \Delta x' \; , \tag{4.19b}$$

where it is implicit that the events are all indpendent. In deriving equations (4.19) we have taken the mean number of events per second as $pA\Delta x'/\overline{\tau}_f$ and, from equation (2.57), have employed the condition $D = \mu k\theta/q = \overline{l_f^2}/2\overline{\tau}_f$. By integrating these equations we obtain

$$\overline{S'_{i_0}(\omega)} = \frac{4A}{D} \int_0^W p \left| \frac{k_0 k_2}{k_1 + k_2} \right|^2 dx' \tag{4.20a}$$

and

$$\overline{S'_{iw}(\omega)} = \frac{4A}{D} \int_0^W p \left| \frac{k_1 k_w}{k_1 + k_2} \right|^2 dx' \tag{4.20b}$$

for the power spectra of the current fluctuations at the boundaries due to all the events in the n-region.

For the moment we shall not evaluate the integrals in equations (4.20) but will return to them shortly in connection with the discussion of the total noise in the ideal diode. Meanwhile, the significance of the two expressions should be borne in mind.

It is apparent from equation (4.16) that the thermal motion causes a current flow *through* the material, rather than a fluctuation in carrier concentration. Equivalently, we could say that there is a fluctuation in the *difference*, $\Delta p(t)$, in carrier concentration at the planes $x = x'$ and $x = x' + \Delta x'$, for, from equations (4.13) and (4.16) we have

$$\overline{\Delta p^2 (j\omega)} = \frac{2p\overline{l_x^2}}{D^2 \overline{\tau}_f} \Delta x' = \frac{4p}{AD} \Delta x' \tag{4.21}$$

as $\Delta x'$ approaches zero.

A similar but not identical formula to equation (4.21) was introduced by van der Ziel (1955) in his discussion based on the transmission line analogy. In his version he had on the left-hand side the mean-square value of the (transform of the) carrier concentration itself at the plane $x = x'$. It is difficult to reconcile van der Ziel's formulation with the diffusion analysis since, according to the latter, the mean-square fluctuation in minority carrier concentration at any point in the n-region can only be obtained by an integration across the whole region, taking into account the boundary conditions at $x = 0$ and $x = W$. The result of this integration would be a function of W and the boundary concentrations p_0 and p_W, none of which appear on the right-hand side of equation (4.21).

4.2.2 Bulk region recombination–generation noise
When a recombination or generation event occurs there is no change in the overall charge distribution and hence there is no majority carrier relaxation. There is, however, a perturbation in the minority carrier distribution, which is removed by minority carrier flows away from the vicinity of the disturbance.

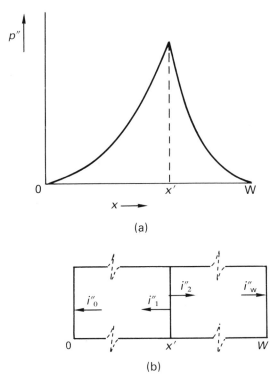

(a)

(b)

Fig. 4.4 – (a) Departure from the equilibrium minority carrier concentration due to a recombination event at the $x = x'$ plane, and (b) the corresponding relaxation current flows.

The perturbation, or initial action, is the instantaneous appearance (or disappearance) of a minority carrier, which is equivalent to a flow $q\delta(t)$ from 'nowhere' into the x' plane. This causes a departure from the unperturbed hole distribution, as sketched in Fig. 4.4(a) for a hole generation event. Figure 4.4(b) shows the minority carrier currents, i''_1 and i''_2, which flow towards the boundaries at $x = 0$ and $x = W$, respectively, as a result of the event. These currents are

$$i''_1(j\omega) = -A k_1(j\omega)p''(j\omega) \tag{4.22a}$$

and

$$i''_2(j\omega) = A k_2(j\omega)p''(j\omega) \quad, \tag{4.22b}$$

where p'' is the excess hole concentration at $x = x'$. Like equations (4.13), these expressions are derived from the time-dependent diffusion equation. Since there can be no accumulation of charge in the material, we must have current continuity at $x = x'$, which requires that

$$i''_1(j\omega) - i''_2(j\omega) + q = 0 \;. \tag{4.23}$$

From the three equations given above the excess hole concentration at $x = x'$ is

$$p''(j\omega) = q/A(k_1 + k_2) \;, \tag{4.24}$$

and hence in an analogous fashion to equations (4.17) we have for the minority flows across the boundaries

$$i''_0(j\omega) = -A k_0(j\omega)p''(j\omega) = -q \frac{k_0}{(k_1 + k_2)} \tag{4.25a}$$

and

$$i''_W(j\omega) = -A k_W(j\omega)p''(j\omega) = -\frac{q\, k_W}{(k_1 + k_2)} \;. \tag{4.25b}$$

As in equations (4.17), the signs in equations (4.25) have been chosen to conform with the convention that currents flowing into the n-region are positive.

If τ_R is the minority carrier lifetime in the n-region, then the number of recombination events per second in a volume element $A\Delta x'$ is $pA\Delta x'/\tau_R$; and the number of generation events per second is $p_n A\Delta x'/\tau_R$ — which of course equals the recombination rate when p takes its equilibrium value of p_n. When these recombination/generation rates and equations (4.25) are substituted into Carson's theorem we find that the power spectra of the noise currents at the boundaries from an element of length $\Delta x'$ are

$$\Delta \overline{S''_{i_0}(\omega)} = \frac{2(p + p_n)q^2 A}{\tau_R} \left| \frac{k_0}{k_1 + k_2} \right|^2 \Delta x' \tag{4.26a}$$

and

$$\overline{\Delta S_{i_W}''(\omega)} = \frac{2(p + p_n)q^2 A}{\tau_R} \left| \frac{k_W}{k_1 + k_2} \right|^2 \Delta x' \, . \tag{4.26b}$$

The total recombination–generation noise found by integrating these expressions over the whole region is

$$\overline{S_{i_0}''(\omega)} = \frac{2q^2 A}{\tau_R} \int_0^W (p + p_n) \left| \frac{k_0}{k_1 + k_2} \right|^2 dx' \tag{4.27a}$$

and

$$\overline{S_{i_W}''(\omega)} = \frac{2q^2 A}{\tau_R} \int_0^W (p + p_n) \left| \frac{k_W}{k_1 + k_2} \right|^2 dx' \, . \tag{4.27b}$$

4.2.3 Total noise in the junction
The spectra of the total noise fluctuations at the boundaries of an ideal diode are sums of the expressions in equations (4.20) and (4.27):

$$\overline{S_{i_0}(\omega)} = \overline{S_{i_0}'(\omega)} + \overline{S_{i_0}''(\omega)} \tag{4.28a}$$

and

$$\overline{S_{i_W}(\omega)} = \overline{S_{i_W}'(\omega)} + \overline{S_{i_W}''(\omega)} \, . \tag{4.28b}$$

Straightforward summing of the power spectra is permissible here since recombination–generation events and thermal flow events are independent.

The noise current in the external circuit of the ideal diode equals the minority carrier noise current at the $x = 0$ plane; that is, its power spectrum is given by the expression in equation (4.28a). The noise at $x = W$, given by equation (4.28b), is not in general the same, the difference being made up by a majority carrier flow through the contact.

In general, the integrands in equations (4.28) involve combinations of hyperbolic functions, making the integrations laborious and not very edifying. Instead of evaluating these power spectra for the general case, we now examine two special cases of practical importance, namely the long diode ($W \gg L_0$) and the short diode ($W \ll L_0$).

4.2.4 The long diode
From equations (4.20a), (4.27a) and (4.28a) the total noise in the long diode is

$$\overline{S_{i_0}(\omega)} = \frac{4A}{D} \int_0^\infty p \left| \frac{k_0 k_2}{k_1 + k_2} \right|^2 dx' + \frac{2q^2 AD}{L_0^2} \int_0^\infty (p + p_n) \left| \frac{k_0}{k_1 + k_2} \right|^2 dx' \, , \tag{4.29}$$

where the upper limit of infinity on the integrals is consistent with the condition $W \gg L_0$. The hole distribution in the long n-region, obtained from a solution of the diffusion equation, takes the form of an exponential decay:

$$(p - p_n) = (p_0 - p_n) \exp(-x'/L_0) \tag{4.30a}$$

and likewise, according to the definitions of the k functions, the remaining position dependent terms in the integrands of equation (4.29) can also be expressed as decaying exponentials:

$$\frac{k_0 k_2}{(k_1 + k_2)} = \frac{qD}{L} \exp(-x'/L) \tag{4.30b}$$

and

$$\frac{k_0}{(k_1 + k_2)} = \exp(-x'/L) . \tag{4.30c}$$

On substituting these three expressions into equation (4.29) and performing the integrations the spectral density of the noise is found to be

$$\overline{S_{i_0}(\omega)} = \frac{4q^2 AD}{L_0}(a^2 + b^2)\left[\frac{(p_0 - p_n)}{(2a + 1)} + \frac{p_n}{2a}\right] + \frac{2q^2 AD}{L_0}\left[\frac{(p_0 - p_n)}{(2a + 1)} + \frac{p_n}{a}\right], \tag{4.31}$$

where a and b are frequency dependent terms defined through equations (4.10).

Now the steady terminal current in the long diode, obtained from the diffusion equation, is

$$I_D = \frac{qAD}{L_0}(p_0 - p_n) \tag{4.32a}$$

and the reverse saturation current is

$$I_s = \frac{qAD}{L_0}p_n . \tag{4.32b}$$

The low frequency conductance of the junction is given in terms of I_D and I_s as

$$G_0 = \frac{q}{k\theta}(I_D + I_s) \tag{4.33a}$$

and, from the time-dependent diffusion equation, the conductance of the junction is

$$G_j = a G_0 . \tag{4.33b}$$

When equations (4.32) and (4.33) are combined with (4.31) we find, after some simple algebraic manipulation and bearing in mind that $b^2 = a^2 - 1$, that

$$\overline{S_{i_0}(\omega)} = 2q\,(I_D + 2I_s) + 4k\theta\,(G_j - G_0) \ . \tag{4.34}$$

This is van der Ziel's expression for the total noise in the ideal diode. It has been derived here for the special case of the long diode, but it can be shown (Buckingham and Faulkner, 1974) that the same result holds for the general case when W/L_0 is finite. The calculation is then rather tedious because the integrals, though still tractable, lead to lengthy algebraic expressions which do little to aid in understanding the problem and which ultimately reduce to the formulation in equation (4.34).

4.2.5 The short diode

Noise in the short diode is not in itself a subject of overwhelming practical significance. It is nevertheless of considerable importance in connection with the bipolar transistor, where the base region is very narrow compared with the diffusion length of the minority carriers. There are of course differences between the situations in the diode and the transistor. In the diode the metal contact maintains the boundary condition $p = p_W$ at $x = W$, and also acts as a source of majority carriers for neutralizing potentials in the n-region. These two functions are split in the transistor between the collector junction, which maintains the boundary condition, and the base contact which provides the majority carriers. However, the noise mechanisms in the two cases are the same, and from a discussion of the short diode it is possible to draw certain conclusions about the noise properties of the transistor.

Consider first the recombination–generation noise in the diode. At low frequencies, where carrier storage effects are negligible, the condition $W \ll |L|$ holds and it is easily shown that

$$\frac{k_0}{(k_1 + k_2)} \simeq (1 - x'/W) \tag{4.35a}$$

and

$$\frac{k_W}{(k_1 + k_2)} \simeq x'/W \ , \tag{4.35b}$$

from which it follows that the minority carrier outflows from a single event, given by equations (4.25), are

$$i_0''(j\omega) \simeq -q(1 - x'/W) \tag{4.36a}$$

and

$$i_W''(j\omega) \simeq -qx'/W \ . \tag{4.36b}$$

These equations show that the *total* outflow from the event is equal to the charge on one carrier, as would be expected. This outflow causes a departure from charge equilibrium, which is restored by a majority carrier flow q through the metal contact in the case of the diode, or through the base lead in the transistor. It follows immediately that when $\omega\tau_R \ll 1$ the component of the base current arising from bulk region recombination–generation shows full shot noise.

With modern high gain transistors, the effects of bulk region recombination on the collector/emitter current noise can usually be safely ignored. The noise is then due entirely to thermal fluctuations in the minority carrier flow (bearing in mind that we are still considering an ideal device in which there is negligible depletion-layer recombination). For this situation we again employ the low-frequency condition $W \ll |L|$, to give

$$\frac{k_0 k_2}{(k_1 + k_2)} \simeq \frac{k_1 k_W}{(k_1 + k_2)} \simeq \frac{qD}{W} , \tag{4.37}$$

from which it follows that

$$\overline{S_{i_0}(\omega)} \simeq \overline{S_{i_W}(\omega)} \simeq \frac{4q^2 AD}{W^2} \int_0^W p\,dx' . \tag{4.38}$$

Evidently in this case, the spectral densities of the current fluctuations at $x = 0$ and $x = W$ are equal — which, as we have already noted, is not in general true, the difference being accounted for by a majority carrier flow at $x = W$.

The integration of the hole distribution in equation (4.38) is particularly simple to perform, since p shows a linear dependence on position: when $W \ll L_0$ we have from the diffusion equation

$$p = p_0 - (p_0 - p_W)(x'/W) \tag{4.39}$$

The integral of this expression taken over the n-region is $(p_0 + p_W)/2$ which, from equation (4.38), gives

$$\overline{S_{i_0}(\omega)} \simeq \overline{S_{i_W}(\omega)} \simeq \frac{2q^2 AD}{W}(p_0 + p_W) , \tag{4.40a}$$

or alternatively, in terms of the diffusion current I_D,

$$\overline{S_{i_0}(\omega)} \simeq \overline{S_{i_W}(\omega)} \simeq 2qI_D \frac{(p_0 + p_W)}{(p_0 - p_W)} . \tag{4.40b}$$

As a check on equation (4.40a) it is easily shown that, when $p_W = p_n$, the right-hand side is equal to the frequency independent term in equation (4.34), as we would expect.

In a bipolar transistor with a reverse biased collector junction, the minority carrier concentration at $x = W$ is $p_W = 0$. It is clear from equation (4.40b) that in this case the noise at $x = W$ is simply shot noise associated with the collector current. This is obviously a physically reasonable conclusion since it is what would have been predicted on the assumption of independent carrier motion across the collector junction. Another interesting aspect of equation (4.40b) is that it correctly describes the high level of noise in a saturated transistor where $p_W \simeq p_0$.

4.2.6 Summarizing comments

The physical mechanisms responsible for the noise in an ideal p–n junction are *thermal fluctuations in minority carrier flow* and *generation–recombination events* in the bulk regions. Both processes give rise to disturbances in the minority carrier distribution, and these disturbances produce diffusion currents through the bulk material which tend to restore the distribution to its undisturbed condition. The total minority carrier diffusion current fluctuation at the edge of the depletion layer due to all the events throughout the bulk region constitutes the noise in the terminal current, I_D.

This description of junction noise differs from the commonly held view that there are independent currents $(I_D + I_s)$ and I_s flowing in opposite directions across the depletion layer, each of which shows full shot noise. In fact the currents that actually flow across the junction are not $(I_D + I_s)$ and I_s, but I_F and I_R, both of which are enormously greater than I_D or I_s; I_D is just the small difference between I_F and I_R. When the applied voltage is zero these large currents are determined solely by the carrier concentrations on either side of the depletion layer. In contrast, I_D and I_s, which are functions of the geometry of the base region and of the carrier lifetime τ_R, are controlled by the rate at which carriers can diffuse through the bulk material. Thus diffusion is the mechanism controlling I_D and, according to the present theory of noise, it is also the mechanism underlying the fluctuations in I_D. As pointed out by Robinson (1974), shot noise in I_F and I_R is negligible because of a very efficient smoothing process in which charge accumulation on either side of the junction alters the electric field across the depletion layer which in turn effectively nullifies the fluctuations in the external circuit (see Appendix 4).

The diffusion theory of junction noise leads to exactly the same expression for the spectral density of the current fluctuations as that derived by van der Ziel on the basis of his transmission line analogy. In view of the considerable weight of experimental evidence supporting van der Ziel's result, this is a prerequisite for any satisfactory theory of junction noise. The advantage of the diffusion theory is that it provides a description of the fluctuations which is consistent with the accepted physics of junction action and which does not rely on arbitrary assumptions – such as those made by van der Ziel and Becking (1958) – concerning the nature of the currents crossing the depletion layer.

4.3 DEPLETION-LAYER RECOMBINATION NOISE

The current/voltage characteristics of most germanium p–n junction diodes show close agreement with Shockley's formula for an ideal diode (equation (4.1)). This is not usually the case for silicon junctions at room temperature, the departure from the 'ideal' characteristic being due to carrier recombination and generation in the depletion layer. The statistics of recombination–generation through single energy level centres in the forbidden energy gap of the semiconductor were formulated by Hall (1952) and Shockley and Read (1952), and such centres are nowadays commonly referred to as HSR centres. Hall–Shockley–Read statistics were employed by Sah, Noyce and Shockley (1957) as a basis for discussing the current/voltage relation in silicon junctions. Their theory shows that in such devices, over a considerable range of bias conditions and temperature, the component of the steady terminal current arising from recombination and generation in the depletion layer predominates over the diffusion component.

Obviously the fluctuations in the recombination current cannot be explained in terms of the diffusion theory of noise discussed in § 4.2. In the early days of silicon technology, several theoretical and experimental investigations of depletion-layer recombination noise appeared in the literature (Schneider and Strutt (1959), Chenette (1960), van der Ziel (1960), Chenette and van der Ziel (1962)), though no entirely satisfactory theory of the phenomenon was forthcoming at that time, and the experimental data that were reported showed a certain amount of variability. The situation was to improve when Scott and Strutt (1966) published measurements on the depletion-layer generation noise in a reverse biased, large area silicon diode which had been constructed for use as a nuclear radiation detector. For frequencies in the range 28 kHz to 91.2 kHz they found the noise to be two-thirds of full shot noise. Shortly after Scott and Strutt's paper, Lauritzen (1968) presented a general theory, based on HSR statistics, for the current fluctuations associated with single level recombination–generation centres in forward or reverse biased junctions. For a reverse biased junction he derived a high-frequency value of two-thirds shot noise, in agreement with Scott and Strutt's measurements, whilst for low frequencies and widely differing hole and electron emission coefficients (a situation which is likely to be found in practice) he predicted full shot noise. Under forward bias, Lauritzen concluded that the noise level at low frequencies is always between three-quarters and full shot noise. A natural conclusion from these results is that depletion layer recombination–generation noise is not very sensitive to the bias condition of the junction or the frequency; it always lies between one half and full shot noise.

4.3.1 The noise mechanism

When a carrier diffuses from one or other of the bulk regions into the depletion region, one of three things can happen. It can cross the depletion layer and enter the bulk region on the far side, in which case it becomes one of the many contributors to the large fluxes of carriers crossing the depletion layer, discussed in § 4.2 and Appendix 4; or it may enter the depletion layer, not have sufficient energy to cross it, and simply return to the region from whence it came, giving a net current in the external circuit of zero on a time scale appropriate to semiconductor devices; or it may fall into an HSR centre within the depletion layer, where it will remain for a time which depends on the dynamics of the centre. In the latter case a current pulse is produced in the external circuit, and the sum of all such pulses, due to all the centres in the depletion layer, constitutes the recombination current in the circuit. This current consists of a steady-state component, superimposed upon which are random, recombination-noise fluctuations. Similarly, when a generation event occurs at a centre, the generated carrier is swept through the transition region by the electric field, towards the bulk region where it will become a majority carrier. A current pulse (opposite in sign to that from a recombination event) is produced in the external circuit, and the sum of all such pulses constitutes the generation current in the circuit. This current also consists of a steady-state component, superimposed upon which are random, generation-noise fluctuations (Van Vliet, 1970).

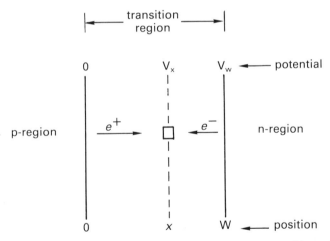

Fig. 4.5 – Sketch of a single HSR recombination centre, □, at position x in a one-dimensional transition region.

The transit of a hole between the p-region and an HSR centre located at x (see Fig. 4.5) produces a transfer of charge around the external circuit equal to $q_p(x)$; and likewise, the transit of an electron between the n-region and the same centre produces a transfer of charge around the external circuit equal to $q_n(x)$. Since we must have

$$q = q_n(x) + q_p(x) , \qquad (4.41)$$

it is easily shown that

$$q_p(x) = -\frac{q}{V_W} \int_0^x E(x)dx = \frac{qV_x}{V_W} \tag{4.42a}$$

and

$$q_n(x) = -\frac{q}{V_W} \int_x^W E(x)dx = q\left(1 - \frac{V_x}{V_W}\right), \tag{4.42b}$$

where W and V_W are the width of and the potential drop across the transition region, respectively, $E(x)$ is the field within the same region, and V_x is the potential at x. Assuming the transit times to be infinitesimally small, the currents in the external circuit associated with q_p and q_n can be written as

$$i_p(x, t) = q_p\delta(t - t_p) = q\frac{V_x}{V_W}\delta(t - t_p) \tag{4.43a}$$

and

$$i_n(x, t) = q_n\delta(t - t_n) = q\left(1 - \frac{V_x}{V_W}\right)\delta(t - t_n), \tag{4.43b}$$

where t_p and t_n are the times at which the hole and electron transits occur.

The individual HSR centres through which recombination and generation occur, exist in one of two possible charge states. The average times spent by a centre in each of these states are not likely to be equal, the more highly charged state being less probable due to Coulombic interactions with carriers of opposite sign (assuming an abundance of both types of carrier). Thus, under forward bias, a centre which can exist in a neutral or negatively charged state will spend most of its time in the neutral state because an electron capture event, causing the centre to go negative, will be followed almost immediately by a hole capture event which will return it to the neutral condition where it is likely to remain for some relatively long time until the next electron capture event occurs. This asymmetry between the capture cross-sections of the centre means that at low (compared with the reciprocal of the time spent in the less probable state) frequencies the hole and electron transfers giving rise to equations (4.43) can be regarded as simultaneous events, in which case instead of a pair of independent current pulses occurring for each full recombination cycle, there is just one, given by

$$i(x, t) = i_p(x, t) + i_n(x, t) = q\delta(t), \tag{4.44}$$

where we have set $t_p = t_n = 0$. An analogous argument applies to the generation of carriers, since it is most unlikely that the emission coefficients of the centre should be the same or even similar.

Suppose that the recombination and generation rates per unit volume at position x in the depletion layer, are $R(x)$ and $G(x)$, respectively. The recombination and generation currents from the HRS centres distributed throughout the transition region are then

$$I_R = qA \int_0^W R(x)\,dx \qquad (4.45a)$$

and

$$I_G = qA \int_0^W G(x)\,dx \ , \qquad (4.45b)$$

where A is the area of the junction. From equation (4.44) in conjunction with Carson's theorem the recombination and generation noise power spectra are

$$\overline{S_R(\omega)} = 2q^2A \int_0^W R(x)\,dx \qquad (4.46a)$$

and

$$\overline{S_G(\omega)} = 2q^2A \int_0^W G(x)\,dx \ , \qquad (4.46b)$$

where the integrations over x take account of the noise contributions from all regions of the depletion layer. It is implicit in equations (4.46) that the HRS centres act independently of one another, which is reasonable provided their concentration is not too high (Sah, 1967). From equations (4.45) and (4.46) it is easily seen that

$$\overline{S_R(\omega)} = 2qI_R \qquad (4.47a)$$

and

$$\overline{S_G(\omega)} = 2qI_G \ . \qquad (4.47b)$$

Thus, at low frequencies such that equation (4.44) holds, the forward-bias recombination current and reverse-bias generation current both show full shot noise.

For zero bias, and again for low frequencies so that equation (4.44) applies, the spectral density of the noise is

$$\overline{S_0(\omega)} = 2q^2A \int_0^W [R_0(x) + G_0(x)]\,dx \ , \qquad (4.48)$$

where $R_0(x)$ and $G_0(x)$ are the equilibrium recombination and generation rates, which are of course equal. Now the conductance of the junction is

$$G_{RG} = \frac{\partial I}{\partial V}\bigg|_{V=0} = qA \int_0^W \frac{\partial\{R(x)-G(x)\}}{\partial V}\bigg|_{V=0} dx \quad , \tag{4.49}$$

where, from the Hall–Shockley–Read theory, the net combination rate is

$$R(x)-G(x) = \frac{pn-n_i^2}{(p+p_1)\tau_n + (n+n_1)\tau_p} \quad . \tag{4.50}$$

In this expression p_1, n_1, τ_p, and τ_n are the familiar terms defined by Shockley and Read. On differentiating equation (4.50) with respect to the applied voltage – bearing in mind that $pn = n_i^2 \exp(qV/k\theta)$ across the depletion layer – and letting $V \to 0$, we find from equation (4.49) that

$$G_{RG} = \frac{q^2 A}{k\theta} \int_0^W R_0(x)\,dx \quad . \tag{4.51}$$

It follows from equation (4.48) that

$$\overline{S_0(\omega)} = 4k\theta G_{RG} \quad , \tag{4.52}$$

which is exactly as we would expect for the spectral density of the equilibrium noise, since the right-hand side of equation (4.52) is just the Nyquist expression for the thermal noise in the conductance G_{RG}.

At higher frequencies than the reciprocal of the average time spent in the less probable of the two charge states of the HRS centres, the current pulses in equations (4.43) may be treated as independent, random events. In this case equation (4.44) does not apply, and instead of equation (4.46) we have

$$\overline{S_R(\omega)} = 2A \int_0^W R(x)\,(q_n^2 + q_p^2)\,dx \tag{4.53a}$$

and

$$\overline{S_G(\omega)} = 2A \int_0^W G(x)\,(q_n^2 + q_p^2)\,dx \quad . \tag{4.53b}$$

For large reverse bias the carriers are swept out of the transition region, so that $p \simeq n \simeq 0$ for all x, and the generation rate from equation (4.50) is

$$G = -\frac{n_i^2}{p_1\tau_n + n_1\tau_p} \quad , \tag{4.54}$$

which is independent of x. Therefore the spectral density of the noise from equations (4.43) and (4.53b) is

$$\overline{S_G(\omega)} = 2q^2 AG \int_0^W \left\{ \left(\frac{V_x}{V_W}\right)^2 + \left(1 - \frac{V_x}{V_W}\right)^2 \right\} dx .$$

(4.55)

Assuming a linear potential distribution across the junction of the form $V_x = V_W(x/W)$, it follows that

$$\overline{S_G(\omega)} = 2q^2 AG \int_0^W \left(1 - \frac{2x}{W} + \frac{2x^2}{W^2}\right) dx$$

$$= \frac{4}{3} q^2 AGW = \frac{4}{3} q I_G ,$$

(4.56)

where the last expression has been derived with the aid of equation (4.45b). Evidently, for high frequencies and reverse bias the generation noise is two-thirds full shot noise. This result was derived by Lauritzen (1968), and shows good agreement with the data of Scott and Strutt (1966).

Van der Ziel (1975a, b) evaluated the integral in equation (4.45) for a linear, rather than a constant, field profile and found noise suppression factors of 11/15 for a highly asymmetrical junction and 23/30 for a symmetrical diode. These values are so close to the factor 2/3 derived by Lauritzen that it seems doubtful whether the presently available data are sufficiently accurate to distinguish between them.

Under conditions of moderate forward bias the recombination rate in equation (4.50),

$$R(x) = \frac{pn}{(p + p_1)\tau_n + (n + n_1)\tau_p} ,$$

(4.57)

shows a sharp maximum at the plane where $p\tau_n = n\tau_p$. If we let $x = x_m$ be the position of this maximum, then from equations (4.45a) and (4.53a) we have for the high-frequency recombination noise spectrum

$$\overline{S_R(\omega)} = \frac{2I_R}{q} \{q_n^2(x_m) + q_p^2(x_m)\} = 2q I_R \left(1 - 2\frac{V_{x_m}}{V_W} + 2\frac{V_{x_m}^2}{V_W^2}\right) ,$$

(4.58)

where the integrals have been taken as proportional (with the same constant of proportionality) to their integrands evaluated at $x = x_m$. In a symmetrical junction the plane of maximum recombination is at the centre of the transition region, hence $V_{x_m}/V_W = \frac{1}{2}$, and

$$\overline{S_R(\omega)} = q I_R ,$$

(4.59a)

corresponding to a suppression factor equal to $1/2$. On the other hand, when the junction is highly asymmetrical the plane $x = x_m$ is close to the edge of the transition region on the low-doped side of the junction, V_{x_m}/V_W is close to zero or close to unity, and in either case

$$\overline{S_R(\omega)} \simeq 2qI_R \ . \tag{4.59b}$$

Thus, in this case the suppression factor is close to but less than unity.

The suppression factors characterizing the recombination–generation noise currents for the various conditions discussed above, are summarized in Table 4.1. It should be borne in mind that the results shown in the table have been derived on the assumption that the recombination–generation events responsible for the noise occur either in pairs, giving a single current pulse $q\delta(t)$ in the external circuit (low frequencies), or are completely independent (high frequencies). These simplifying assumptions are not essential to the theory, as exemplified by Lauritzen's treatment of the problem, involving a statistical analysis of the recombination–generation current waveforms associated with individual HSR centres. However, the quantitative results of Lauritzen's analysis do not differ significantly from those presented above, as we would expect, since the recombination–generation noise is not very sensitive to the conditions in the junction. In view of the relative complexity of Lauritzen's approach, it has not been reproduced here.

Table 4.1

Summary of depletion-layer recombination–generation noise suppression factors

		Symmetrical junction	Asymmetrical junction
Forward bias	low frequency	1	1
	high frequency	0.5	$\simeq 1$
Reverse bias	low frequency	1	1
	high frequency	$\simeq \frac{2}{3}$	$\simeq \frac{2}{3}$

4.4 THE BIPOLAR TRANSISTOR

As we have already seen in § 4.2.5 in the discussion of the short diode, the noise in a bipolar transistor can be treated in much the same way as noise in a p–n junction diode. Naturally, in discussing noise in the transistor, the role of the base contact must be taken into account, particularly the fact that all the depletion-layer recombination current from the emitter-base junction flows through the base lead and none flows through the collector lead. The depletion-layer recombination component of the base current shows full shot noise, corresponding to a suppression factor of unity (see Table 4.1 for the case of a forward biased, asymmetrical junction).

In the absence of emitter-base depletion-layer recombination, the transistor may be regarded as 'ideal' in the sense that its current/voltage behaviour is governed by Shockley's diffusion theory of junction action. The effect of emitter-base depletion-layer recombination on the noise in the leads of the device is merely to add an independent component to the 'ideal' emitter and base noise currents.

In the following discussion we shall use the term 'ideal' in connection with certain transistor parameters. For example, Y_E is the 'ideal' admittance of the emitter-base junction, that is, the admittance that would be observed in the absence of depletion-layer recombination. Shockley's theory gives an expression for Y_E in terms of the geometry of the base, the minority carrier recombination time in the base, etc.

The power spectral densities of the noise fluctuations in the emitter and collector currents can be written down immediately from equations (4.20) and (4.27):

$$\overline{S_{i_E}(\omega)} = \frac{4A}{D} I_1 + \frac{2q^2 A}{\tau_R} I_2 + 2q I_{BR} \tag{4.60}$$

and

$$\overline{S_{i_C}(\omega)} = \frac{4A}{D} I_3 + \frac{2q^2 A}{\tau_R} I_4 \quad , \tag{4.61}$$

where I_{BR} is the depletion-layer recombination component of the base (and also the emitter) current. The integrals over the base region in equations (4.60) and (4.61) take the same form as the integrals over the n-region in equations (4.20) and (4.27):

$$I_1 = \int_0^W p \left| \frac{k_0 k_2}{k_1 + k_2} \right|^2 dx' \tag{4.62a}$$

$$I_2 = \int_0^W (p + p_n) \left| \frac{k_0}{k_1 + k_2} \right|^2 dx' \tag{4.62b}$$

$$I_3 = \int_0^W p \left| \frac{k_1 k_W}{k_1 + k_2} \right|^2 dx' \tag{4.62c}$$

and

$$I_4 = \int_0^W (p + p_n) \left| \frac{k_W}{k_1 + k_2} \right|^2 dx' , \qquad (4.62d)$$

where W is now the width of the base region and the k parameters are exactly as defined previously. By analogy with equations (4.20) and (4.27) it is clear that the integrals in equations (4.62) are associated with thermal fluctuations (I_1 and I_3) and recombination–generation (I_2 and I_4) in the base region.

On evaluating the integrals in equations (4.62) it can be shown that equations (4.60) and (4.61) reduce to the forms (Buckingham and Faulkner, 1974)

$$\overline{S_{i_E}(\omega)} = 4qI_E \left(\frac{G_E}{G_{E_0}} - \frac{1}{2} \right) - 4qI_{BR} \left(\frac{G_E}{G_{E_0}} - 1 \right)$$

$$= 4qI_{ES} \left(\frac{G_E}{G_{E0}} - \frac{1}{2} \right) + 2qI_{BR} \qquad (4.63)$$

and

$$\overline{S_{i_C}(\omega)} = 2qI_C , \qquad (4.64)$$

where G_E is the 'ideal' conductance of the emitter-base junction, G_{EO} is the low frequency value of G_E, I_E and I_C are the steady emitter and collector currents, and $I_{ES} = I_E - I_{BR}$ is the 'ideal' or diffusion component of the steady emitter current. The form of the power spectrum in equation (4.63) shows that the fluctuations in the emitter current contain a shot noise component due to depletion-layer recombination and a further component due to thermal fluctuations and recombination–generation in the base. At low frequencies, when $G_E \simeq G_{EO}$, the latter component reduces to simple shot noise in I_{ES}, from which it follows that, at these frequencies, the total emitter current shows full shot noise. Equation (4.64) shows that at all frequencies the collector current exhibits full shot noise, which is consistent with the idea of a flux of independent carriers crossing the collector junction.

As well as the spectral densities of the current fluctuations in the transistor, we are also interested in the cross-spectral densities between the fluctuations in different leads of the device. Since the same physical mechanisms are responsible for the fluctuations at each of the terminals it is reasonable to expect some degree of correlation to exist between these fluctuations.

The cross-spectral density between the emitter and collector noise currents is obtained from the corollary to Carson's theorem in § 2.14, equation (2.99). Bearing in mind that the noise contributions from thermal fluctuations, bulk region recombination–generation and depletion-layer recombination–generation are independent, we have from equations (4.17) and (4.25), after integrating over the base region,

$$S_{CE}(\omega) = -\frac{4A}{D} I_5 + \frac{2q^2 A}{\tau_R} I_6 , \tag{4.65}$$

where I_5 and I_6 are the following integrals:

$$I_5 = \int_0^W p \frac{k_1 k_W k_0^* k_2^*}{|k_1 + k_2|^2} \, dx' \tag{4.66a}$$

and

$$I_6 = \int_0^W (p + p_n) \frac{k_W k_0^*}{|k_1 + k_2|^2} \, dx' . \tag{4.66b}$$

The first of these integrals, I_5, is associated with thermal fluctuations in the base and the second, I_6, with recombination–generation in the same region. Notice that depletion-layer recombination does not affect equation (4.65) because the collector current fluctuations are independent of recombination events in the emitter-base junction.

Equation (4.65) is derived from an analogous argument to those giving the spectral densities in equations (4.20) and (4.27), except that *cross-products* between collector and emitter noise pulses have been formed above to give the cross-spectral density, as opposed to *squaring* pulses to obtain power spectra.

On evaluating the integrals in equation (4.66) and comparing the results with the expressions for the appropriate transistor parameters, it is readily shown that

$$\overline{S_{CE}(\omega)} = -2q I_C \frac{\alpha_S}{\alpha_{0S}} \frac{Y_E}{G_{E0}} , \tag{4.67}$$

where Y_E is the 'ideal' admittance of the emitter-base junction, α_S is the ratio of the alternating collector current to the 'ideal' alternating emitter current, and α_{0S} is the low frequency value of α_S. For low frequencies, when $\alpha_S \simeq \alpha_{0S}$ and $Y_E \simeq G_{E0}$, the cross-spectral density in equation (4.67) reduces to the frequency-independent form

$$\overline{S_{CE}(\omega)} \simeq -2q I_C . \tag{4.68}$$

The normalized cross-spectral density, or coherence function, between the fluctuations in the collector and emitter currents is defined as (see § 2.14, equation (2.81))

$$\Gamma_{CE} = \frac{\overline{S_{CE}(\omega)}}{[\overline{S_{i_C}(\omega)} \, \overline{S_{i_E}(\omega)}]^{\frac{1}{2}}} = -\frac{2q I_C \dfrac{\alpha_s}{\alpha_{0S}} \dfrac{Y_E}{G_{E0}}}{[\overline{S_{i_C}(\omega)} \, \overline{S_{i_E}(\omega)}]^{\frac{1}{2}}} . \tag{4.69}$$

For low frequencies, when $Y_E \simeq G_E \simeq G_{E0}$ and $\alpha_S = \alpha_{0S}$, we have from equation (4.63) that $\overline{S_{i_E}(\omega)} \simeq 2qI_E = 2q\alpha_0 I_C$, where I_E is the *total* steady emitter current, and hence

$$\Gamma_{CE} \simeq -\alpha_0^{\frac{1}{2}} . \tag{4.70}$$

Since, in most modern transistors, α_0 is very close to unity we see that, over the frequency range where equation (4.70) is valid, the fluctuations in the collector and emitter currents are highly correlated. The negative sign in equation (4.70) is a result of the sign convention adopted here, in which currents flowing into the base are positive; this gives rise to a negative correlation between the low-frequency fluctuations in the in-going emitter current and the out-going collector current.

The power spectrum of the fluctuations in the base current is given by the following relationship:

$$\overline{S_{i_B}(\omega)} = \overline{S_{i_E}(\omega)} + \overline{S_{i_C}(\omega)} + 2\mathrm{Re}\,\overline{S_{C_E}(\omega)} . \tag{4.71}$$

This expression is obtained by setting $i_B(t) = -\{i_E(t) + i_C(t)\}$, forming the transform, $i_B(j\omega)$, taking the square of the modulus and averaging to obtain the power spectrum. On substituting from equations (4.67)–(4.69) we find, after some algebraic manipulation, that

$$\overline{S_{i_B}(\omega)} = 2qI_C \left[\frac{1}{\beta_0} + \frac{2G_E - (\alpha_S Y_E + \alpha_S^* Y_E^*)}{\alpha_{0S} G_{E0}} - \frac{2(1 - \alpha_{0S})}{\alpha_{0S}} \right], \tag{4.72}$$

where $\beta_0 = I_C/I_B$ is the ratio of the steady collector current to the steady base current (including the recombination component). At low frequencies the spectrum in equation (4.72) reduces to $2qI_B$, that is the base current shows full shot noise, as we would expect.

From an argument similar to that giving equation (4.71), the cross-spectral density between the current fluctuations in the collector and base leads is as follows:

$$\overline{S_{CB}(\omega)} = -\overline{S_{CE}(\omega)} - \overline{S_{i_C}(\omega)} , \tag{4.73}$$

and hence, from equations (4.64) and (4.67),

$$\overline{S_{CB}(\omega)} = -2qI_C \left[1 - \frac{\alpha_S}{\alpha_{0S}} \frac{Y_E}{G_{E0}} \right] . \tag{4.74}$$

If $\alpha_S Y_E$ is expanded in a Taylor series, we find that to first order in frequency

$$\alpha_S Y_E \simeq \alpha_{0S} G_{E0} \left(1 - \frac{j\omega\tau_j}{3} \right) , \tag{4.75}$$

where $\tau_j = W^2/2D$ and D is the diffusion constant of the minority carriers in the base. The diffusion constant for holes in silicon approximately equals $12\,\text{cm}^2/\text{sec}$ which, for a base region $2\,\mu\text{m}$ thick, gives $\tau_j = 1.7$ nsec. It follows from equation (4.75) that, to first order in frequency, the cross-spectral density in equation (4.74) is

$$\overline{S_{CB}(\omega)} \simeq -2qI_C\,(j\omega\tau_j/3)\ , \tag{4.76}$$

which is purely imaginary. The normalized cross-spectral density between the current fluctuations in the collector and base leads is defined as

$$\Gamma_{CB} = \frac{\overline{S_{CB}(\omega)}}{[\overline{S_{i_B}(\omega)}\,\overline{S_{i_C}(\omega)}]^{\frac{1}{2}}}\ , \tag{4.77}$$

which at low frequencies takes the form

$$\Gamma_{CB} \simeq -\beta_0^{\frac{1}{2}}(j\omega\tau_j/3)\ . \tag{4.78}$$

Finally, we are interested in the cross-spectral density between the current fluctuations in the emitter and base leads:

$$\overline{S_{EB}(\omega)} = -\overline{S_{i_E}(\omega)} - \overline{S_{EC}(\omega)}$$

$$= -2qI_C\left[\frac{2\,G_E}{\alpha_{0S}\,G_{E0}} - \frac{2}{\alpha_{0S}} + \frac{1}{\beta_0} + 1 - \frac{\alpha_S^*\,Y_E^*}{\alpha_{0S}\,G_{E0}}\right]\ . \tag{4.79}$$

At low frequencies this expression reduces to

$$\overline{S_{EB}(\omega)} \simeq -2qI_C/\beta_0 \tag{4.80}$$

Table 4.2

Summary of the spectral densities and the coherence functions for the current fluctuations in a bipolar transistor

Spectral densities	Coherence functions
$S_{i_E}(\omega) \simeq 2qI_E$	$\Gamma_{CE} \simeq -\alpha_0^{\frac{1}{2}}$
$S_{i_C}(\omega) = 2qI_C$	$\Gamma_{CB} \simeq -\beta_0^{\frac{1}{2}}\left(\dfrac{j\omega\tau_j}{3}\right)$
$S_{i_B}(\omega) \simeq 2qI_B$	$\Gamma_{EB} \simeq -\left(\dfrac{\alpha_0}{\beta_0}\right)^{\frac{1}{2}}$

and the corresponding normalized cross-spectral density is

$$\Gamma_{EB} = \frac{\overline{S_{EB}(\omega)}}{[\overline{S_{i_E}(\omega)}\,\overline{S_{i_B}(\omega)}]^{\frac{1}{2}}} \simeq - \left(\frac{\alpha_0}{\beta_0}\right)^{\frac{1}{2}}. \tag{4.81}$$

A summary of the low-frequency expressions derived above for the various spectral and cross-spectral densities of the current fluctuations in the leads of the transistor is shown in Table 4.2.

4.5 NOISE FIGURE OF THE 'INTRINSIC' TRANSISTOR

We have seen in the previous chapter that it is convenient to express the noise characteristics of a linear two-port in terms of a noise figure. This quantity is a figure of merit which can be determined directly by measurements made at the terminals of the two-port; and obviously it depends on the nature of the circuitry within the system. Thus, in order to derive the noise figure of a bipolar transistor it is necessary first to introduce an appropriate equivalent circuit for the device.

For small signal conditions the transistor is essentially a linear device, and nowadays it is usually represented by the so-called 'hybrid-π' circuit shown in Fig. 4.6, which contains only resistors, capacitors and a single current generator controlled by the voltage between the emitter terminal and the fictitious 'internal base terminal' B'. In this circuit the base resistance, r_b, is treated as an external component connected in series with the external impedance between the base-emitter terminals. The 'intrinsic transistor', with terminals B', E, C is clearly the appropriate model to use in conjunction with the noise calculations of § 4.4, since it is implicit in these calculations that the base resistance is zero. As base resistance effects can be included in the equivalent circuit so readily, we shall for the moment ignore them and concentrate on the intrinsic transistor B', E, C.

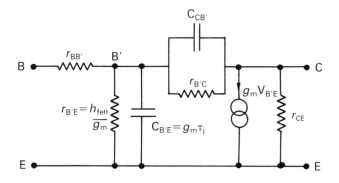

Fig. 4.6 – 'Hybrid-π' equivalent circuit of the transistor.

The admittance matrix of the intrinsic transistor is

$$\mathbf{Y} = \begin{bmatrix} \dfrac{1}{r_{\mathrm{B'E}}} + \dfrac{1}{r_{\mathrm{B'C}}} + j\omega(C_{\mathrm{B'E}} + C_{\mathrm{CB'}}) & -\dfrac{1}{r_{\mathrm{B'C}}} - j\omega C_{\mathrm{CB'}} \\[3mm] g_{\mathrm{m}} - \dfrac{1}{r_{\mathrm{B'C}}} - j\omega\, C_{\mathrm{CB'}} & \dfrac{1}{r_{\mathrm{B'C}}} + \dfrac{1}{r_{\mathrm{CE}}} + j\omega C_{\mathrm{CB'}} \end{bmatrix}, \qquad (4.82)$$

where g_{m} is the mutual conductance of the transistor. For most design purposes this matrix can be satisfactorily approximated as follows:

$$\mathbf{Y} \simeq \begin{bmatrix} g_{\mathrm{m}}\left(\dfrac{1}{h_{\mathrm{feo}}} + j\omega\tau_{\mathrm{j}}\right) & -j\omega C_{\mathrm{CB'}} \\[3mm] g_{\mathrm{m}} & 0 \end{bmatrix}, \qquad (4.83)$$

where h_{feo} is the low frequency current gain and, as defined previously, $\tau_{\mathrm{j}} = W^2/2D$.

The noise sources in the transistor can now be represented by external noise generators connected to the terminals of the 'noiseless' equivalent circuit. This is illustrated in Fig. 4.7 which shows current generators $i_{\mathrm{GB}}(t)$ and $i_{\mathrm{GC}}(t)$ connected across the emitter-base and emitter-collector terminals, respectively. The spectral densities of these generators and the cross-spectral density between them are given by equations (4.72), (4.64) and (4.74).

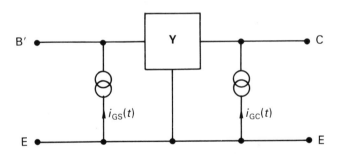

Fig. 4.7 – External noise generators connected across the emitter-base and emitter-collector terminals of the intrinsic transistor.

The circuit in Fig. 4.7 is not in a convenient form for calculating the noise figure of the transistor because the noise generators are not referred to the input port. On transferring the output generator to the input we obtain the two new generators $i_{\mathrm{na}}(t)$ and $v_{\mathrm{na}}(t)$ at the input, as illustrated in Fig. 4.8. This circuit is

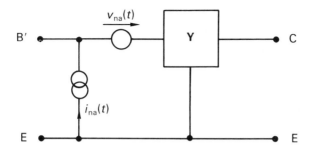

Fig. 4.8 – Equivalent circuit of that in Fig. 4.7 with both generators referred to the input.

valid for calculating the noise current and voltage at the output, but not for calculating the noise current at the input (see page 62, Chapter 3). In transform terminology the generators in Fig. 4.8 are

$$V_{na} = -I_{GC}/Y_{21} \tag{4.84a}$$

and

$$I_{na} = I_{GB} - (Y_{11}/Y_{21})I_{GC} , \tag{4.84b}$$

where the Y parameters are the components of the admittance matrix in equation (4.83). It follows from equations (4.72), (4.64) and (4.74) that the spectral densities of the two generators are

$$\overline{S_{va}(\omega)} \simeq 2qI_C/G_{E0}^2 \tag{4.85a}$$

and

$$\overline{S_{ia}(\omega)} \simeq 2qI_B + \frac{4}{3}qI_C\omega^2\tau_j^2 , \tag{4.85b}$$

and the cross-spectral density between them is

$$\overline{S_{vi}(\omega)} \simeq \frac{2qI_C}{G_{E0}}\left(\frac{1}{h_{feo}} - \frac{2j\omega\tau_j}{3}\right) , \tag{4.85c}$$

where G_{E0} is the low-frequency value of the emitter-base conductance. The following relationships have been employed in deriving equations (4.85):

$$\alpha_{0s} \simeq 1$$

$$h_{feo}^2 \gg \beta_0 \gg 1 \tag{4.86}$$

$$\omega^2\tau_j^2 \ll 1 .$$

The noise figure of the intrinsic transistor is now determined from the circuit shown in Fig. 4.9. In this circuit $i_s(t)$ is the signal current generator, $Y_s = G_s + jB_s$ is the source admittance, $i_{ns}(t)$ is the source noise current generator with spectral density equal to $4k\theta G_s$, and $i_{nb}(t)$ is the equivalent current generator to $v_{na}(t)$ in Fig. 4.8. By Norton's theorem this generator is given in transform terminology as

$$I_{nb} = Y_s V_{na} . \tag{4.87}$$

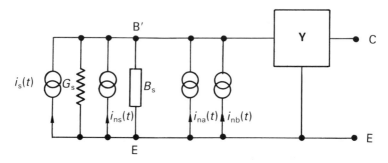

Fig. 4.9 – Circuit for deriving the noise figure of the intrinsic transistor.

It follows immediately from the circuit in Fig. 4.9 and equations (4.85) that the noise figure is

$$F' = 1 + \frac{2qI_C}{4k\theta G_s} \left[\left(\frac{B_s^2}{G_{E0}^2} + \frac{4}{3} \frac{B_s \omega \tau_j}{G_{E0}} + \frac{2}{3} \omega^2 \tau_j^2 \right) + \frac{1}{\beta_0} + \frac{G_s^2}{G_{E0}^2} + \frac{2G_s}{G_{E0} h_{fe0}} \right] , \tag{4.88a}$$

which, by completing the square, can be expressed in the form

$$F' = 1 + \frac{2qI_C}{4k\theta G_s} \left[\left(\frac{B_s}{G_{E0}} + \frac{2}{3} \omega \tau_j \right)^2 + \frac{2}{9} \omega^2 \tau_j^2 + \frac{1}{\beta_0} + \frac{G_s^2}{G_{E0}^2} + \frac{2G_s}{G_{E0} h_{fe0}} \right] . \tag{4.88b}$$

The primes in these equations indicate an affiliation with the intrinsic transistor.

The noise tuning condition is achieved when the source susceptance takes the value

$$B_{s0} = -\frac{2}{3} G_{E0} \omega \tau_j \tag{4.89a}$$

corresponding to a source inductance $3/(2G_{E0}\omega^2\tau_j)$. When B_{s0} is substituted for B_s in equation (4.88b), the noise matching condition can be determined by minimizing the resultant noise figure with respect to the source conductance.

This procedure gives the optimum source conductance as

$$G_{s0} = G_{E0} \left(\frac{1}{\beta_0} + \frac{2}{9} \omega^2 \tau_j^2 \right)^{\frac{1}{2}} \tag{4.89b}$$

and it follows that the minimum noise figure is

$$F_0' \simeq 1 + \left(\frac{1}{\beta_0} + \frac{2}{9} \omega^2 \tau_j^2 \right)^{\frac{1}{2}}, \tag{4.90}$$

where the term in $1/h_{fe0}$ has been neglected and we have used the relationship $G_{E0} \simeq qI_C/k\theta$. Note that according to equation (4.90) the minimum noise figure increases with increasing frequency, in agreement with the observed behaviour of actual transistors.

Equation (4.90) represents a physical limitation to the performance that can be expected of a transistor. At low frequencies, when the frequency-dependent term is negligible, F_0' reduces to $1 + 1/\sqrt{\beta_0}$, which for $\beta_0 = 100$ is equal to 0.4 dB. This value is in good agreement with the measurements of Faulkner and Harding (1968), indicating that the noise performance of modern silicon planar transistors can come close to the optimum, and that the effects of a non-zero base resistance and excess noise, neither of which have been included in the above analysis, can be reduced to a negligibly low level.

If for some reason the source admittance has to be real, the noise tuning condition cannot be satisfied – that is, it is not possible to tune out inductively part of the frequency-dependent contribution to the noise figure. By setting $B_s = 0$ and minimizing the resultant expression for F' in equation (4.88b) with respect to G_s, the minimum noise figure under these conditions is found to be

$$F_0' \simeq 1 + \left(\frac{1}{\beta_0} + \frac{2}{3} \omega^2 \tau_j^2 \right)^{\frac{1}{2}}. \tag{4.91}$$

Clearly, this expression differs from that in equation (4.90) only at the higher frequencies when the frequency-dependent term is significant. Like equation (4.90), it represents a physical limitation to performance, which may be approached in practice, but which will not actually be reached due to base resistance effects and excess noise in the transistor.

4.6 THE LOW-FREQUENCY EQUIVALENT CIRCUIT INCLUDING BASE RESISTANCE

Although the equivalent circuit in Fig. 4.8 accurately represents the noise in the intrinsic transistor over a wide range of frequency, it is unnecessarily complicated for many circuit design purposes. For frequencies well below $1/(\tau_j \sqrt{\beta_0})$ a simplification can be achieved by recognizing that the voltage and current generators in Fig. 4.8 are essentially 'white' and uncorrelated; that is (cf. equations (4.85a) and (4.85b)),

$$\overline{S_{va}(\omega)} \simeq 2qI_C/G_{E0}^2 \simeq 2k\theta/G_{E0} \tag{4.92a}$$

$$\overline{S_{ia}(\omega)} \simeq 2qI_B \simeq 2k\theta/\beta_0 G_{E0} \tag{4.92b}$$

and

$$\Gamma_{vi} \simeq 0 \ , \tag{4.92c}$$

where Γ_{vi} is the normalized cross-spectral density between the generators.

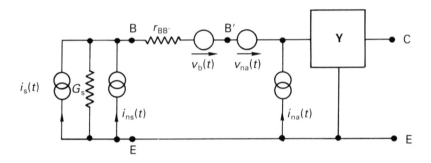

Fig. 4.10 — Equivalent circuit of the transistor, including the base resistance.

At this point in the discussion it is conveneient to take the base resistance into account; it is included in the equivalent circuit in Fig. 4.10, where $v_b(t)$ is the thermal noise voltage generator associated with r_b. The low-frequency noise figure of the complete transistor follows directly from this circuit in conjunction with equations (4.92). Assuming a real source admittance, the result is

$$F \simeq \left[1 + \frac{(r_b + r_e/2)}{R_s} + \frac{(R_s + r_b)^2}{2r_e\beta_0 R_s} \right] \ , \tag{4.93}$$

where $r_e = 1/G_{E0}$. Provided the conditions

$$\beta_0 \gg 1 \quad \text{and} \quad r_e \geqslant r_b \tag{4.94}$$

are satisfied the expression for the noise figure reduces to

$$F \simeq \left[1 + \frac{(r_b + r_e/2)}{R_s} + \frac{R_s}{2r_e\beta_0} \right] \ . \tag{4.95}$$

Now this expression can be written as

$$F \simeq 1 + \frac{R_{nv}}{R_s} + \frac{R_s}{R_{ni}} , \qquad (4.96)$$

where

$$R_{nv} = r_b + r_e/2 \qquad (4.97a)$$

and

$$R_{ni} = 2r_e\beta_0 . \qquad (4.97b)$$

It is apparent from this formulation of the noise figure that R_{nv} and R_{ni} may be interpreted as the noise resistances of the uncorrelated series voltage and parallel current noise generators at the emitter-base terminals. Notice that the effect of the base resistance on the noise figure is included simply by adding r_b to the series noise resistance of the intrinsic transistor.

The noise matching condition is achieved when the source resistance takes the optimum value

$$R_{s0} = \sqrt{(R_{nv}R_{ni})} = \{2r_e\beta_0(r_b + r_e/2)\}^{\frac{1}{2}} , \qquad (4.98)$$

giving a minimum noise figure

$$F_0 \simeq 1 + 2 \left\{ \frac{(r_b + r_e/2)}{2r_e\beta_0} \right\}^{\frac{1}{2}} . \qquad (4.99)$$

Clearly, when $r_e \gg r_b$, so that base resistance effects are negligible, the expression for the minimum noise figure reduces to

$$F_0 \simeq 1 + 1/\sqrt{\beta_0} , \qquad (4.100)$$

as we would expect on the basis of equations (4.90) and (4.91).

The operating range over which the second inequality in equation (4.94) is valid is best illustrated with a specific example (Faulkner, 1968). We recall that $r_e = k\theta/qI_C$, and hence when $I_C = 1$ mA, $r_e \simeq 25$ Ω at room temperature. Now a low-noise silicon planar transistor designed for audio frequency applications may typically show a value of $r_b = 200$ Ω. In this case $r_e \gg r_b$ for values of I_C up to 125 μA. At this operating current the series and parallel noise resistances in equations (4.97) are $R_{nv} = 300$ Ω and, with $\beta_0 = 100$, $R_{ni} = 40$ kΩ, giving from equation (4.98) an optimum source resistance $R_{s0} = 3.5$ kΩ. The corresponding minimum noise figure, from equation (4.99), is $F_0 \simeq 0.68$ dB.

An improvement on this value of F_0 can be achieved by reducing I_C by a factor of 10 or more, in which case r_b in equation (4.99) is negligible compared with the term $r_e/2$ and the minimum noise figure is as given in equation (4.100). Again with $\beta_0 = 100$, we find for this case that $F_0 \simeq 0.43$ dB.

An interesting implication of the low-frequency analysis given above has been pointed out by Faulkner (1968). He observes that the low-frequency common-emitter input resistance of the transistor, $(r_b + h_{feo}r_e)$, is approximately equal to the parallel noise resistance, R_{ni}, since h_{feo} generally lies somewhere between β_0 and $2\,\beta_0$. This means that a bipolar transistor in a common-emitter configuration will not show a good noise figure unless voltage-amplifier conditions prevail; that is, unless its input resistance is substantially greater than the source resistance. This contradicts the assertion frequently made in the literature that the bipolar transistor is basically a current amplifier, and which Faulkner claims has led to a great deal of misunderstanding.

4.7 THE EFFECT OF $1/f$ NOISE ON THE NOISE FIGURE

The discussion given above neglects entirely the effect of $1/f$ noise on the noise figure of the transistor. Faulkner (1968) has shown that $1/f$ noise may be taken into account simply by replacing β_0 everywhere in the equations of § 4.6 by an effective current gain, β_0', defined as

$$\beta_0' = \frac{\beta_0}{(1 + \omega_F/\omega)} , \qquad (4.101)$$

where ω_F is the $1/f$ noise characteristic angular frequency. A good low-noise silicon planar transistor shows a value of $\omega_F/2\pi$ less than about 1 kHz (Faulkner and Harding, 1968). For frequencies substantially greater than this $\beta_0' \simeq \beta_0$ and the expressions for the noise figure given above hold. When $\omega \ll \omega_F$, however, $1/f$ noise is a significant contributor to the overall noise of the transistor, and by the time β_0' has fallen to unity the device can no longer be said to have a low-noise capability. In this frequency range the analysis of the previous sections is obviously not applicable.

4.8 THE PROBLEM OF LOW SOURCE RESISTANCE

According to equations (4.94) and (4.98) the optimum source resistance takes its *minimum* value for fixed values of β_0 and r_b, when $r_e = r_b$. Thus the minimum value of the optimum source resistance is $r_b\sqrt{3\beta_0}$ which, with $\beta_0 = 100$ and $r_b = 200\ \Omega$, is equal to 3.5 kΩ; and as we would rarely expect to find r_b much less than 100 Ω, it follows that the optimum source resistance will always be greater than about 2 kΩ.

Now, an obvious difficulty arises with sources whose resistance is less than 2 kΩ: such a source cannot satisfy directly the noise matching condition necessary for achieving the minimum noise figure. One solution to the problem is to match the low impedance source to the transistor by means of a transformer. However, the use of an input transformer in a low-noise audio frequency amplifier is usually highly undesirable, and in many cases impossible.

An alternative solution has been proposed by Faulkner (1966) in which several transistors are connected in parallel, as illustrated in Fig. 4.11. The important point concerning this technique is that, when n identical amplifiers are connected in parallel, the series and parallel noise resistances of the combination are both reduced by a factor n relative to the corresponding values for one of the amplifiers in the circuit. Thus, the optimum source resistance is also reduced by a factor n compared with that for a single transistor. For example, with $n = 4$ and the base resistance of each transistor equal to $100\,\Omega$, the optimum source resistance is $500\,\Omega$, compared with $2\,k\Omega$ for a single transistor cited above.

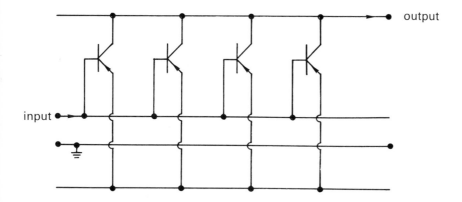

Fig. 4.11 – Parallel connection of input transistors. (All components except the transistors themselves have been omitted from the diagram.)

Even when the noise matching condition is not satisfied, the technique of connecting transistors in parallel can still significantly improve the noise figure. For instance, with a 30 Ω source resistance, $r_e = r_b = 100\,\Omega$ and $\beta_0 = 100$, the noise figure from equation (4.95) for a single transistor is 7.8 dB. On connecting four transistors in parallel the noise figure decreases to 3.5 dB, representing a rather dramatic improvement in performance.

REFERENCES

M. J. Buckingham and E. A. Faulkner (1974), The theory of inherent noise in p–n junction diodes and bipolar transistors, *The Radio and Elect. Eng.*, **44**, 125–140.

E. R. Chenette (1960), Frequency dependence of the noise and the current amplification factor of silicon transistors, *Proc. IRE* (correspondence), **48**, 111-112.

E. R. Chenette and A. van der Ziel (1962), Accurate noise measurements on transistors, *IRE Trans. on Elect. Dev.*, **ED–9**, 123–128.

E. A. Faulkner (1966), Optimum design of low-noise amplifiers, *Elect. Lett.*, **2**, 426–427.

E. A. Faulkner (1968), The design of low-noise audio frequency amplifiers, *The Radio and Elect. Eng.*, **36**, 17–30.

E. A. Faulkner and D. W. Harding (1968), Some measurements on low-noise transistors for audio frequency applications, *The Radio and Elect. Eng.*, **36**, 31-33.

W. Guggenbuehl and M. J. O. Strutt (1957), Theory and experiments on shot noise in semiconductor junction diodes and transistors, *Proc. IRE*, **45**, 839–854.

R. N. Hall (1952), Electron-hole recombination in germanium, *Phys. Rev.*, **87**, 387.

G. H. Hanson and A. van der Ziel (1957), Shot noise in transistors, *Proc. IRE*, **45**, 1538–1542.

P. O. Lauritzen, (1968), Noise due to generation and recombination of carriers in p–n junction transition regions, *IEEE Trans. Elect. Dev.*, **ED–15**, 770–776.

H. C. Montgomery (1952a), Transistor noise in circuit applications, *Proc. IRE*, **40**, 1461–1471.

H. C. Montgomery (1952b), Electrical noise in semiconductors, *Bell Syst. Tech. J.*, **31**, 950–975.

H. C. Montgomery and M. A. Clark (1953), Shot noise in junction transistors, *J. Appl. Phys.*, **24**, 1337–1338.

F. N. H. Robinson (1974), *Noise and Fluctuations in Electronic Devices and Circuits*, Clarendon Press, Oxford, pp. 99 and 237–239.

R. M. Ryder and R. J. Kircher (1949), Some circuit aspects of the transistor, *Bell Syst. Tech. J.*, **28**, 367–400.

C. T. Sah (1967), The equivalent circuit model in solid-state electronics, part 1: The single energy level defect centres, *Proc. IEEE*, **55**, 654–671.

C. T. Sah, R. N. Noyce and W. Shockley (1957), Carrier generation and recombination in p–n junctions and p–n junction characteristics, *Proc. IRE*, **45**, 1228–1243.

B. Schneider and M. J. O. Strutt (1959), Theory and experiments on shot noise in silicon p–n junction diodes and transistors, *Proc. IRE*, **47**, 546–554.

L. Scott and M. J. O. Strutt (1966), Spontaneous fluctuations in the leakage current due to charge generation and recombination in semiconductor diodes, *Solid State Elect.*, **9**, 1067–1073.

W. Shockley (1949), The theory of p–n junctions in semiconductors and p–n junction transistors, *Bell Syst. Tech. J.*, **28**, 435–489.

W. Shockley and W. T. Read Jr. (1952), Statistics of recombination of holes and electrons, *Phys. Rev.*, **87**, 835–842.

A. van der Ziel (1955), Theory of shot noise in junction diodes and junction transistors, *Proc. IRE,* **43**, 1639–1646.

A. van der Ziel (1957), Theory of shot noise in junction diodes and junction transistors, *Proc. IRE,* **45**, 1011.

A. van der Ziel (1960), Shot noise in transistors, *Proc. IRE* (correspondence), **48**, 114–115.

A. van der Ziel (1975a), The state of solid state device noise research, Invited paper at the Fourth Conference on Physical Aspects of Noise in Solid State Devices, Noordwijkerhout, The Netherlands, September 9–11.

A. van der Ziel (1975b), Shot noise in back biased p–n silicon diodes, *Solid State Elect.,* **18**, 969–970.

A. van der Ziel and A. G. Th. Becking (1958), Theory of junction diode and junction transistor noise, *Proc. IRE,* **46**, 589–594.

A. van der Ziel and E. R. Chenette (1978), Noise in solid state devices, *Advances in Electronics and Electron Physics,* **46**, 313–383.

K. M. Van Vliet, (1970), Noise sources in transport equations associated with ambipolar diffusion and Shockley–Read recombination, *Solid State Elect.,* **13**, 649–657.

R. L. Wallace and W. J. Pietenpol (1951), Some circuit properties and applications of n–p–n transistors, *Bell Syst. Tech. J.,* **30**, 530–563.

5

Noise in JFETs and MOSFETs

5.1 INTRODUCTION

The noise sources in JFETs and MOSFETs are similar except for $1/f$ noise, which is almost absent from the former but predominates in the latter at low frequencies. This behaviour strongly suggests that the $1/f$ noise in this case is a surface-related effect.

An important noise mechanism in FETs is thermal fluctuations amongst the carrier population in the channel. These conductivity fluctuations produce thermal noise in the drain and gate currents. A further significant source of noise is carrier generation in the channel-gate depletion region; and in JFETs this same mechanism gives rise to shot noise in the leakage current flowing through the gate. Other sources of noise in FETs, including carrier concentration fluctuations in the channel due to recombination–generation through HSR centres located in the channel or through partially ionized donors (n-type channel) or acceptors (p-type channel), are generally negligible.

5.2 OPERATING CHARACTERISTICS OF THE JFET

Apart from their opposite polarities, p-channel and n-channel JFETs operate in much the same way. In order to avoid unnecessary duplication, only the former is considered in the following discussion.

A cross-section through a planar, p-channel JFET is illustrated schematically in Fig. 5.1. The terminals at either end of the p-region are the source and the drain, and are usually bonded to inserts of heavily doped p-type material. The current, I_D, through the device flows via these two terminals, as shown in the figure, and a third terminal, the gate, is connected to the n⁺-type layers sandwiching the p-type material.

The JFET operates with a reverse bias applied across the p–n junctions. Since the n-regions are heavily doped in relation to the p-region, the space-charge regions of the junctions fall almost entirely within the p-region, as indicated by the hatching in the figure. With different voltages at the source

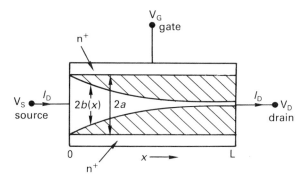

Fig. 5.1 — Cross-section through a p-channel JFET. The hatched areas are space-charge regions.

and the drain, the reverse bias across the p—n junctions varies with x, the position coordinate along the device, and consequently the thickness of the space-charge regions also depends on x. As the carrier concentration in the space-charge regions is negligible, the current I_D flows in a channel of p-type material whose boundaries are formed by the edges of the space-charge regions. Note that the current in the device is carried almost exclusively by majority carriers (holes in the case of the p-channel device), a fact which led Shockley (1952) to describe JFETs as unipolar in order to distinguish them from point contact and junction transistors, which are bipolar.

Shockley originally derived the small signal characteristics of the JFET by considering the geometry of the conducting channel. Grove (1967) discusses Shockley's analysis in a general account of the operating characteristics of JFETs. We give below a brief discussion of those aspects of performance which are pertinent to the noise behaviour of these devices.

Suppose that the potential at a point x in the channel is $\phi(x)$, then the potential drop across the junctions at that point is†

$$V(x) = V_B + V_G - \phi(x) , \tag{5.1}$$

where V_B (which approximately equals 1 volt for silicon) is the built-in voltage of the junction and $V_G \geqslant 0$ is the reverse voltage on the gate terminal. By solving Poisson's equation for a highly asymmetrical step junction, the potential $V(x)$ can be expressed in terms of the thickness of the space-charge regions as follows:

$$V(x) = \frac{q N_A}{2 \epsilon_r \epsilon_0} \left[a - b(x) \right]^2 , \tag{5.2}$$

---

† It is implicit in equation (5.1) that $\phi(x)$ is measured relative to the same reference potential as the terminal potentials.

where q is the magnitude of the electronic charge, N_A is the density of acceptors in the p-type channel, ϵ_r is the dielectric constant and ϵ_0 is the permittivity of free space. If at any point x the width of the channel is zero, i.e. $b = 0$, the channel is said to be 'pinched-off'. This is the normal operating condition for the JFET. By setting $b(x) = 0$ in equation (5.2), a pinch-off voltage can be defined as

$$V_p = \frac{qN_Aa^2}{2\,\epsilon_r\epsilon_0}\,, \tag{5.3}$$

and hence the channel thickness in equation (5.2) can be expressed in the form

$$b(x) = a\,[1 - \{V(x)/V_p\}^{\frac{1}{2}}]\,, \tag{5.4}$$

where $V(x) \leqslant V_p$.

The potential $V(x)$ increases between the source and the drain, and the pinched-off region of the channel, if it exists, normally falls close to the drain extending over a length which is a small but not insignificant fraction of the total length of the channel. Most of the voltage drop along the channel then falls across the pinched region, in which case the current through the device is essentially independent of the drain voltage. When these conditions prevail the JFET is said to be operating in the saturated region.

The current flowing through the channel is independent of x and is given by the expression

$$I_D = 2\,b(x)w\sigma\,\frac{dV(x)}{dx}\,, \tag{5.5}$$

where σ is the conductivity of the channel and w is the thickness of the JFET, i.e. the dimension normal to the plane of the paper in Fig. 5.1. On integrating the left-hand side of equation (5.5) over x between the limits x_1 and x_2 and the right-hand side over potential between the corresponding limits V_1 and V_2, the current can be expressed as

$$I_D = \frac{2w\sigma a}{(x_2 - x_1)}\,[(V_2 - V_1) - (2/3)\,V_p^{-\frac{1}{2}}\,(V_2^{\frac{3}{2}} - V_1^{\frac{3}{2}})]\,. \tag{5.6}$$

Assuming that the channel is not pinched off, we may set $x_1 = 0$ and $x_2 = L$, in which case $V_1 = (V_B + V_G - V_S)$ and $V_2 = (V_B + V_G - V_D)$, where V_s and V_D are the source and drain voltages, respectively. Then

$$I_D = -g_0\,[V_D + (2/3)\,V_p^{-\frac{1}{2}}\,\{(V_B + V_G - V_D)^{\frac{3}{2}} - (V_B + V_G)^{\frac{3}{2}}\}]\,, \tag{5.7}$$

where V_s has been set equal to zero and

$$g_0 = \frac{2\sigma a w}{L} \tag{5.8}$$

is the conductance of the metallurgical channel of uniform width $2a$. Note that in order to ensure the junctions are reverse-biased along the whole length of the channel, V_D must be zero or negative.

Equation (5.7) is the fundamental equation in the theory of JFETs. It applies provided $(V_B + V_G - V_D) < V_p$. If this inequality is not satisfied, the channel is pinched-off and the device is operating in the saturation region. The onset of saturation occurs when

$$V_D = V_{Dsat} = -(V_p - V_B - V_G) \ ,$$ (5.9)

corresponding to a saturation current

$$I_D = I_{Dsat} = \frac{g_0 V_p}{3} \left[1 - 3 \frac{(V_B + V_G)}{V_p} + 2 \left(\frac{V_B + V_G}{V_p} \right)^{\frac{3}{2}} \right] \ .$$ (5.10)

Below saturation, when V_D is relatively small and the channel is far from being pinched-off, i.e. the channel width shows only a weak dependence on x, the JFET behaves like a resistor. This regime is usually referred to as the linear region of operation. The ohmic relationship between I_D and V_D in the linear region can be derived from equation (5.7) by expanding the right-hand side in a Taylor series to first order in V_D, on the assumption that $|V_D| \ll (V_B + V_G)$. The result is

$$I_D \simeq -g_0 \left[1 - \left(\frac{V_B + V_G}{V_p} \right)^{\frac{1}{2}} \right] V_D \ .$$ (5.11)

Note that the resistance of the channel, given by the reciprocal of the coefficient multiplying V_D, increases with increasing V_G. This resistance becomes extremely large when V_G approaches $(V_p - V_B)$, which is called the turn-off voltage. Turn-off occurs when the reverse bias on the gate is such that the channel is completely depleted of carriers along its length.

The conductance of the channel is defined as

$$g = -\frac{\partial I_D}{\partial V_D} \bigg|_{V_G} \ .$$ (5.12)

Hence it follows immediately from equation (5.11) that

$$g = g_0 \left[1 - \left(\frac{V_B + V_G}{V_p} \right)^{\frac{1}{2}} \right] \text{ (linear region) } .$$ (5.13)

The transconductance of the FET is defined as

$$g_m = -\frac{\partial I_D}{\partial V_G} \bigg|_{V_D} \ ,$$ (5.14)

and again from equation (5.11) we have

$$g_m = \frac{-g_0 V_D}{2\sqrt{[V_p(V_B + V_G)]}} \text{(linear region)} . \tag{5.15a}$$

In the saturation region it follows from equation (5.10) that the transconductance is

$$g_{m\text{sat}} = g_0 \left[1 - \left(\frac{V_B + V_G}{V_p}\right)^{\frac{1}{2}}\right] \text{(saturation region)} . \tag{5.15b}$$

On comparing equations (5.13) and (5.15b) it can be seen that the channel conductance in the linear region is equal to the transconductance in saturation.

Finally, before going on to discuss noise in the JFET, we derive the input capacitance, C, of the device. The argument concerns the variation in the space charge in the junctions and the way in which it varies with V_G.

The space charge in the depletion layers between x and $x + dx$ is

$$dQ = -2qN_A \{a - b(x)\} w \, dx , \tag{5.16}$$

and hence the total charge over the length of the JFET, assuming that the channel is not pinched-off, is

$$Q = -2q N_A w \int_0^L \{a - b(x)\} dx$$

$$= -2q N_A w \left\{aL - \int_0^L b(x) dx\right\} . \tag{5.17}$$

Since the charge induced on the gate is equal to $-Q$, the input capacitance is

$$C = -\frac{\partial Q}{\partial V_G} = -2q N_A w \frac{\partial}{\partial V_G} \left\{\int_0^L b(x) dx\right\} . \tag{5.18}$$

The integral in the expressions for Q and C can be evaluated with the aid of equations (5.4) and (5.5), bearing in mind that I_D is independent of x:

$$\int_0^L b(x) dx = \left(\frac{2w\sigma a^2}{g_0}\right) \frac{[(z_D - z_S) - (4/3)(z_D^{\frac{3}{2}} - z_S^{\frac{3}{2}}) + (1/2)(z_D^2 - z_S^2)]}{[(z_D - z_S) - (2/3)(z_D^{\frac{3}{2}} - z_S^{\frac{3}{2}})]}, \tag{5.19}$$

where we have set

$$z_D = z_S - (V_D/V_p) = (V_B + V_G - V_D)/V_p . \tag{5.20}$$

It is implicit that equation (5.19) applies below saturation. On differentiating equation (5.19) with respect to V_G and substituting the result into equation (5.18), the input capacitance below saturation is found to be

$$C = -\frac{2q N_A w a L}{V_p} (z_D - z_S) \times$$

$$\times \frac{[(z_D - z_S) - (z_D^{\frac{1}{2}} - z_S^{\frac{1}{2}}) - (2/3)(z_D^{\frac{3}{2}} - z_S^{\frac{3}{2}}) + (1/2)(z_D + z_S)(z_D^{\frac{1}{2}} - z_S^{\frac{1}{2}})]}{[(z_D - z_S) - (2/3)(z_D^{\frac{3}{2}} - z_S^{\frac{3}{2}})]^2} .$$

$$5.21$$

When the channel is pinched-off, the formulation in equation (5.17) for the total charge over the length of the FET can be retained except that, strictly, the upper limit on the integral should be replaced with x_0, the position coordinate at the source end of the pinched region. However, Robinson (1969) has shown that, as most of the voltage drop along the channel occurs over the pinched region, x_0 can be safely approximated by L as the upper limit on the integral. Hence, in the saturation region, when pinch-off obtains, it follows from equation (5.19) on setting $z_D = 1$ that

$$\int_0^{x_0} b(x)\,dx \simeq \int_0^L b(x)\,dx = \frac{w\sigma a^2}{g_0}\frac{(1 - 6z_S + 8 z_S^{\frac{3}{2}} - 4 z_S^2)}{(1 - 3z_S + 2z_S^{\frac{3}{2}})}$$

$$= \frac{w\sigma a^2}{g_0}\frac{(1 - z_S^{\frac{1}{2}})(1 + 3z_S^{\frac{1}{2}})}{(1 + 2z_S^{\frac{1}{2}})} , \quad (5.22)$$

corresponding to an input capacitance

$$C = \frac{3 q N_A w a L}{V_p}\frac{(1 + z_S^{\frac{1}{2}})}{(1 + 2z_S^{\frac{1}{2}})} \quad \text{(saturation region)} . \quad (5.23)$$

5.3 NOISE SOURCES IN THE JFET

One of the principal noise mechanisms in the JFET, and the first to be treated analytically (van der Ziel, 1962), is thermal fluctuations amongst the current carriers in the channel. These fluctuations produce thermal noise in the drain current and also, because of the capacitive coupling between the gate and the channel, in the gate current. A partial correlation exists between the thermal noise currents in the drain and the gate, but it is not large enough to have a significant effect on the optimum noise figure.

A second important source of noise in the JFET is carrier generation through centres located in the space-charge region of the channel-gate junctions.

Under normal operating conditions, the junctions are reverse-biased and a HSR centre in the depletion region generates alternately a hole and an electron, which are immediately swept out of the region by the strong electric field. In silicon devices at room temperature, these generated carriers constitute the main part of the leakage current in the gate lead. There is therefore a shot noise component in the gate current which, at low frequencies, predominates over the thermal noise component.

Depletion-layer generation also produces noise in the output, or drain, current. This occurs because a given centre is continually changing its charge state, thereby causing a local variation in the depletion layer width, and hence the channel width, which in turn produces a noise current in the external circuit.

Fluctuations in the carrier concentration in the channel can constitute a further source of noise in the FET (van der Ziel, 1963b). At room temperature, such fluctuations may occur as a result of recombination–generation of carriers through HSR centres located in the channel; whilst at low temperatures a similar recombination–generation process but involving the partially ionized donor population (n-type channel) or acceptor population (p-type channel) may give rise to noise. But in either case, the fluctuations in the external circuit are usually negligible compared with the generation noise produced by the HSR centres in the space-charge regions.

5.4 THERMAL NOISE IN THE JFET

5.4.1 Thermal noise in the output current
For the purpose of calculating the thermal noise in the output current of the JFET, the channel may be treated as a resistor whose incremental resistance between x and $x + dx$ is a function of the position coordinate x. Then the thermal fluctuations in the channel can be analysed on the basis of the same model as applied to thermal noise in a resistor (Chapter 2, § 2.8) except that, in the JFET, the mean number of elementary events per second, ν, is no longer uniform but depends on x as $[2\,N_A w b(x)/\overline{\tau}_F]\,dx$, where $\overline{\tau}_F$ is the mean free time of the carriers.

By analogy with the analysis of thermal noise in a resistance in § 2.8, the power spectrum of the noise current produced between x and $x + dx$ is

$$d\overline{S}_i = 4\frac{N_A w b(x)}{\overline{\tau}_F}\frac{q^2\overline{l_f^2}}{L^2}\,dx\ ,\tag{5.24}$$

where $\overline{l_f^2}$, the mean-square free path length, can be expressed in terms of the carrier mobility as follows (see Appendix 3):

$$\overline{l_f^2} = 2\frac{\overline{\tau}_F k\theta\mu}{q}\ .\tag{5.25}$$

Hence,

$$dS_i = 8 \frac{N_A w b(x)}{L^2} q \mu k \theta \, dx , \qquad (5.26)$$

and on integrating over the length of the channel, on the assumption that the random 'events' in each increment of length dx are independent, we find that

$$\overline{S_i} = 8 \frac{N_A w q \mu k \theta}{L^2} \int_0^L b(x) \, dx , \qquad (5.27)$$

which is the power spectrum of the thermal noise current in the channel.

The integral here is given in equation (5.19) for the case when the channel is not pinched-off. On making the substitution, and remembering that $\sigma = N_A q \mu$, we find that

$$\overline{S_i} = 4 k \theta \, g_0 \frac{[(z_D - z_S) - (4/3)(z_D^{\frac{3}{2}} - z_S^{\frac{3}{2}}) + (1/2)(z_D^2 - z_S^2)]}{[(z_D - z_S) - (2/3)(z_D^{\frac{3}{2}} - z_S^{\frac{3}{2}})]} . \qquad (5.28a)$$

This expression was derived originally by van der Ziel (1962), and has since been discussed by Haslett and Trofimenkoff (1969a) in connection with a model based on the analysis of an equivalent circuit of the FET. In the limit as $z_D \rightarrow z_S$, corresponding to a drain voltage of zero, equation (5.28a) reduces to the form

$$\overline{S_i} = 4 k \theta g , \qquad (5.28b)$$

where g is the conductance of the channel, given in equation (5.13). Thus, in thermal equilibrium the thermal noise in the channel obeys the Nyquist law, as we should expect.

In the saturation region, the integral in equation (5.27) can be safely approximated by the expression in equation (5.22). Then the power spectrum of the channel noise current is given as

$$\overline{S_i} \simeq 4 k \theta g_{msat} \left[\frac{(1 + 3 z_S^{\frac{1}{2}})}{2(1 + 2 z_S^{\frac{1}{2}})} \right] . \qquad (5.29a)$$

As discussed by Robinson (1969), over the usual range of biasing conditions the quantity $z_S = (V_B + V_G)/V_p$ lies between 0.1 and 1, corresponding to a variation of the term in square brackets from 0.6 to 0.67. Thus, to a good approximation,

$$\overline{S_i} \simeq 4 k \theta (2 g_{msat}/3) ; \qquad (5.29b)$$

that is, in the saturation region, the thermal noise current in the channel is approximately equal to Johnson noise in a conductance equal to 2/3 times the transconductance.

5.4.2 Thermal noise in the gate

The thermal motion of the carriers in the channel causes a random variation in potential along the length of the channel, which in turn produces a fluctuation in the channel width or, equivalently, in the width of the depletion layers bordering the channel. Thus, the total charge in the depletion regions also fluctuates randomly and hence, through the capacitive coupling between the gate and the channel, a noise current, $i_g(t)$, flows in the gate terminal.

If $\bar{q}_T$ is the total fixed charge in the depletion regions between the channel and the gate, then

$$\bar{q}_T = -2qN_A w \int_0^L \{a - b(x)\}\,\mathrm{d}x \ , \tag{5.30}$$

where $b(x)$ is now a (random) function of time as well as a function of position. The gate current is

$$i_g(t) = -\frac{\mathrm{d}\bar{q}_T(t)}{\mathrm{d}t} \ , \tag{5.31}$$

or in transform terminology,

$$I_g(j\omega) = -j\omega\,\overline{Q}_T(j\omega) \ , \tag{5.32}$$

where $I_g(j\omega)$ and $\overline{Q}_T(j\omega)$ are the Fourier transforms of $i_g(t)$ and $\bar{q}_T(t)$, respectively.

The power spectrum of i_g may be derived using a similar procedure to that applied to noise in a resistor: the noise is assumed to be due to a random succession of independent, microscopic events, each of which consists of a carrier motion between collisions followed by a relaxation of the system back towards the equilibrium state. Each event gives rise to a potential fluctuation along the length of the channel, thereby producing a current pulse in the gate, given by equation (5.31). Once the shape of each current pulse has been determined, by solving for the fluctuation in the depletion layer widths, the power spectrum of i_g can be obtained from Carson's theorem followed by an integration over the length of the channel. It should be borne in mind, however, that (as in the preceding discussion of thermal noise in the channel) the mean number of events per second is a function of position in the channel, a consequence of the fact that the channel width itself depends on x.

It can be seen immediately from equation (5.32), without the need to perform the calculation, that through the capacitive coupling between the gate and the channel the power spectrum of i_g varies with frequency as ω^2. Thus, at low frequencies, below about 100 kHz, thermal noise in the gate is usually negligible compared with the shot noise in the leakage current flowing through the reverse-biased junctions; but at higher frequencies, the thermal noise component may be expected to predominate.

The power spectrum of the gate current has been derived by van der Ziel (1963a) for the case when the JFET is operating below saturation. This treatment has been extended by Robinson (1969) to include the saturation region. The algebra in the derivation is rather lengthy, involving integrals of the form

$$\int_0^L b^n(x) \, dx \ ,$$

where $n = 1$, 2 and 3. Instead of repeating the analysis here, we merely quote Robinson's result. He finds that the power spectrum of the thermal noise in the gate is

$$\overline{S_{i_g}(\omega)} = \frac{\omega^2 C^2}{g_{msat}} 4k\theta \left[\frac{(1 + 7z_S^{\frac{1}{2}})}{10(1 + 2z_S^{\frac{1}{2}})} \right] \ , \tag{5.33a}$$

where g_{msat} is the transconductance in saturation given in equation (5.15b) and C is the input capacitance in equation (5.23).

The normalized potential z_S, defined in equation (5.20) as $(V_B + V_G)/V_p$, takes values between 0.1 and 1 over the usual range of bias conditions. Over this range, the expression in square brackets in equation (5.33a) varies slowly between 0.2 and 0.27. Thus, to a good approximation we have

$$\overline{S_{i_g}(\omega)} \simeq \frac{\omega^2 C^2}{4g_{msat}} 4k\theta \ . \tag{5.33b}$$

Since the same physical processes are responsible for the thermal noise currents in the gate and the channel, some degree of correlation may be expected between these current fluctuations. Robinson has shown that the cross-spectral density between the channel and the gate thermal noise currents is

$$\overline{S_{i_g i}(\omega)} = -j\omega C \, 4k\theta \left[\frac{1 + 6z_S^{\frac{1}{2}} + 3z_S}{10(1 + z_S)^{\frac{1}{2}} (1 + 2z_S^{\frac{1}{2}})} \right] \ . \tag{5.34a}$$

For $0.1 < z_S < 1$ the term in square brackets is essentially constant, lying between 0.15 and 0.17, and hence

$$\overline{S_{i_g i}(\omega)} \simeq -0.16j\omega C \, 4k\theta \ . \tag{5.34b}$$

From equations (5.29b), (5.33b) and (5.34b) the normalized cross-spectral density is

$$\Gamma_{i_g i} \simeq -0.16\sqrt{6}j \ ; \tag{5.35}$$

that is, $\Gamma_{i_g i}$ is purely imaginary with $|\Gamma_{i_g i}|^2 \simeq 0.16$. According to Takagi et al. (1976), this rather small value is reduced even further if the electric field dependence of the mobility is taken into account. For the purpose of circuit design, it is sufficient to approximate $\Gamma_{i_g i}$ by zero, which is equivalent to treating the thermal noise currents in the gate and the channel as independent fluctuations.

5.4.3 Series resistance in the channel

The gate contacts do not usually extend over the whole length of the channel. The unmodulated, or non-active, regions appearing at the source and drain ends of the channel act as series resistances, r_S and r_D, in the source and drain leads. These resistances modify the transconductance, g_m, and the channel conductance, g (van der Ziel, 1962), which become g'_m and g', where

$$g'_m = g_m / (1 + r_S g_{msat} + r_D g) \tag{5.36a}$$

and

$$g' = g / (1 + r_S g_{msat} + r_D g) . \tag{5.36b}$$

In the saturation region $g' = g = 0$ and

$$g'_{msat} = g_{msat} / (1 + r_S g_{msat}) . \tag{5.37}$$

Notice that this expression does not contain r_D, and that g' (linear region) is now no longer equal to g'_{msat}. Van der Ziel (1962) has suggested that this lack of equality might be used to identify series resistance effects.

It is apparent from equation (5.37) that, provided $r_S g_{msat}$ is small compared with unity, which is usually the case, the effect of the series resistances on the noise is insignificant. For practical purposes they may be safely neglected.

Brunke (1963) has reported noise measurements on field effect transistors which demonstrate the effect of r_S and r_D on the noise. His results are shown in Fig. 5.2, in which I_{eq} is the equivalent saturated diode current for the noise. Curve B in the figure was calculated with $r_S = 960\ \Omega$, $r_D = 2440\ \Omega$ and $g_{msat} = 330\ \mu$mho.

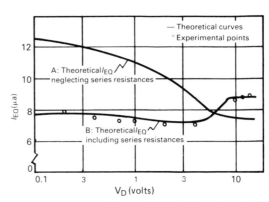

Fig. 5.2 — I_{eq} as a function of V_D at 25 kHz (from Brunke (1963), by kind permission, © 1963 IEEE).

5.4.4 High-frequency effects

At high frequencies, thermal noise in FETs usually predominates over other types of noise. Van der Ziel and Ero (1964) have presented a small-signal, high-frequency theory of a FET in which the channel is treated as an active, non-uniform transmission line. They derive the wave equation for the line and, from an approximate treatment, determine the high-frequency admittance matrix of the device. Hauser (1965) has also discussed the wave equation—transmission line representation of the JFET.

Van der Ziel and Ero derive an expression for the cut-off frequency of the JFET. For the special case of zero gate voltage they find, after substituting reasonable numerical values for the carrier mobility, the pinch-off voltage and the length of the channel, a cut-off frequency equal to 40 MHz. With the gate biases likely to be used in practice, the cut-off frequency is less than this, and may typically be in the region of 10 MHz.

The approximations employed by van der Ziel and Ero include an expansion of the transconductance which is terminated beyond the first power of the frequency. This puts an upper limit on the frequency range over which the theory applies. A more general analysis has been presented by Geurst (1965), whose results are applicable at very high frequencies. He shows that the elements of the admittance matrix of the JFET can be represented in terms of certain known special functions of mathematics, namely Weber's parabolic cylinder functions, $D_n(x)$.

Geurst's differential equation, modified slightly to include a thermal noise current source, has been used by Klaassen (1967) in connection with an analysis of the high-frequency noise in a JFET. Klaassen's results show that at low frequencies the spectral density of the noise current in the channel is independent of frequency and very much greater than that of the gate noise current (in agreement with van der Ziel, 1962); but at high frequencies, the noise spectra of the gate and drain currents both vary as ω^2 and are similar in magnitude. Moreover, Klaassen finds that the high-frequency coupling between the gate and the channel leads to a small but nevertheless non-zero value (approximately equal to $0.13\omega C/g_m$, where C is the input capacitance) for the real part of the normalized cross-spectral density between the gate and the channel noise currents.

Measurements of the high-frequency behaviour of noise in JFETs have been reported by Brunke and van der Ziel (1966). They present the spectral densities of the drain and gate noise currents in terms of the equivalent saturated diode currents I_{nd} and I_{ng}; that is, they set

$$\overline{S_i(\omega)} = 2qI_{nd} \tag{5.38}$$

and

$$\overline{S_{i_g}(\omega)} = 2qI_{ng} . \tag{5.39}$$

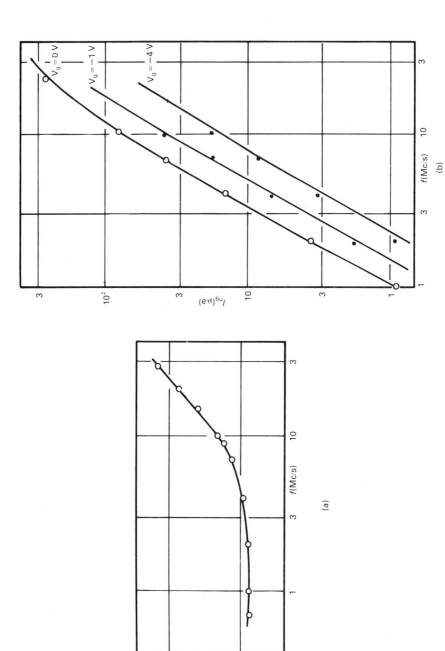
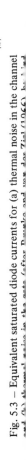

Fig. 5.3 – Equivalent saturated diode currents for (a) thermal noise in the channel and (b) thermal noise in the gate (after Bruncke and van der Ziel (1966) by kind

Figure 5.3, after Brunke and van der Ziel, shows I_{nd} and I_{ng} plotted as functions of frequency for the saturated region of operation. The points in the figure represent experimental measurements and the full lines are theoretical curves of the form

$$I_{nd} = I_{nd0} + (2k\theta/q)|g_{12}| \tag{5.40}$$

and

$$I_{ng} = (2k\theta/q)g_{11} , \tag{5.41}$$

where g_{11} and g_{12} are elements of the conductance matrix of the FET, both of which are proportional to ω^2, and I_{nd0} is the low-frequency value of I_{nd}. The agreement between theory and experiment is very good, and the high-frequency dependence of I_{nd} on ω is clearly demonstrated. However, as this dependence appears at frequencies greater than about 10 MHz, which is close to the cut-off frequency of the FET, it may be safely ignored in connection with most practical applications; that is, I_{nd} may be approximated as I_{nd0}. This greatly facilitates the calculation of the noise figure.

It has been reported recently (van der Ziel and Chenette, 1978) that, owing to the availability of improved materials and advances in device fabrication, it is possible to produce GaAs FETs with superior high-frequency, low-noise characteristics. They quote tuned-noise figures of between 1.2 dB and 4 dB over the frequency range 4 GHz–18 GHz. The same authors also suggest that further improvements may be achieved with FETs fabricated from ternary compounds such as InGaAs.

5.5 GENERATION–RECOMBINATION NOISE

5.5.1 Depletion-layer generation noise

An important component of the noise at low frequencies in a silicon JFET originates in the fluctuations in the charge states of HSR centres located in the depletion layers of the channel-gate junctions. These centres are also responsible for the main component of the leakage current across the junctions.

The problem of depletion-layer generation noise in silicon FETs was originally addressed by Lauritzen and Sah (1963), who later presented detailed discussions of the mechanism in separate publications (Sah, 1964, and Lauritzen, 1965). Their model is supported by experimental measurements of noise made on conventional and gold-doped silicon FETs over a wide range of bias conditions. Gold-doped silicon JFETs have also been discussed by Fu and Sah (1969) in connection with a lumped equivalent circuit analysis of generation–recombination noise due to charge fluctuations at impurity centres in the depletion layers of the channel-gate junctions.

Depletion-layer generation noise occurs because each of the HSR centres emits alternately a hole and an electron, which are then swept out of the junction by the high electric field. This fluctuation in the charge state of a centre produces a local modulation of the depletion-layer width, and hence the channel width, which in turn produces fluctuations in the current flowing in the external circuit.

The noise can be represented by a voltage generator in the input to the JFET, as shown in Fig. 5.4. The equivalent noise resistance of this generator is of the form

$$R_n = \frac{F(V_G, V_D)\tau_t}{(1 + \omega^2\tau_t^2)} \,, \tag{5.42}$$

where $F(V_G, V_D)$ gives the functional dependence of R_n on the gate and drain bias levels, and

$$\tau_t \simeq (c_p p_1 + c_n n_1)^{-1} \tag{5.43}$$

is the time constant of the HSR centres. The approximation here is appropriate to conditions obtaining in a reverse-biased p–n junction. In equation (5.43), c_p and c_n are the hole and electron capture probabilities, and p_1 and n_1 are defined in terms of the trap level, E_T, and the intrinsic Fermi level, E_i, as follows (Shockley and Read, 1952):

$$p_1 = n_i \exp\left[(E_i - E_T)/k\theta\right] \tag{5.44a}$$

and

$$n_1 = n_i \exp\left[(E_T - E_i)/k\theta\right] \,, \tag{5.44b}$$

where n_i is the intrinsic carrier concentration.

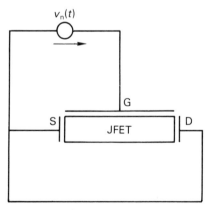

Fig. 5.4 – Schematic showing noise voltage generator in the gate lead representing depletion-layer generation noise.

The magnitude of R_n depends strongly on temperature, through the parameter τ_t in the numerator of equation (5.42). This is illustrated most easily for the case where the trap level falls at the intrinsic Fermi level. The temperature dependence of τ_t is then contained entirely in the expression

$$\tau_t \propto n_i^{-1} \propto \theta^{-\frac{3}{2}} \exp\left(E_{G0}/2k\theta\right) \ ,$$

where E_{G0} is the bandgap of the semiconductor at $\theta = 0$ K. The factor $\theta^{-\frac{3}{2}}$ in this expression has only a minor effect compared with the exponential term, which governs the temperature dependence of τ_t. As the temperature increases, τ_t decreases and hence R_n also decreases. This variation in τ_t has the further effect of increasing the bandwidth of R_n as the temperature is raised. The temperature dependence of R_n is illustrated for the case of a silicon JFET in Fig. 5.5. The curves in the figure were calculated with $E_{G0} = 1.2$ eV and a room temperature value of $\tau_t = 5$ msec.

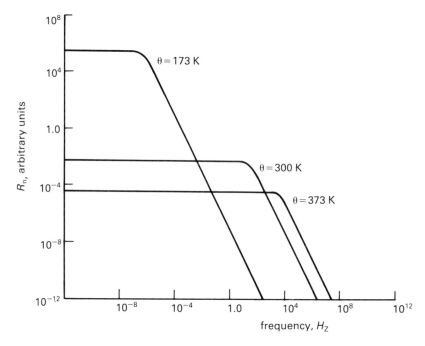

Fig. 5.5 – The temperature dependence of R_n in equation (5.42) for a silicon JFET.

The density of generation centres appears in R_n only through a scaling factor in the function $F(V_G, V_D)$. Thus, the magnitude of the noise depends on the trap concentration, but the frequency dependence does not.

The function $F(V_G, V_D)$ in equation (5.42) diverges logarithmically as the JFET is biased into saturation. This difficulty was recognized by Lauritzen (1965), who avoided it in his two-dimensional model by allowing the channel width at the drain to remain finite at pinch-off. The problem has also been treated by Haslett and Trofimenkoff (1969b) in an analysis based on Trofimenkoff and Nordquist's (1968) model for pinch-off operation.

5.5.2 Channel noise

The noise due to carrier density fluctuations in the channel has been calculated by van der Ziel (1963b) and more recently by van Vliet and Hiatt (1975). Such fluctuations could occur if HSR recombination–generation centres were present in the channel or, at low temperatures, if the donor or acceptor population in the channel were only partially ionized.

The spectrum of the noise can be expressed in terms of an equivalent noise resistance, R_n, which is similar in form to the expression in equation (5.42):

$$R_n \propto \frac{\tau}{(1 + \omega^2 \tau^2)} , \qquad (5.45)$$

where τ is the time constant of the fluctuations. This spectral shape has been observed experimentally by Halladay and Brunke (1963), who made measurements on silicon and germanium FETs. However, it is difficult to say whether the noise they measured was due to carrier density fluctuations in the channel or to depletion-layer generation processes. According to Sah (1964), the former noise component should be negligible at room temperature.

At lower temperatures, however, carrier density fluctuations in the channel may make a significant contribution to the noise. Churchill and Lauritzen (1971) have made measurements of the noise in silicon JFETs at temperatures below 125 K. They found noise levels up to 23 dB higher than that expected of thermal noise in the channel. Their experimental results are in good agreement with the carrier density fluctuation hypothesis when the field dependence of the carrier mobility is included in the expressions.

In general, the low-temperature noise of JFETs cannot be explained in terms of just a single mechanism. Hiatt et al. (1975) measured the noise spectra in the low temperature range 80–200 K in a number of devices, and found evidence for the presence of several types of generation–recombination process, two of which they attributed to traps in the channel. The activation energies found for these two processes are 0.17–0.19 eV and 0.34–0.36 eV. The shallower trap could be the 0.16 eV level often found in silicon and usually ascribed to oxygen (Klaassen and Robinson, 1970, and Courvoisier et al. 1963); and the deeper level could be due to nickel, which has an acceptor level in silicon at 0.35 eV below the conduction band.

5.5.3 Leakage current noise

In addition to thermal noise in the gate lead, there is also noise associated with the leakage current, I_g, across the gate-channel junctions. The spectral density of the latter noise component is

$$\overline{S_{i_{gl}}(\omega)} = 2qI_g , \qquad (5.46)$$

that is, it behaves like shot noise. Since $\overline{S_{i_{gl}}(\omega)}$ is independent of frequency and the thermal noise varies as ω^2, there is a cross-over frequency below which the leakage current noise component is dominant. This frequency, found by comparing equations (5.33b) and (5.46), is

$$f_c = f_T \left\{ \frac{2qI_g}{k\theta \, g_{msat}} \right\}^{\frac{1}{2}} , \qquad (5.47)$$

where $f_T = g_m/2\pi C$ is the cut-off frequency of the JFET. With $f_T = 10$ MHz, $I_g = 10^{-9}$ A and $g_{msat} = 10^{-3}$ mho, we find from equation (5.47) that $f_c \simeq 100$ kHz.

There is also a noise component at the drain due to leakage across the junctions in a JFET which is strongly correlated with the leakage current noise in the gate. However, at room temperatures, in modern devices, the former noise component is negligible; but at high temperatures it becomes significant and can impose a high temperature limit on the low-noise performance of a JFET. The effect has been analysed by van der Ziel (1969). He includes MOSFETs in his treatment, even though the phenomenon in these devices is much less pronounced than in JFETs.

5.6 GENERATION–RECOMBINATION NOISE EXPERIMENTS

The equivalent noise resistance of the voltage fluctuations due to generation through HSR centres in the depletion layers of the gate-channel junctions is of the form (see equation (5.42))

$$R_n \propto \tau_t/(1 + \omega^2 \tau_t^2) , \qquad (5.48)$$

where τ_t is the time constant of the centres (equation (5.43)) and the constant of proportionality is a function of the gate and drain voltages. Under fixed bias conditions and at a given frequency ω_1, R_n shows a maximum when $\tau_t = 1/\omega_1$. As τ_t varies rapidly with temperature and the constant of proportionality shows only a weak dependence on temperature (at least, when the trap level falls at the intrinsic Fermi level), it follows that R_n should pass through one or more maxima, depending on the number of different trapping centre species present, as the temperature is varied.

Such behaviour has indeed been observed by Haslett and Kendall (1972).

They made measurements of the noise in many n-channel JFETs from different manufacturers, and found surpisingly consistent results among the specimens. An example of one of their measurements of the equivalent noise voltage versus temperature is shown in Fig. 5.6. The device used in the measurement was selected for its low noise at room temperature. Almost identical curves were obtained for many other devices of the same type number, the only differences being minor ones in the magnitude of the peaks and the temperatures at which they occurred. Even devices from different manufacturers showed similar behaviour, in all cases displaying prominent peaks like those in Fig. 5.6. It was found that the temperature at which a given peak occurred decreased as the frequency of the measurement was reduced. This behavious is consistent with equation (5.48).

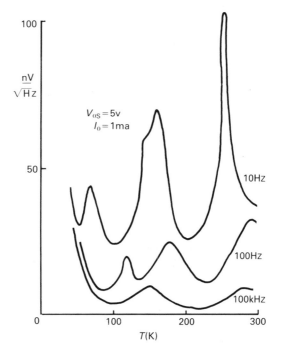

Fig. 5.6 – Equivalent noise voltage as a function of temperature at constant frequency (after Haslett and Kendall (1972), by kind permission © 1972 IEEE).

As gold is used in the fabrication of commercial JFETs, it is likely that some gold is present in the silicon host lattice of such devices. Now gold in silicon is a well-known amphoteric centre, with an acceptor level very close to the intrinsic Fermi level at the middle of the bandgap and a donor level about 0.4 eV above the valence band edge. Haslett and Kendall suggest that the gold acceptor level is responsible for the peak in the noise at temperatures in the vicinity of 300 K (see Fig. 5.6). This is a prominent peak and the most consistent one observed. The peak between 100 K and 200 K is more difficult to explain, though it is safe to assert that it is not due to the gold donor level because this is too far below the middle of the bandgap to produce noise comparable in magnitude with that of the gold acceptor level. Furthermore, oxygen is unlikely to be responsible for the peak, as suggested by Klaassen and Robinson (1970), because the capture cross-sections required are unreasonably high. Haslett and Kendall suggest that some interchange of carriers between dislocations – which act as traps with similar energy and capture cross-sections as the gold acceptor level – and a gold level which is split due to lattice strain, may be the mechanism responsible for this peak. The third peak, which at 10 Hz always occurred near 80 K, was the sharpest and in some cases the largest observed. The origin of this peak is also uncertain, but according to Haslett and Kendall it may be due to some odd coupling between trap levels which alters the noise contribution from the edges of the transition region.

An interesting study of low-frequency excess noise in JFETs has recently been conducted by Kandiah and Whiting (1978). Their ultimate aim was to improve the performance of JFETs in amplifiers used for spectrometry with nuclear radiation detectors, an application in which the requirements of low noise are very exacting. They made their measurements on n-channel four-terminal JFETs in which the extra terminal was connected to the p^+ substrate and acted as a second gate. This second gate terminal was used solely to determine the bias conditions whilst the top gate was used as the signal gate. The noise was measured as a function of the second gate bias, with temperature and drain current as additional variables. As with Haslett and Kendall, peaks in the noise were observed to appear and disappear as a function of temperature.

But the fascinating aspect of Kandiah and Whiting's experimental technique is that it permits the observation of unit electronic charge fluctuations at *individual* HSR centres. This is achieved by sweeping the channel through the device by varying the second gate bias. Thus, a centre which at the beginning of the sweep is located in the depletion layer above the channel, finds itself in the channel as the second gate is increased to a moderate level, and then in the depletion layer below the channel as the bias is swept to its maximum value. If the charge fluctuations at the HSR centre depend on the nature of its local environment, showing large differences between, say, neutral and depleted junction regions, then the noise as a function of the second-gate bias should show a profile which could be used to investigate the sources of excess noise in JFETs.

Kandiah and Whiting found that most of the excess noise is due to HSR centres located in a narrow transition region† about 1000 Å thick between the channel and the fully depleted region of the gate-channel junction. Figure 5.7 is

---

† Note that the term 'transition region' in this context differs in meaning from the more conventional usage as a synonym for the depletion region of a p–n junction.

an example of their observations, indicating the presence of two centres, designated A and B. These centres produce relatively large amounts of noise when they are either just above or just below the channel. The larger differences in V_{SS} between the peaks A–A and B–B at the higher drain current is a measure of the increased thickness of the channel in this case. In general, the higher the number of active centres in the volume swept out by the channel, the greater the number of peaks there will be in noise plots like that in Fig. 5.7; and, as Kandiah and Whiting point out, it would be a mistake to assume that all situations are of such simplicity as the case shown here.

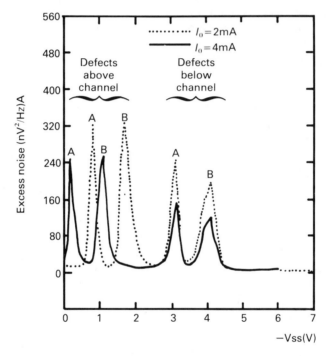

Fig. 5.7 – Excess noise of a JFET as a function of second gate bias V_{SS} at two drain currents (after Kandiah and Whiting (1978), by kind permission of Pergammon Press).

Kandiah and Whiting attribute the noise peaks to HSR centres with shallow energy levels in the bandgap of the semiconductor. At low temperatures, significant noise from such centres has been reported by Haslett and Kendall (1972) and Wang *et al.* (1975). If charge emission were the only mechanism operating at a shallow centre, the time spent in one of the charge states would be extremely long compared with that spent in the other, due to the asymmetry of the energy level of the centre in the energy gap. The noise from the centre would then be negligible. However, if some free charge were available in the vicinity of the centre, the fluctuation time would be considerably reduced as a result of charge capture processes, and the noise would be appreciable. This mechanism has been proposed by Kandiah and Whiting to account for the excess noise observed from the very narrow transition region (see footnote on page 131) between the channel and the fully depleted junction. It appears that the alternate emission of electrons and holes from mid-band centres located throughout the depletion layer — the mechanism discussed by Sah (1964) — makes a negligible contribution to the low-frequency noise at temperatures below 200 K.

According to Kandiah and Whiting, the number of active HSR centres with shallow energy levels in the transition region (see footnote on page 131) of good JFETs with channel widths in the region of 1000 μm and channel lengths of about 2 μm, is in the range 3 to 10. Since these isolated centres are largely responsible for the noise, it should be possible, at least in a four-terminal JFET, to choose a set of operating conditions — temperature, drain current and substrate bias — which minimizes the noise over a useful range of frequencies. This technique could yield substantial benefits in applications such as X-ray spectrometry.

5.7 HIGH FIELD EFFECTS

In JFETs and MOSFETs with very short channels, the electric field in the channel can become very large. Above a threshold field strength, E_0, the mobility decreases appreciably with increasing field, E, obeying a law which can be approximated as (Trofimenkoff, 1965)

$$\mu \simeq \mu_0 \left(1 + E/E_0\right)^{-\frac{1}{2}}, \tag{5.49}$$

where μ_0 is the low-field mobility and $E_0 \simeq 5000$ v/cm for electrons in silicon. Alternatives to equation (5.49) have often appeared in the literature, most notably that due to Dacey and Ross (1953) and employed by Halladay and van der Ziel (1968a), in which μ varies as $E^{-\frac{1}{2}}$.

As well as influencing the carrier mobility, the high electric field has the further effect of increasing the free-carrier temperature above the lattice temperature: it creates a population of hot carriers. The effective temperature of the carriers, θ_e, is approximately given by the relation†

$$\theta_e \simeq \theta_0 \left\{1 + \beta(E/E_0)\right\}, \tag{5.50}$$

† The first reported measurements of θ_e, by Erlbach and Gunn (1962), were made on single crystal n-type germanium, and showed $(\theta_e - \theta_0)$ varying quadratically with E at low fields; but at higher fields, above about 900 v/cm, $(\theta_e - \theta_0)$ showed a more nearly linear dependence on E, as in equation (5.50). Takagi et al. (1976) claim to have demonstrated experimentally the relationship $(\theta_e - \theta_0) \propto E^2$ for a silicon epitaxial layer. The linear relationship in equation (5.50) is used here because it is a reasonable approximation over a large range of measurements of θ_e; and it is the approximation which is perhaps most commonly found in the literature.

where the parameter β depends on the lattice temperature θ_0. In germanium, $\beta \simeq 1$ at $\theta_0 = 300$ K and $10 < \beta < 15$ at $\theta_0 = 77$ K. Klaassen (1970), whilst acknowledging that no data were available for β in silicon, suggests that it takes much the same values as in germanium and accordingly used the values cited above.

The high-field dependences expressed through equations (5.49) and (5.50) are both responsible for an increase in the thermal noise in the channel. If the equivalent thermal noise resistance in the saturation region is expressed as

$$R_n = \alpha/g_{m\,sat} , \qquad (5.51)$$

then, in the absence of high-field effects, $\alpha \simeq 2/3$ (see equation (5.29b)). Klaassen (1970) has shown that the field dependent mobility and the effect of the hot carriers can lead to values of α in excess of unity, with $\alpha \simeq 5$ being possible, depending on bias conditions, the geometry of the device and most notably on the temperature. At room temperature, $\alpha \simeq 1$ but at liquid nitrogen temperature α can be considerably greater than unity. This is principally due to the hot carrier effect, which alone is usually sufficient to produce higher thermal noise at 77 K than at room temperature (Radeka, 1969; Klaassen and Robinson, 1970). For optimum low-noise performance, the operating temperature of high-frequency JFETs is usually somewhat higher than 77 K.

In the operating region well beyond pinch-off, charge multiplication (avalanching) associated with the extremely high field in the vicinity of the drain can also contribute significantly to the output noise of a FET, especially at low temperatures.

When avalanching occurs, hole–electron pairs are formed as a result of impact ionization, and the minority carriers are immediately swept into the gate, thereby increasing the gate current, whilst the majority carriers proceed to the drain. The increased gate current has been observed by Ryan (1969) and an analogous effect in n-channel MOS devices has been reported by Nakahara *et al.* (1968). The increased noise due to the mechanism has been observed in germanium JFETs by Radeka (1967) and in silicon JFETs by Nakahara and Kobayashi (1970) and Klaassen and Robinson (1970). A brief theoretical treatment of the phenomenon is given by van der Ziel and Chenette (1978).

5.8 1/f NOISE IN JFETs

1/f noise in good low-noise silicon JFETs is usually negligible at room temperature. The virtual absence of 1/f noise in these devices distinguishes them from almost all other solid state devices. It also leads naturally to the inference that 1/f noise is not a bulk effect but is a phenomenon associated with the semiconductor–oxide interface, possibly due to fluctuations in the occupancy of interface states; for interface effects are absent from the JFET since the channel is modulated by a depletion layer located in the bulk of the device.

Surprisingly, GaAs FETs show a considerable amount of $1/f$ noise (Takagi and van der Ziel, 1979). This has been explained by van der Ziel (1979), who points out that in such devices the width of the gate is much less than the length of the metallurgical channel. There are, therefore, large areas of semiconductor–oxide interface between the source and the gate and between the gate and the drain, and these regions produce a substantial amount of $1/f$ noise.

At low temperatures, below 200 K, the noise spectra of silicon JFETs show evidence of several types of generation–recombination process, but there is no $1/f$ noise component (Hiatt *et al.*, 1975). This is in agreement with earlier observations of Klaassen and Robinson (1970). At temperatures above 200 K, however, there is a $1/f$ noise contribution, according to Klaassen and Robinson, which decreases with increasing temperature. There is no obvious explanation for this temperature dependence.

5.9 EQUIVALENT CIRCUITS FOR THE JFET

Fortunately, the various noise sources in a JFET are not all present at the same time. At room temperature under normal operating conditions, thermal noise in the channel and shot noise associated with the leakage current in the gate are usually the most important types of noise. These can be represented by two current generators, $i_1(t)$ and $i_2(t)$, connected between the source and the drain, and the gate and the source, respectively, as illustrated in Fig. 5.8(a). By a simple circuit transformation, the output generator can be transferred to the input as a series voltage generator, $v_n(t)$, as shown in Fig. 5.8(b).

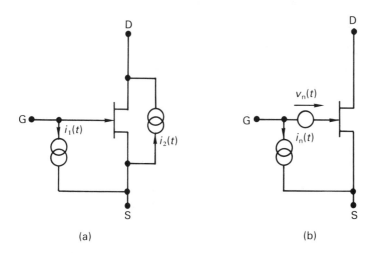

(a)   (b)

Fig. 5.8 – (a) Noise current generators connected across the input and output of a JFET and (b) an equivalent circuit in which the output current generator has been transferred to the input as a series voltage generator.

The power spectral densities of the current and voltage input noise generators can be written immediately from equations (5.29b), (5.33b) and (5.46) as

$$\overline{S_{i_n}(\omega)} \simeq \frac{\omega^2 C^2}{g_{msat}} k\theta + 2qI_g \tag{5.52}$$

and

$$\overline{S_{v_n}(\omega)} \simeq (2/3) \left(\frac{4k\theta}{g_{msat}} \right) , \tag{5.53}$$

and from equation (5.35) the normalized cross-spectral density between the thermal noise in the gate and the drain is

$$\Gamma_{i_n v_n} \simeq -0.4j . \tag{5.54}$$

The corresponding quantity for the noise in the gate leakage current and the thermal noise in the channel is zero.

When the signal source is resistive, the noise figure of the JFET does not involve $\Gamma_{i_n v_n}$ at all, since the latter is purely imaginary. Assuming the source resistance is R_s, the noise figure is

$$F = 1 + \frac{\overline{S_{i_n}} R_s}{4k\theta} + \frac{\overline{S_{v_n}}}{4k\theta R_s} . \tag{5.55}$$

The condition on R_s to minimize F is

$$R_s = \sqrt{(\overline{S_{v_n}}/\overline{S_{i_n}})} , \tag{5.56}$$

giving the minimum noise figure

$$F_0 = 1 + \frac{2\sqrt{(S_{v_n} S_{i_n})}}{4k\theta} . \tag{5.57}$$

For high frequencies, when the first term on the right of equation (5.52) is predominant, the minimum noise figure from equation (5.57) is

$$F_0 = 1 + (2/3)^{\frac{1}{2}} \frac{\omega C}{g_{msat}} . \tag{5.58}$$

At a frequency equal to $g_{msat}/2\pi C$ (the gain-bandwidth product), the minimum noise figure from this expression is 1.82, or 2.6 dB.

A marginal improvement can be achieved by noise tuning, using a reactive source. In this case, the minimum noise figure is

$$F_0 = 1 + \frac{\omega C}{g_{msat}} (2/3)^{\frac{1}{2}} (1 + \Gamma^2_{i_n v_n})^{\frac{1}{2}}$$

$$= 1 + \frac{\omega C}{g_{msat}} (2/3)^{\frac{1}{2}} (1 - 0.16)^{\frac{1}{2}} . \tag{5.59}$$

When $\omega C/g_{msat} = 1$, the value of this expression is 1.75, or 2.4 dB.

5.10 NOISE IN MOSFETS

The noise mechanisms in MOSFETs are essentially the same as those in JFETs, the main exception being $1/f$ noise which, at low frequencies, is the dominant source of noise in a MOSFET (Jordan and Jordan 1965, Yau and Sah 1969b). The fact that $1/f$ noise is present in MOSFETs but is almost entirely absent from good low-noise JFETs, strongly suggests that $1/f$ noise is a surface effect rather than a bulk effect. Whatever its origin, $1/f$ noise prevents MOSFETs from being low-noise devices.

The phenomenon of $1/f$ noise in MOS devices has been investigated by numerous authors, including Abowitz et al. (1967), Christenson et al. (1968), Hsu et al. (1968), Berz (1970), Klaassen (1970), Klaassen (1971), Nicollian and Melchior (1967), Fu and Sah (1972), Hsu (1970), Katto et al. (1974, 1975), and Katto et al. (1977); and related papers have also been published by Leventhal (1968) and Leuenberger (1968). Most of the theoretical models of $1/f$ fluctuations in MOS devices treat the noise as a surface-related effect, although the details of the mechanism involved are still not entirely understood. One of the most recent of these models (van der Ziel, 1980), which incorporates several of the features of previous theories, involves the interaction of carriers in the channel with surface states at the semiconductor/oxide interface. It is then postulated that a further interaction occurs, via tunnelling, between the carriers in the surface states and traps in the oxide layer. This mechanism results in a modulation of the number of free carriers in the channel, and it also modulates the surface potential, thereby producing a fluctuation in the surface mobility. These two effects, the number fluctuation and the mobility fluctuation, form the basis of van der Ziel's treatment. An alternative model has also been proposed recently, by Vandamme (1980), in which the $1/f$ noise is treated as a true mobility fluctuation in the channel. Now this is a bulk effect and hence should also give rise to $1/f$ noise in JFETs. Since $1/f$ noise is absent from JFETs, it would appear that the bulk model does not apply to these devices; and if it does not apply to JFETs, then presumably it is not a valid model for $1/f$ noise in other silicon devices either.

Apart from $1/f$ noise, the main sources of noise in MOSFETs are thermal noise in the channel (Klaassen and Prins, 1967) and generation–recombination noise in the space charge regions (Yau and Sah, 1969a, 1969c). Both can be treated in much the same way as the corresponding noise mechanisms in JFETs. At one time it was thought by van der Ziel and his coworkers (Halladay and van der Ziel, 1968b, 1969; Takagi and van der Ziel, 1969) that an excess 'white' noise component existed in MOSFETs, in addition to the thermal noise in the channel. This suggestion has since been refuted by Yau and Sah (1969b), who claim that the observed excess component of noise is in fact a $1/f$ noise component, and that the misinterpretation on the part of the previous investigators was due to noise measurements made at insufficiently high frequencies on devices showing very high $1/f$ noise.

High frequency noise in MOSFETs has been studied by Shoji (1966), Klaassen and Prins (1968, 1969), and Leupp and Strutt (1968, 1969).

The simple theory of thermal noise fluctuations in the channel must be modified if the bulk charge associated with ionized impurities in the substrate is taken into account. Sah *et al.* (1966) have investigated the effect of this bulk charge on the drain current noise, and have found that it can increase the thermal noise level by a factor of up to four above the level predicted by the simple theory. The gate current noise is essentially unaffected by the bulk charge (Rao, 1969).

REFERENCES

G. Abowitz, E. Arnold and E. A. Leventhal (1967), Surface states and $1/f$ noise in MOS transistors, *IEEE Trans. Elect. Dev.*, **ED–14**, 775–777.

F. Berz (1970), Theory of low frequency noise in Si MOSTs, *Solid State Elect.*, **13**, 631–647.

W. C. Brunke (1963), Noise measurements in field effect transistors, *Proc. IEEE (Correspondence)*, **51**, 378–379.

W. C. Brunke and A. van der Ziel (1966), Thermal noise in junction-gate field-effect transistors, *IEEE Trans. Elect. Dev.*, **ED–13**, 323–329.

S. Christenson, I. Lundstrom and C. Svensson (1968), Low frequency noise in MOS transistors. I: theory, *Solid State Elect.*, **11**, 797–812; II: experiment, *Solid State Elect.*, **11**, 812–820.

M. J. Churchill and P. O. Lauritzen (1971), Carrier density fluctuation noise in silicon junction field effect transistors at low temperatures, *Solid State Elect.*, **14**, 985–993.

J. C. Courvoisier, W. Haidinger, P. J. W. Jochems and L. J. Tummers (1963), Evaporation–condensation method for making germanium layers for transistor purposes, *Solid State Elect.*, **6**, 265–270.

G. C. Dacey and I. M. Ross (1953), Unipolar 'field effect' transistor, *Proc. IRE*, **41**, 970–979.

E. Erlbach and J. B. Gunn (1962), Noise temperature of hot electrons in germanium, *Phys. Rev. Lett.*, **8**, 280–282.

H. S. Fu and C. T. Sah (1969), Lumped model analysis of the low frequency generation noise in gold-doped silicon junction-gate field-effect transistors, *Solid State Elect.*, **12**, 605–618.

H. S. Fu and C. T. Sah (1972), Theory and experiments on surface $1/f$ noise, *IEEE Trans. Elect. Dev.*, **ED–19**, 273–285.

A. S. Grove (1967), *Physics and Technology of Semiconductor Devices*, John Wiley, Chapter 8.

J. A. Geurst (1965), Calculation of the high-frequency characteristics of field-effect transistors, *Solid State Elect.*, **8**, 563–566.

H. E. Halladay and W. C. Brunke (1963), Excess noise in field-effect transistors, *Proc. IEEE*, **51**, 1671.

H. E. Halladay and A. van der Ziel (1968a), Field-dependent mobility effects in the excess noise of junction-gate field-effect transistors, *IEEE Trans. Elect. Dev. (Correspondence)*, **ED–14**, 110–111.

H. E. Halladay and A. van der Ziel (1968b), Test of the thermal noise hypothesis in MOSFETs, *Elect. Lett.*, **4**, 366–367.

H. E. Halladay and A. van der Ziel (1969), On the high frequency excess noise and equivalent circuit representation of the MOS–FET with n-type channel, *Solid State Elect.*, **12**, 161–176.

J. W. Haslett and J. M. Kendall (1972), Temperature dependence of low-frequency excess noise in junction-gate FETs, *IEEE Trans. Elect. Dev.*, **ED–19**, 943–950.

J. W. Haslett and F. N. Trofimenkoff (1969a), Thermal noise in field-effect devices, *Proc. IEE*, **116**, 1863–1868.

J. W. Haslett and F. N. Trofimenkoff (1969b), Generation noise resistance in junction field effect transistors at pinch-off, *Solid State Elect.*, **12**, 747–750.

J. R. Hauser (1965), Small signal properties of field effect devices, *IEEE Trans. Elect. Dev.*, **ED–22**, 614–616.

C. F. Hiatt, A. van der Ziel and K. M. van Vliet (1975), Generation–recombination noise produced in the channel of JFETs, *IEEE Trans. Elect. Dev.*, **ED–22**, 614–616.

S. T. Hsu (1970), Surface state related $1/f$ noise in MOS transistors, *Solid State Elect.*, **13**, 1451–1459.

S. T. Hsu, D. J. Fitzgerald and A. S. Grove (1968), Surface-related $1/f$ noise in p–n junctions and MOS transistors, *Appl. Phys. Lett.*, **12**, 287–289.

A. G. Jordan and N. A. Jordan (1965), Theory of noise in metal oxide semiconductor devices, *IEEE Trans. Elect. Dev.*, **ED–12**, 148–156.

K. Kandiah and F. B. Whiting (1978), Low frequency noise in junction field effect transistors, *Solid State Elect.*, **21**, 1079–1088.

H. Katto, M. Aoki and E. Yamada (1977), *Proc. Symposium on $1/f$ fluctuations, Tokyo, July 11–13;* (1977) *Conf. Rep.*, 148–153.

H. Katto, Y. Kamigaki and Y. Itoh (1974) *Proc. 6th Conf. on Solid State Devices, Tokyo;* (1975), *Supplement to the J. of Japan Soc. of Applied Physics,* **44,** 243–248.

F. M. Klaassen (1967), High-frequency noise of the junction field effect transistor, *IEEE Trans. Elect. Dev.,* **ED–14,** 368–373.

F. M. Klaassen (1970), On the geometrical dependence of $1/f$ noise in MOS transistors, *Philips Research Reports,* **25,** 171–174.

F. M. Klaassen (1970), On the influence of hot carrier effects on the thermal noise of field-effect transistors, *IEEE Trans Elect. Dev.,* **ED–17,** 858–862.

F. M. Klaassen and J. Prins (1967), Thermal noise of MOS transistors, *Philips Research Reports,* **22,** 505–514.

F. M. Klaassen and J. Prins (1968), Noise in VHF and UHF MOS tetrodes, *Philips Research Reports,* **23,** 478–484.

F. M. Klaassen and J. Prins (1969), Noise of field-effect transistors at very high frequencies, *IEEE Trans. Elect. Dev.,* **ED–16,** 952–957.

F. M. Klaassen and J. R. Robinson (1970), Anomalous noise behaviour of the junction gate field effect transistor at low temperatures, *IEEE Trans. Elect. Dev.,* **ED–17,** 852–857.

P. O. Lauritzen (1965), Low-frequency generation noise in junction field effect transistors, *Solid State Elect.,* **8,** 41–58.

P. O. Lauritzen and C. T. Sah (1963), Low frequency recombination–generation noise in silicon FETs, presented at 1963 IEEE Solid State Device Research Conference, Lansing, Mich.; abstract, *IEEE Trans. Elect. Dev.,* **ED–10,** 334–335.

F. Leuenberger (1968), $1/f$ noise in gate-controlled planar silicon diodes, *Elect. Lett.,* **4,** 280.

A. Leupp and M. J. O. Strutt (1968), Noise behaviour of the MOSFET at VHF and UHF, *Elect. Lett.,* **4,** 313–314.

A. Leupp and M. J. O. Strutt (1969), High-frequency FET noise parameters and approximation of the optimum source admittance, *IEEE Trans. Elect. Dev.,* **ED–16,** 428–431.

E. A. Leventhal (1968), Derivation of $1/f$ noise in silicon inversion layers from carrier motion in a surface band, *Solid State Elect.,* **11,** 621–627.

M. Nakahara, H. Iwasawa and K. Yasutake (1968), Anomalous enhancement of substrate terminal current beyond pinch-off in silicon n-channel MOS transistors and its related phenomena, *Proc. IEEE (letters),* **56,** 2088–2090.

M. Nakahara and I. Kobayashi (1970), On the gate current and noise behaviour in pinched-off silicon junction field-effect transistors, *Proc. IEEE (letters),* **58,** 1158–1159.

E. M. Nicollian and H. Melchior (1967), A quantitative theory of $1/f$ type noise due to interface states in thermally oxidized silicon, *Bell Syst. Tech. J.,* **46,** 2019–2033.

V. Radeka (1967), Field effect transistor noise as a function of temperature and frequency, *Conference on Semiconductor Radiation Detectors and Circuits, Gatlinburg, Tenn., May 1967.*

V. Radeka (1969), FET noise as a function of temperature and frequency, in *Semiconductor Nuclear Particle Detectors,* Washington, DC, National Academy of Sciences, publ. 1593.

P. S. Rao (1969), The effect of the substrate upon the gate and drain noise of MOSFETs, *Solid State Elect.,* 12, 549–555.

F. N. H. Robinson (1969), Noise in field-effect transistors at moderately high frequencies, *Elect. Eng.,* 41, 353–355.

R. D. Ryan (1969), The gate currents of junction field-effect transistors at low temperatures, *Proc. IEEE (letters),* 57, 1225–1226.

C. T. Sah (1964), Theory of low-frequency generation noise in junction-gate field-effect transistors, *Proc. IEEE,* 52, 795–814.

C. T. Sah, S. Y. Wu and F. H. Hielscher (1966), The effects of fixed bulk charge on the thermal noise in metal-oxide semiconductor transistors, *IEEE Trans. Elect. Dev.,* ED–12, 148–156.

W. Shockley (1952), A unipolar 'field-effect' transistor, *Proc. IRE,* 40, 1365–1376.

W. Shockley and W. T. Read Jr. (1952), Statistics of recombination of holes and electrons, *Phys. Rev.,* 87, 835–842.

M. Shoji (1966), Analysis of high-frequency thermal noise of enhancement mode MOS field-effect transistors, *IEEE Trans. Elect. Dev.,* ED–13, 520–524.

K. Takagi, Y. Sumino and K. Tabata (1976), Correlation coefficient of gate and drain noise at high electric field, *Solid State Elect.,* 19, 1043–1045.

K. Takagi and A. van der Ziel (1969), Non-thermal noise in MOS FETs and MOS tetrodes, *Solid State Elect.,* 12, 907–913.

K. Takagi and A. van der Ziel (1979), High frequency excess noise and flicker noise in GaAs FETs, *Solid State Elect.,* 22, 285–287.

F. N. Trofimenkoff (1965), Field-dependent mobility analysis of the field-effect transistor, *Proc. IEEE (Correspondence),* 53, 1765–1766.

F. N. Trofimenkoff and A. Nordquist (1968), FET operation in the pinch-off mode, *Proc. IEE,* 115, 496–502.

A. van der Ziel (1962), Thermal noise in field-effect transistors, *Proc. IRE,* 50, 1808–1812.

A. van der Ziel (1963a), Gate noise in field-effect transistors at moderately high frequencies, *Proc. IEEE,* 51, 461–467.

A. van der Ziel (1963b), Carrier density fluctuation noise in field effect transistors, *Proc. IEEE (Correspondence),* 51, 1670–1671.

A. van der Ziel (1969), Noise in junction and MOS–FETs at high temperatures, *Solid State Elect.,* 12, 861–866.

A. van der Ziel (1979), Flicker noise in electronic devices, *Advances in Electronics and Electron Physics,* 49, 225–297.

A. van Ziel (1980), The oxide trap model of $1/f$ noise in MOSFETs, *Proc. Symp. on 1/f fluctuations, Orlando, Florida.*

A. van der Ziel and E. R. Chenette (1978), Noise in solid state devices, *Advances in Electronics and Electron Physics,* **46**, 313–383.

A. van der Ziel and J. W. Ero (1964), Small signal, high frequency theory of field-effect transistors, *IEEE Trans. Elect. Dev.,* **ED–11**, 128–135.

L. K. J. Vandamme (1980), Model for $1/f$ noise in MOS transistors biased in the linear region, *Solid State Elect.,* **23**, 317–323; $1/f$ noise model for MOSTs biased in nonohmic region, *Solid State Elect.,* **23**, 325–329.

K. M. van Vliet and C. F. Hiatt (1975), Theory of generation–recombination noise in the channel of junction field effect transistors, *IEEE Trans. Elect. Dev.,* **ED–22**, 616–617.

K. K. Wang, A. van der Ziel and E. R. Chenette (1975), Neutron-induced noise in junction field effect transistors, *IEEE Trans. Elect. Dev.,* **ED–22**, 591–593.

L. D. Yau and C. T. Sah (1969a), Geometrical dependences of the low-frequency generation–recombination noise in MOS transistors, *Solid State Elect.,* **12**, 903–905.

L. D. Yau and C. T. Sah (1969b), On the 'excess white noise' in MOS transistors, *Solid State Elect.,* **12**, 927–936.

L. D. Yau and C. T. Sah (1969c), Theory and experiment of low-frequency generation–recombination noise in MOS transistors, *IEEE Trans. Elect. Dev.,* **ED–16**, 170–177.

6

$1/f$ Noise

6.1 INTRODUCTION

When a constant voltage is applied across a resistor, a fluctuating component is observed in the current in addition to the thermal noise which is also present. Similarly, when a constant current flows through a resistor, an excess random fluctuation is observed in the voltage. These excess noise components, which are observed in most resistors in the presence of a dc current or voltage, show a power spectral density which varies as $|f|^{-\alpha}$, where α is more or less constant and usually lies between 0.8 and 1.4. This spectral shape has been observed over a wide frequency range, spanning a dozen decades or so, from 10^{-6} Hz to 10^6 Hz or higher in some microwave devices. In fact, noise obeying the inverse frequency power law is known nowadays to exist in practically all electronic materials and devices, including homogeneous semiconductors and junction devices, metal films and whiskers, liquid metals, electrolytic solutions, thermionic tubes, superconductors and Josephson junctions; and usually, irrespective of where the phenomenon occurs it goes under the generic name of $1/f$ noise.

In the past, $1/f$ noise has been variously called current noise, excess noise, flicker noise (usually in connection with the fluctuations in electron emission from a thermionic cathode), semiconductor noise (before it was appreciated that it also appears in metals and aqueous electrolytes) and contact noise (although it is well known that $1/f$ noise is not in general a contact effect). On a cautionary note, it is perhaps worth commenting that the name $1/f$ noise, common to all manifestations of the phenomenon, should not be taken to imply the existence of a common physical mechanism giving rise to them all. Indeed, the available evidence seems to suggest that the origins of $1/f$ noise in different types of device may be quite different.

The first observations of $1/f$ noise in an electronic system were made over fifty years ago (Johnson, 1925) and since then the subject has developed extensively. Much of the early work on the phenomenon has been described by Bell (1960), who has also published a review article which includes more recent developments (Bell, 1980). Some interesting and novel aspects of $1/f$ noise are discussed by van der Ziel (1979) and an up-to-date survey of $1/f$ noise theories has been produced by Weissman (1981). Some of the mathematical and empirical models that have been proposed for $1/f$ noise are described by Hooge (1976). All of these reviews contain numerous references to relevant papers in the field.

$1/f$ noise is a ubiquitous type of fluctuation, appearing not only in measurements on electronic systems but also in a diverse range of observations elsewhere. It has been reported, for example, in connection with earthquakes and thunderstorms (Machlup, 1981) and the height of the floods of the river Nile (Gardner, 1978), although the spectra involved here are not power spectra in the usual sense of the term. Biological systems also exhibit $1/f$ noise: the normal human heartbeat period shows a fluctuation whose power spectral density varies approximately as $1/|f|$ at frequencies below 0.3 Hz, and a similar spectral shape is observed in connection with brain-wave fluctuations, in particular with the alpha wave component of the electroencephalogram (EEG). Both of these observations are reported by Musha (1981). Neuro-membranes are well known to exhibit $1/f$ fluctuations, and a number of references on this particular manifestation of $1/f$ noise are cited by Hooge (1976). Another area where $1/f$ noise is encountered is music. Voss and Clarke (1978) found that the relationship between intensity and pitch in classical music (Mozart, Bach, Beethoven, Debussy), western music, the Beatles and jazz, as well as music from a variety of different cultures, is $1/|f|$ in character. What is perhaps more surprising is that the individual's perception of music is strongly influenced by its spectral shape: three pieces of music 'composed' from random numbers with spectral densities varying as $1/|f|^2$, $1/|f|$ and $1/|f|^0$ (white noise) were judged to be boring ($1/|f|^2$) irritating (white noise) and pleasing ($1/|f|$). It appears that 'good' music has a $1/|f|$ spectrum, presumably because its correlation time is neither so short as to make it disturbingly irregular nor so long as to make it predictable.

There are many other examples of the $1/|f|$ law appearing in non-electronic systems, but we shall not discuss them here. The remainder of the chapter is devoted to $1/f$ noise in connection with electronic devices and conductors.

An enormous pool of data has been accumulating on $1/f$ noise in various electronic devices over the past two decades. The experimental results have often raised more questions than they have answered, and in certain cases they appear to be contradictory. The physical origin of $1/f$ noise is not understood, except perhaps in a few specific cases, and indeed it is not possible even now to say with certainty whether $1/f$ noise originates at the surface or within the volume of a specimen. Most of the evidence suggests that in some types of device it is a surface effect, as in the case of a MOSFET where the semiconductor/oxide interface plays an important role; but in other devices, such as a homogeneous resistor, $1/f$ noise is thought to be a bulk effect associated with a random modulation of the resistance, implying a fluctuation in either the number or the mobility of the charge carriers. Somewhat puzzlingly, there is experimental evidence which appears to support both the number and mobility fluctuation hypotheses.

In contrast with many of the experimental data on the phenomenon, the power spectral density of $1/f$ noise in homogeneous materials behaves more or less systematically. Hooge (1969) has formulated an empirical law, in which the spectral density varies inversely with the total number of carriers in the specimen, and although it may not be of universal validity the law appears to be representative of many observed $1/f$ noise spectra. However, Hooge's law has not led to the emergence of a physical mechanism for $1/f$ noise, and indeed the explanation for the phenomenon seems to be as remote as ever. Several theories have been proposed, the most widely discussed being surface trapping and equilibrium thermal energy exchange with the environment. Although these theories may individually account for certain specific manifestations of the phenomenon, no general theory of $1/f$ noise exists at present. It was the absence of such a theory that led to the appearance in the literature of an alternative approach to the problem, based on the fractional-order integration of white noise. The procedure results in a mathematical construction which gives rise to the required spectral form but which sheds little light on the physical mechanism responsible for the noise.

Before describing some of the recently acquired experimental data on $1/f$ noise and discussing the more tenable theories of the phenomenon, we examine some of the implications and properties of a power spectrum which varies as $1/|f|^{\alpha}$.

6.2 SCALE INVARIANCE

A 'true' $1/f$ noise waveform, $x(t)$, is characterized by a power spectral density function in which α is identically unity:

$$\overline{S_x(\omega)} = c/|\omega| \, , \tag{6.1}$$

where c is independent of frequency. This spectral form has been the subject of extensive discussions in the literature, often in connection with philosophical arguments concerning a hypothetical low-frequency limit below which it may no longer apply. Much effort has been expended in the search for such a limit, but none has been found. Mansour et al. (1968) made $1/f$ noise measurements on MOSFETs down to a frequency of 5×10^{-5} Hz and similar measurements were performed by Caloyanides (1974) on semiconductors down to a frequency of 5×10^{-7} Hz but no significant departure from the $1/|f|$ law was observed in either case.

A low frequency limit was postulated because in the absence of such a limit the total power in a spectrum having the form in equation (6.1) is infinite. In fact, both the high and low frequency extremes of the spectrum give rise to an infinity, but in the case of the former the difficulty is easily eliminated on recognizing that a high-frequency limit will exist — associated with the inherent response time of either the mechanism producing the noise or the system in which it is produced — beyond which the law in equation (6.1) will no longer hold; instead the spectrum will decay at least as fast as $1/f^2$. By way of contrast, the infinity associated with the low frequency end of the $1/|f|$ spectrum cannot be removed on the basis of any *physical* argument, at least not one that is known at present. It is important, however, not to overstate the significance of this infinity, which is logarithmic in character; as discussed below, there is no *practical* reason for expecting it to cause embarrassment, nor is there any reason for believing a low-frequency limit to the $1/|f|$ law will occur within the observable range of frequencies.

Before pursuing the question of the low-frequency behaviour of $1/f$ noise, it is expedient to examine the integrated power in the spectrum between the (positive) angular frequencies ω_1 and ω_2:

$$P_x(\omega_1, \omega_2) = \frac{1}{2\pi} \int_{\omega_1}^{\omega_2} \overline{S_x(\omega)} \, d\omega$$

$$= (c/2\pi) \ln (\omega_2/\omega_1) \ . \tag{6.2}$$

This simple result shows that for a fixed frequency ratio ω_2/ω_1, the integrated power is a constant. Thus the total noise power between, say, 0.1 Hz and 1 Hz is the same as that between 1 Hz and 10 Hz, or 10 Hz and 100 Hz, or in any other decade of frequency. This property of $1/f$ noise is known as *scale invariance*.

If ω_2 in equation (6.2) is allowed to become indefinitely large, with ω_1 held constant and finite, it is apparent that the total power shows a logarithmic infinity. For the reason given above, there is no serious difficulty here since the $1/|f|$ law cannot extend to indefinitely high frequencies. When ω_1 in equation (6.2) goes to zero with ω_2 held constant and finite the total power again shows a logarithmic infinity, but now the implications of such behaviour cannot be dismissed quite so lightly.

Consider the following question: does the low-frequency infinity in the total power militate against the existence of the $1/|f|$ law down to zero frequency? It may be argued that it does not, because any *measurement* on the noise waveform will be performed in a time which, however long, will always be finite in duration. This means that the lowest frequency measured, however small it may be, will always be greater than zero, and hence the total noise power measured will always be finite. The unavoidable conclusion is that there is no reason in principle why $1/f$ noise should show a flattening of the spectrum below some low-frequency limit; an infinity which appears only after looking for an infinite time is not cause for concern.

A slightly different perspective on the question of the extent of the $1/|f|$ law has been presented by Flinn (1968), who attempted to quantify the maximum number of decades in frequency over which a $1/|f|$ spectrum could conceivably exist. He set the upper limit at 10^{23} Hz, corresponding to the time it takes light to traverse the classical radius of an electron, and the lower limit at 10^{-17} Hz, based on the estimated age of the universe. This is a span of 40 decades, which is an enormous range and well beyond experimental investigation. Yet the total r.m.s. value of the $1/f$ fluctuation is merely $\sqrt{40} \simeq 6$ times that in one decade. Flinn employed an experimental result reported by Brophy (1968) to obtain a figure of 3.5×10^{-7} W for the total power in the 40 decades, which is six orders of magnitude less than the dc input power of 0.39 W used in Brophy's experiment. Clearly, even in this extreme calculation the total noise power is insignificant in terms, for example, of the power rating of the dc supply used in the experiment. This would seem to eliminate any practical reason for expecting the $1/|f|$ dependence to cease at some low-frequency limit falling within the accessible range of frequencies.

6.3 STATIONARITY

A debate on whether $1/f$ noise is statistically stationary has been argued in the literature for a number of years. Statements to the effect that $1/f$ noise is a 'stationary' fluctuation are as common as those proclaiming that it exhibits some degree of 'non-stationarity'. Usually the precise meaning of these terms is left unspecified, with the result that the whole issue is somewhat confused. The problems arise, of course, in connection with the divergence of the spectrum in the low-frequency limit.

In order to clarify the situation two specific waveforms are discussed below, namely band-limited $1/f$ noise (in which the low-frequency components are absent) and low-pass filtered $1/f$ noise (in which the low-frequency components down to zero frequency are present). The band-limited fluctuation corresponds to the actual waveforms investigated in experimental measurements of $1/f$ noise (since all observed $1/f$ noise is band-limited, either directly by filtering or indirectly by a limited observation time), and such a process is statistically stationary. In contrast, low-pass filtered $1/f$ noise is a theoretical abstraction and is non-stationary.

6.3.1 Band-limited 1/f noise

Consider a $1/f$ noise process, $x(t)$, which is band-pass filtered so that its power spectral density is

$$\overline{S_x(\omega)} = \begin{cases} c/|\omega| & \text{for } \omega_1 \leqslant \omega \leqslant \omega_2 \\ 0 & \text{otherwise} \end{cases}, \tag{6.3}$$

where ω_2 and ω_1 are the upper and lower angular frequencies of the passband.

It follows from the Wiener–Khintchine theorem in conjunction with equation (6.3) that the autocorrelation function of $x(t)$ is

$$\overline{\phi_x(\tau)} = c/2\pi \int_{\omega_1}^{\omega_2} \frac{\cos\omega\tau}{\omega} \, d\omega \; , \tag{6.4}$$

which, by making a simple change of variable and rearranging the limits on the integral, can be expressed in the form

$$\overline{\phi_x(\tau)} = \frac{c}{2\pi} \left\{ Ci\,(\omega_2\tau) - Ci\,(\omega_1\tau) \right\} \; , \tag{6.5}$$

where

$$Ci(z) = \int_\infty^z \frac{\cos y}{y} \, dy \tag{6.6}$$

is the cosine integral. The series expansion of the cosine integral is (e.g. see Lebedev (1965))

$$Ci(z) = \gamma + \ln(z) + \sum_{k=1}^\infty \frac{(-1)^k z^{2k}}{(2k)!\,2k} \; , \tag{6.7}$$

where $\gamma = 0.5772 \ldots$ is Euler's constant; and thus when $z \to 0$ the function $Ci(z)$ behaves as $\ln(z)$. It follows that the mean-square value of $x(t)$, obtained from equation (6.5) by taking the limit as $\tau \to 0$, is

$$\overline{\phi_x(0)} = \frac{c}{2\pi} \ln\,(\omega_2/\omega_1) \; , \tag{6.8}$$

in agreement with equation (6.2).

It is evident from equations (6.5) and (6.8) that, for a fixed passband with $\omega_1 > 0$ the autocorrelation function and the mean-square value of the process whose power spectral density is given in equation (6.3) both converge to unique limiting forms. This implies that the statistical measures (i.e. probability density functions) underlying these second-order quantities depend only on the delay time τ and not on the absolute times at which the ensemble averaging is performed – which is the condition for wide-sense stationarity. Thus, band-limited $1/f$ noise is at least wide-sense stationary.

The question of the stationarity of $1/f$ noise was originally raised by Brophy (1968, 1970), who made measurements of the 'variance of the variance', as it has come to be known. He took samples of $1/f$ noise in a large number of epochs and evaluated the variance of each sample. These variances themselves fluctuated, however, because each was measured in an interval of finite duration. The degree of this fluctuation is expressed as the variance of the variance. Brophy found that the variance of the variance was greater for band-limited $1/f$ noise than for stationary, white, thermal noise, which accounts for his reference to $1/f$ noise as 'noisy noise'; and he concluded that $1/f$ noise 'possesses some form of conditional stationarity'.

Qualitatively at least, Brophy's results are consistent with the conclusion reached above that band-limited $1/f$ noise is wide-sense stationary. Moreover, since the correlation time of $1/f$ noise is much greater than that of white noise, the relatively high variance of the variance shown by the former is to be expected, and should not be construed as evidence supporting the view that band-limited $1/f$ noise is non-stationary.

Further experiments investigating the stationarity of band-limited $1/f$ noise have been performed by Stoisiek and Wolf (1976), who measured fluctuations in the variance of the noise from two types of physical source (carbon resistors and bipolar transistors) and compared them with similar measurements made on 'artificial' $1/f$ noise, produced by shaping stationary gaussian noise with a $1/|f|$ power spectrum. They found no reason to doubt that band-limited $1/f$ noise is statistically stationary. Strasilla and Strutt (1974) reached a similar conclusion.

6.3.2 Low-pass filtered $1/f$ noise

When the lower limit of the pass band, ω_1, is zero the spectrum in equation (6.3) diverges, behaving as $1/|f|$ down to zero frequency. It is interesting to examine the stationarity of the $1/f$ noise waveform $x(t)$ in this case, even though such a spectrum would not be observed in practice. The following argument is based on establishing a condition which must be satisfied by any stationary process.

If a random fluctuation, $x(t)$, is statistically stationary, the Wiener–Khintchine theorem applies, allowing the power spectral density to be expressed as

$$\overline{S_x(\omega)} = 4 \int_0^\infty \overline{\phi_x(\tau)} \cos \omega\tau \, d\tau \, , \tag{6.9}$$

where $\overline{\phi_x(\tau)}$ is the autocorrelation function of $x(t)$. By differentiating with respect to ω, we find that

$$\lim_{\omega \to 0} \frac{d\,\overline{S_x(\omega)}}{d\omega} = 4 \lim_{\omega \to 0} \int_0^\infty \frac{\partial}{\partial \omega} \left[\overline{\phi_x(\tau)} \cos \omega\tau \right] d\tau$$

$$= -4 \lim_{\omega \to 0} \int_0^\infty \tau \, \overline{\phi_x(\tau)} \sin \omega\tau \, d\tau$$

$$= 0 \, . \tag{6.10}$$

Thus, the slope of the power spectral density of a stationary process at zero frequency is zero, or in other words the spectrum is flat in the limit of low frequency.

This condition is obviously not satisfied by a spectrum which keeps rising as the zero frequency limit is approached, which establishes low-pass filtered $1/f$ noise as a non-stationary process. Taken at face value this would seem to imply serious difficulties, associated with intractable mathematics (divergent integrals), in the construction of theories of $1/f$ noise, but in reality this is not the case, or at least need not be so, as discussed below.

6.3.3 Stationarity and theoretical modelling

As we have already mentioned, the experimentally accessible part of the spectrum is limited at the low-frequency end by the observation time, T, of the measurement. Since T must be finite, there is always a low-frequency region of the spectrum which cannot be observed. It is easy to see that if the spectrum levelled off at some infinitesimal frequency, well below the observable frequency range, the process would be indistinguishable from one whose spectrum continued rising; moreover, it would then obey the condition arising out of equation (6.10) and accordingly would be designated stationary. Thus, by modifying the spectrum in a minutely detailed, undetectable way, the status of the waveform can be changed from non-stationary to stationary. This suggests that the question of the stationarity of $1/f$ noise is merely one of semantics and is almost immaterial to the physics of the situation. The fact is that a statistically stationary random waveform can be constructed mathematically whose properties over the range of frequencies observable in practice are indistinguishable from those of measured $1/f$ noise processes. The real difficulty lies, not with the mathematics of $1/f$ noise, but in identifying physical mechanisms which might be responsible for generating it.

It is apparent from the above argument that the assumption of wide sense stationarity is not inconsistent with the experimental evidence available on $1/f$ noise. The advantage of adopting this assumption is that mathematical models can then be constructed based on the familiar Wiener—Khintchine and related theorems, which apply to stationary processes. But by assuming stationarity the possibility that $1/f$ noise is an inherently non-stationary waveform is, by implication, discarded. However, it would seem that this is not a serious loss because theoretical constructions purporting to represent a non-stationary process often involve some physically unrealizable feature. This is illustrated in a paper by Tandon and Bilger (1976), where a functional form (their equation (3)) for the statistical expectation $E[y(t)y(t + \tau)]$ of the non-stationary process $y(t)$ is proposed. Inspection of their suggested function reveals that it is non-causal: it contains T_0, the duration of the epoch in which the ensemble averaging is performed, and this implies a 'prediction' of the future. No actual system can behave in this way.

There is evidently no compelling reason for treating $1/f$ noise as non-stationary, and indeed such an approach is positively disadvantageous. On the other hand, by taking the pragmatic view and treating it as wide-sense stationary a degree of clarity is brought to the whole issue. This is apparent from the fact that stationarity demands a low-frequency roll-off, which ensures the convergence of certain integrals, and this in turn facilitates the construction of mathematical models of the noise. It is true that any *measurement* of the power spectrum, autocorrelation function or mean-square value of $1/f$ noise will always depend on the observation time T, but this is simply interpreted as meaning that T is not sufficiently long for these statistical measures to converge to unique limiting forms. Such convergence could only occur if T were to exceed the reciprocal of the (postulated) corner frequency.

The effect of a finite observation time on the measured spectrum is expressed in the empirical formula

$$\overline{S_{obs}(\omega, T)} = \begin{cases} c/|\omega| & \text{for } 2\pi/T \leqslant \omega \leqslant \omega_2 \\ 0 & \text{otherwise} \end{cases} , \qquad (6.11)$$

which is formally similar to equation (6.3) except that ω_1, the lower limit of the band-pass filter, has now been replaced by $2\pi/T$. Notice that the functional dependence on T has been included explicitly on the left of equation (6.11). By analogy with equations (6.5) and (6.8) the autocorrelation function and mean-square value associated with $\overline{S_{obs}}$ can be written, respectively, as

$$\overline{\phi_{obs}(\tau, T)} = \frac{c}{2\pi} \{Ci(\omega_2 \tau) - Ci(2\pi\tau/T)\} \qquad (6.12)$$

and

$$\overline{\phi_{obs}(0, T)} = \frac{c}{2\pi} \ln(f_2 T) \quad , \qquad (6.13)$$

where $f_2 = \omega_2/2\pi$ and $Ci(\ )$ is the cosine integral defined in equations (6.6) and (6.7). The logarithmic dependence of the mean-square value on T, expressed in equation (6.13), has been confirmed experimentally by Brophy (1970).

6.3.4 Comparison with the random walk

The spectra of $1/f$ noise and a Wiener–Lévy process (see § 2.10), which is a version of the random walk, are similar in that both are inverse frequency power laws, the former having a logarithmic slope of approximately -1 whilst that of the latter is -2. Yet the status of the two processes in connection with stationarity is different. The Wiener–Lévy process is a cumulative process and as such the waveform is well-defined. In the limit of low frequency it is guaranteed that the spectral density of such a process continues to vary as $1/\omega^2$, and it follows that the process is categorically non-stationary. There is no question of possibly encountering a corner frequency in the spectrum because, by the very nature of the process, this could not happen (the roll-off shown in Fig. 2.5(b) is not a genuine feature of the Wiener–Levy spectrum, it is an artefact associated with the finite gating of the process). In the case of $1/f$ noise, however, the physical mechanisms responsible for the waveform are not well-defined, a low-frequency roll-off may or may not exist, and the stationarity of the process is open to question, as discussed above.

6.4 $1/f$ WAVEFORMS

It has been mentioned already that, in general, the physical origins of $1/f$ noise are obscure. Some clue to the mechanism responsible for the phenomenon may be contained in the detailed structure of the waveform itself, which suggests that an examination of mathematically constructed processes having the characteristics of $1/f$ noise could provide certain insights into the physics underlying this type of fluctuation.

Two $1/f$ waveforms are discussed below. The first is a random pulse train, which is a construction that in the present context has received surprisingly little attention in the literature. Schönfeld (1955) introduced the idea that $1/f$ noise could be represented by a random succession of similarly shaped pulses and van der Ziel (1979) developed the model in terms of the simplest possible pulse shape function. But otherwise it appears that little interest has been shown in the approach, although a related model has been discussed by Bell (1974).

The second mathematical representation of a $1/f$ waveform described below is based on the superposition of a large number of relaxation processes with a wide spread of time constants (van der Ziel, 1950; McWhorter, 1956). This model has received more acclaim than the random pulse train approach, probably because it relates directly to a surface mechanism for $1/f$ noise, pertinent to MOSFETs, in which carriers tunnel between the semiconductor and traps located in the overlying oxide layer. A generalization of the superposition model has been described by Halford (1968).

6.4.1 A random pulse train model of $1/f$ noise

The power spectral density of a random pulse train, $x(t)$, in which the pulse shape function is $f(t)$, is given by Carson's theorem (see § 2.6):

$$\overline{S_x(\omega)} = 2\nu \, \overline{a^2} \, | \, F(j\omega)|^2 \, , \tag{6.14}$$

where $F(j\omega)$ is the Fourier transform of $f(t)$, $\overline{a^2}$ is the mean-square value of the pulse height and ν is the mean rate of the pulses. An inspection of equation (6.14) reveals that the frequency dependence of $\overline{S_x}$ is determined entirely by the shape of an individual pulse, $f(t)$. Thus, the problem is to specify a pulse shape which will lead to the required $1/|f|^\alpha$ spectral form. (A similar problem has already been encountered in connection with thermal noise and shot noise, where the pulse shape function is a delta function, its transform is unity and the spectra are independent of frequency.)

Consider the shape function

$$f(t) = u(t)t^{-(1-\frac{\alpha}{2})} \exp - \omega_x t \ , \tag{6.15}$$

where α and ω_x are both positive and independent of time, α is in the vicinity of but is not necessarily equal to unity, and $u(t)$ is the unit step function. The Fourier transform of $f(t)$ is

$$F(j\omega) = \int_0^\infty t^{-(1-\frac{\alpha}{2})} \exp - (\omega_x + j\omega)t \ dt$$

$$= \frac{\Gamma(\alpha/2)}{(\omega_x + j\omega)^{\alpha/2}} \ , \tag{6.16}$$

where $\Gamma(\ )$ is the gamma function. The integral in equation (6.16) is a standard form and can be found in any table of integrals, for example, Gradshteyn and Ryzhik (1965). When the result in equation (6.16) is substituted into Carson's theorem, the spectral density of the random pulse train with the pulse shape given by equation (6.15) is found to be

$$\overline{S_x(\omega)} = \frac{2\nu\overline{a^2}\,\Gamma^2(\alpha/2)}{|\omega_x^2 + \omega^2|^{\alpha/2}} \ . \tag{6.17}$$

The function on the right of equation (6.17) is sketched in Fig. 6.1(a). It shows a corner frequency at $\omega = \omega_x$, below which it is essentially flat, and over the frequency range where $\omega \gg \omega_x$ the expression approximates to

$$\overline{S_x(\omega)} \simeq \frac{c}{|\omega|^\alpha} \ , \ \alpha \simeq 1 \tag{6.18}$$

where $c = 2\nu\overline{a^2}\,\Gamma^2(\alpha/2)$. This spectral form is commensurate with $1/f$ noise spectra observed in practice, even to the extent of allowing some variation in the logarithmic slope through the index α. Moreover, the inverse frequency power law can be made to extend over as many decades as required since the roll-off (angular) frequency, ω_x, appearing in equation (6.17) can be made as small as we please. Incidentally, it is clear that however small ω_x may be, provided it is non-zero the function $\overline{S_x(\omega)}$ in equation (6.17) satisfies the condition for (wide-sense) stationarity implicit in equation (6.10).

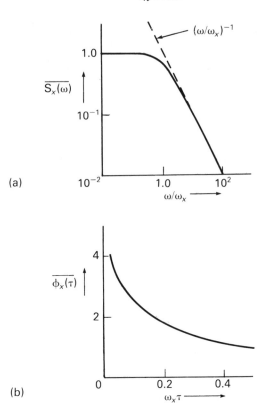

Fig. 6.1 – (a) $\overline{S_x(\omega)}$ from equation (6.17), normalized to the zero frequency value, for $\alpha = 1$. (b) The corresponding autocorrelation function, from equation (6.19) with $c = 2\pi$.

Assuming that ω_x lies well below the observable range of frequencies, the auto-correlation function and mean-square value of $x(t)$ will depend on the observation time, T, as discussed in § 6.3.3. It is, however, of theoretical interest to examine these statistical quantities in the limit as $T \to \infty$. Then the spectral form in equation (6.17) rather than that in (6.18) is the appropriate one to substitute into the Wiener–Khintchine integral for the autocorrelation function:

$$\overline{\phi_x(\tau)} = \frac{1}{2\pi} \int_0^\infty \overline{S_x(\omega)} \cos \omega\tau \, d\omega$$

$$= \frac{c}{2\pi} K_0(\omega_x\tau) \, , \tag{6.19}$$

where for simplicity we have taken α to be identically equal to unity, and $K_0(\ )$ is the modified Bessel function of the second kind of zero order. The function $\overline{\phi_x(\tau)}$ is sketched in Fig. 6.1(b) against $\omega_x \tau$. Perhaps the nature of the auto-correlation function is best appreciated if the Bessel function is approximated by the first few terms in its series expansion:

$$K_0(z) = -\gamma + \ln(2) - \ln(z) + \ldots , \qquad (6.20)$$

where $\gamma = 0.5772\ldots$ is Euler's constant. For small $\omega_x \tau$, which is the case of interest, it is apparent that $\overline{\phi_x(\tau)}$ varies as $\ln(\omega_x \tau)$, which for $\omega_x \neq 0$ is finite everywhere except at the origin in τ, where it shows a logarithmic infinity associated with the high frequency end of the $1/f$ noise spectrum. This is no more embarrassing than the delta function describing the autocorrelation function of white noise (whose spectrum extends uniformly to infinity): in neither case is the infinite mean-square value anything more than a mathematical abstraction, since in reality a bandwidth limitation in both cases ensures finite practical measures. For the case when $\omega_x = 0$, it is evident from equations (6.19) and (6.20) that the autocorrelation function is infinite for all values of τ. This is certainly a strange theoretical attribute, associated intimately with the existence of the $1/|f|$ law down to indefinitely low frequencies; but, while emphasizing the long correlation times to be expected of $1/f$ noise, it represents an extreme (low frequency) condition which for the reasons discussed at length above would never be observed in practice.

It is clear that a random pulse train in which the pulse shape varies approximately as $t^{-\frac{1}{2}}$ shows the same second-order statistical properties as experimentally measured $1/f$ noise. Of course, the physical origins of such a pulse shape, in electronic devices or indeed in the many systems which exhibit non-electronic $1/f$ noise, are not immediately apparent. Thermal diffusion has been proposed as a mechanism for producing a modulation of the electrical resistance of a specimen, which in turn gives rise to $1/f$ noise over a limited range of frequency, and this model can be interpreted in terms of the random pulse train discussed above. Even though thermal fluctuations are no longer thought to be responsible for the observed $1/f$ noise (at least in the majority of cases), the model appeared to hold much promise in the mid-1970s when it was introduced. One appealing feature of the approach is that the equilibrium exchange of thermal energy between a body and its surroundings is a universal phenomenon, and in this sense is very similar to $1/f$ noise itself. Other factors, however, militated against the temperature fluctuation hypothesis and the experimental evidence currently available is such that the model, if not entirely abandoned, has fallen from favour. In view of the importance attached to the mechanism when it was introduced, and the possibility that it may be reinstated in the future in a modified form, its attributes and failings are described later in § 6.7.3.

6.4.2 Superposition of relaxation processes

The power spectral density of a relaxation process, $z(t)$, with relaxation time τ_z can be expressed in the general form

$$S_z(\omega) = \frac{g(\tau_z)}{(1 + \omega^2 \tau_z^2)} ,$$ (6.21)

where $g(\tau_z)$ gives the functional dependence of the numerator on τ_z. The form of $g(\tau_z)$ depends on the physical mechanism responsible for the noise and in some cases, for example thermal noise, $g(\tau_z)$ is independent of τ_z altogether. For our present purpose there is no need to specify $g(\tau_z)$ but later, in connection with McWhorter's model of $1/f$ noise in a MOSFET (§ 6.7.2), the case where $g(\tau_z) \propto \tau_z$ is examined.

Imagine now that a linear superposition, $x(t)$, is constructed from relaxation processes whose time constants are distributed between upper and lower limits τ_2 and τ_1, with a probability density $p(\tau_z)$. The overall power spectral density is then

$$S_x(\omega) = \int_{\tau_1}^{\tau_2} S_z(\omega) p(\tau_z) \, d\tau_z = \int_{\tau_1}^{\tau_2} \frac{p(\tau_z) g(\tau_z)}{(1 + \omega^2 \tau_z^2)} \, d\tau_z .$$ (6.22)

For the case where the product of the two functions appearing in the numerator of the integrand is independent of τ_z and equal to P, say, this integral gives

$$S_x(\omega) = P \left[\tan^{-1}(\omega \tau_2) - \tan^{-1}(\omega \tau_1) \right] / \omega .$$ (6.23)

Note that this is an even function of frequency, and over the frequency range where $\omega \tau_2 \gg 1$ and $0 \leqslant \omega \tau_1 \ll 1$ the two trigonometric functions in the numerator approximate to $\pi/2$ and zero, giving an approximate frequency dependence of the form

$$S_x(\omega) = \frac{\pi P}{2 |\omega|} .$$ (6.24)

Thus, a superposition of relaxation processes can give rise to a spectrum which varies inversely with the frequency. Moreover, a more general form, varying as $|\omega|^{-\alpha}$ where $\alpha \simeq 1$, can also be obtained by setting the numerator in the integrand in equation (6.22) proportional to $\tau_z^{\alpha-1}$.

A difficulty with the model concerns the tremendous spread of time constants required to give the $1/|f|$ law over an extensive range of frequency. With $\tau_2/\tau_1 = 10^6$ the result in equation (6.24) is appropriate over only four decades of frequency, and in order to extend this to ten decades the ratio τ_2/τ_1 must be increased to 10^{12}. In certain specific devices, such as the MOSFET, it is possible to invoke a physical mechanism which could account for relaxation times distributed between, say, 10^{-5} s and 10^8 s, but in general this is not the case, and it seems unlikely that a superposition of relaxation processes is the basis of most observed $1/f$ spectra.

6.5 FRACTIONAL-ORDER INTEGRATION

When it was realized that $1/f$ noise was not going to succumb readily to theoretical analysis, and that physical explanations for most manifestations of the phenomenon were not forthcoming, several papers appeared which adopted an alternative line of attack. They were based on the idea that $1/f$ noise could be produced by half-order integration of white noise (Barnes and Allen, 1966; Mandelbrot, 1967; Mandelbrot and Ness, 1968). A rather formal version of the approach was published recently by Maccone (1981).

In essence the argument is as follows: if a process $x(t)$ has a uniform power spectral density, $\overline{S_x(\omega)} = S_0$ say, then the power spectral density of the process obtained by integrating $x(t)$ m times is

$$\overline{S_x^{(m)}(\omega)} = \frac{S_0}{|\omega|^{2m}} \ . \tag{6.25a}$$

By setting $2m = 1$ this gives

$$\overline{S_x^{(\frac{1}{2})}(\omega)} = \frac{S_0}{|\omega|} \ , \tag{6.25b}$$

which is the required inverse frequency spectrum. The condition $m = 1/2$ corresponds to a half-order integration of $x(t)$.

Although fractional-order integration can produce a $1/|f|$ law, it is such an abstruse concept that one is led to wonder how it could help in constructing a physical model of $1/f$ noise. Radeka (1969) has made some progress in this direction by pointing out that if white noise is passed through a filter whose transfer function is $H(j\omega) = (j\omega)^{-\frac{1}{2}}$, then the output fluctuation shows a $1/|f|$ spectrum. In this case the hypothetical filter is the fractional-order integrator. It is interesting that, since white noise may be represented as a random succession of impulses, the output of Radeka's filter is a random pulse train and the pulse shape is simply the impulse response, $h(t)$, of the filter. Now $h(t)$ is the inverse Fourier transform of $H(j\omega)$, which is zero for $t < 0$ and takes the form $h(t) \propto t^{-\frac{1}{2}}$ when $t \geqslant 0$. But this is exactly the pulse shape discussed by Schönfeld in his random pulse train representation of a $1/f$ noise waveform. Thus, Radeka's approach provides few fresh insights into the structure of $1/f$ waveforms, but his filtering idea could be fruitful in the search for physical mechanisms.

6.6 EXPERIMENTAL EVIDENCE

Perhaps the most impressive feature of $1/f$ noise is its ubiquity: it appears in all types of carbon resistor, in homogeneous single crystal semiconductors including germanium, silicon and III-V compounds, in p–n junction devices, in metal-oxide-semiconductor structures, in continuous and discontinuous metal films, in metallic whiskers and aqueous electrolytes. It is also present in superconductors. On the other hand, a good silicon JFET at room temperature is notable in that $1/f$ noise is essentially absent from the device (Hiatt *et al.,* 1975), which is particularly surprising since GaAs FETs show a considerable amount of $1/f$ noise. Van der Ziel (1978) has pointed out that these two types of device have different structures, the channel in the JFET being confined to the interior of the semiconductor by depletion layers whilst that in the GaAs device, because of the small gate area, is bounded over a relatively extensive region by the semiconductor—oxide interface. Thus, the absence of $1/f$ noise in silicon JFETs is explained if the phenomenon is attributed to a surface mechanism involving traps in the oxide layer, an idea which has already been mentioned in connection with MOSFETs, and which is pursued later in § 6.7.2.

In view of the similarity between the observed spectra from various types of device, it is tempting to think that the same physical mechanism underlies most if not all manifestations of $1/f$ noise. There is a certain amount of experimental support for this line of thought, and this has been embodied in an empirical formula constructed by Hooge (1969) (see § 6.6.4). But, however appealing such a universal explanation may be, the experimental evidence taken overall seems to indicate that at least two and possibly more mechanisms are at work: it appears that $1/f$ noise can be a surface effect and a bulk effect, the physical origins of the noise presumably being different in the two cases.

No attempt is made here to present a comprehensive account of the large number of experiments on $1/f$ noise which have been conducted over the last fifty years or more. Most of these have been surveyed in the review articles by Bell and van der Ziel cited in § 6.1. Instead, just a few experimental facts and findings are described, some of them very recent, which highlight the most significant attributes of $1/f$ noise. Experimental investigations into the low-frequency extent of the $1/f$ law are not included since these have been adequately covered in § 6.2 in the discussion of scale invariance.

6.6.1 Contact noise?
Perhaps the explanation for $1/f$ noise that first comes to mind is that it is merely a spurious effect associated with faulty or loose contacts. In general, this is definitely not the case: modern experiments are carefully designed around four-probe measurements, where the two terminals maintaining the dc level are independent of the pair sensing the fluctuation. In this way, contacts are eliminated as a possible source of the noise. As a further precaution, however, checks are usually run with a control sample, such as a wire wound resistor which is known to be free of $1/f$ noise, to ensure that the experimental arrangement is not itself a significant generator of $1/f$ fluctuations.

6.6.2 The amplitude distribution of $1/f$ noise
The amplitude distribution of $1/f$ noise is gaussian. This has been determined by

Bell (1955) over a noise bandwidth of 40 Hz to 6 kHz, and later by Hooge and Hoppenbrouwers (1969a) in an improved experiment over the frequency range 100 Hz to 100 kHz. A slight departure from a Gaussian was indicated in a measurement in the band 1 to 10 kHz by Bell and Dissanayake (1975) of the ratio of the fourth moment to the second moment of the non-equilibrium noise in various samples; but it transpires (Bell, 1980) that burst noise cannot be eliminated as a possible cause of this. Even if it is a real effect in connection with $1/f$ noise, it represents only a minor distortion in the tails of the distribution which for all practical purposes may be safely ignored.

6.6.3 Resistance fluctuations

The $1/f$ voltage fluctuations observed in homogeneous resistors of various types, including semiconductor, thin-film metal and metal whisker resistors, show a power spectrum which varies as the square of the dc current flowing through the specimen.

Assuming that a constant current source maintains the dc level, the voltage fluctuation $v(t)$ can arise only from a fluctuation $r(t)$ in the resistance of the sample. Since $v(t) = Ir(t)$, where I is the dc current, the power spectral density of the voltage fluctuation is

$$\overline{S_v(\omega)} = I^2 \overline{S_r(\omega)} , \tag{6.26}$$

where $\overline{S_r(\omega)}$ is the power spectral density of the resistance fluctuation. Thus, this simple argument accounts for the observed quadratic dependence on the dc level. Of course, it does not explain the origin of the $1/f$ noise, but merely shifts attention onto the resistance as the source of the fluctuation. As the resistance depends on the density and mobility of the charge carriers, the obvious conclusion is that $1/f$ noise arises either from number or mobility fluctuations.

The square-law dependence on the dc level is not always precisely obeyed (Bell and Chong, 1954). Small departures from it can occur due to joule heating of the sample, for example, which may change the resistance as the current increases.

When an ac current flows through a resistor which shows $1/f$ noise in the presence of a dc current, noise resembling $1/f$ noise is produced in two sidebands either side of the driving frequency, f_0. This is known as $1/\Delta f$ noise because the power spectral density in the sidebands varies as $1/|f_0 - f|$. The $1/\Delta f$ noise scales in proportion to the mean-square level of the ac current, which can also be interpreted as being due to a resistance fluctuation.

The appearance of $1/\Delta f$ noise may be understood by expressing the resistance fluctuation, $r(t)$, as a Fourier integral:

$$r(t) = \frac{1}{2\pi} \int_{-\infty}^{\infty} R(j\omega) \exp j\omega t \, d\omega , \tag{6.27}$$

where $R(j\omega)$ is the Fourier transform of $r(t)$. Now, if the ac current is $I_0 \cos(\omega_0 t)$, the voltage fluctuation is

$$v(t) = \frac{I_0}{4\pi} \int_{-\infty}^{\infty} [R\{j(\omega - \omega_0)\}\exp j\,\omega t + R\{j(\omega + \omega_0)\}\exp j\,\omega t]\,d\omega ,$$

$$(6.28)$$

which describes two noise sidebands each of which shows a power spectral density varying as $1/|f_0 - f|$.

If resistance fluctuations are indeed responsible for the excess non-equilibrium noise in resistors, then a sample in which a dc current and an ac current of frequency f_0 are flowing simultaneously should show a $1/f$ spectral component at frequency f_1 say, and corresponding $1/\Delta f$ components at $f_0 \pm f_1$ and these should be highly correlated. Jones and Francis (1975) conducted an experiment designed to confirm the existence of this expected correlation, and found correlation coefficients lying within 5% of unity. This is convincing evidence that $1/f$ noise in resistors is associated with a fluctuation in the resistance of the sample.

A different type of experiment which led to the same conclusion was performed by Hawkins and Bloodworth (1971). They measured the voltage fluctuations in thick film resistors using a four-probe arrangement, and found $1/f$ voltage fluctuations when the pair of sensing probes was placed perpendicular to the current flow on opposite sides of the film. This transverse noise, as it is called, showed a similar level to that of the $1/f$ noise measured with the sensor probes placed parallel to the current flow and on the same side of the film. Modulation of the conductivity would account for this behaviour.

There is yet another, very important, piece of experimental evidence which eliminates almost any doubt that $1/f$ noise in a homogeneous specimen arises from fluctuations in the resistance. It is implicit that, if such fluctuations exist, they will be present when the sample is in thermal equilibrium with its surroundings as well as when a current is flowing through it. Resistance fluctuations should therefore be detectable in *equilibrium* as a modulation of the thermal noise envelope. Voss and Clarke (1976) were the first to report the observation of a $1/f$-like spectrum associated with the square of the thermal noise voltage, measured in the absent of a current, and similar observations have since been made by Beck and Spruit (1978). The inevitable conclusion is that $1/f$ current or voltage fluctuations are indeed due to $1/f$ fluctuations in the resistance of the sample. It is interesting that, according to this view, the dc current in a specimen does not generate $1/f$ noise, it merely makes apparent the $1/f$ fluctuations that are already present in the resistance in equilibrium.

6.6.4 Hooge's hypothesis
For many years it was generally believed that $1/f$ noise is exhibited by continuous semiconductor materials but not by homogeneous metal films. (It was well

known that inhomogeneous metal layers, consisting of islands or domains with small contacts between them, showed $1/f$ noise.) Then in the late 1960s Hooge and Hoppenbrouwers (1969b) reported observations of $1/f$ noise in continuous thin gold films, and Hooge (1969) proposed that the $1/f$ fluctuations in all homogeneous materials can be represented by the empirical formula

$$\frac{\overline{S_r(\omega)}}{R_0^2} = \frac{\alpha_H}{N_{tot}|f|},$$

(6.29)

where N_{tot} is the *total* number of charge carriers in the specimen, R_0 is the mean resistance, $\overline{S_r(\omega)}$ is the power spectral density of the resistance fluctuations, and $\alpha_H \simeq 2 \times 10^{-3}$ is a 'universal' constant showing only a very weak temperature dependence.

Hooge's law was the first indication that $1/f$ noise in homogeneous materials behaves in a (statistically) systematic way: many homogeneous resistors at room temperature show $1/f$ noise which is satisfactorily described by equation (6.29). The original claim that the law applies to all homogeneous materials has since been modified to include only those cases where lattice scattering predominates over impurity scattering and where boundary scattering is negligible (Hooge et al., 1979). According to Hooge and Vandamme (1978), if impurity scattering is significant then the constant $\alpha_H \simeq 2 \times 10^{-3}$ in equation (6.29) should be reduced by the factor

$$\left(\frac{\mu_{imp}}{\mu_{lat} + \mu_{imp}}\right)^2,$$

where μ_{imp} and μ_{lat} are the mobilities associated with impurity and lattice scattering, respectively.

Despite the considerable success enjoyed by Hooge's law, there is accumulating evidence that α_H, even in the modified form, is not universally representative of the level of $1/f$ noise observed in homogeneous resistors. For example, Dutta et al. (1977) measured $1/f$ noise in copper whiskers and found values of α_H which varied by about a factor of 10 between samples having the same volume, and which were as much as 2×10^3 times larger than Hooge's α_H in equation (6.29). Moreover, Eberhard and Horn (1977) measured $1/f$ noise in silver and copper films and found that the level of noise increased rapidly with increasing temperature, which is equivalent to saying that α_H increased as the temperature rose. No satisfactory explanation for these observations of the variability, high level and temperature dependence of α_H has been offered, though all three are anomalous effects in that they do not accord with the Hooge hypothesis.

Yet another example of the failure of equation (6.29) concerns $1/f$ noise in ionic solutions, where α_H is not constant but is found to scale in proportion to the ionic concentration (Hooge and Gaal, 1971a). However, in view of the lattice scattering mechanism invoked by Hooge and his collaborators, which pertains to solids, it is perhaps unfair to criticize the Hooge law for not satisfactorily describing $1/f$ noise in liquids.

6.6.5 Number fluctuations

The $1/f$ resistance fluctuations observed in homogeneous materials could arise from fluctuations in either the number or mobility of the charge carriers. The inverse dependence on the total number of carriers in Hooge's law would seem to suggest a fluctuation in number as the mechanism responsible for the $1/f$ spectrum. This possibility is examined below.

The number of mobile carriers in a specimen can fluctuate by exchange either with an external bath or with fixed internal energy states such as those associated with trapping or generation–recombination centres. Exchange with an external source is precluded in metals and extrinsic semiconductors because the condition of charge neutrality must be maintained, and hence in these materials trapping is the mechanism that must be invoked if number fluctuations are to occur.

The mean-square resistance fluctuation obtained from Hooge's law by integrating over ten decades of frequency is (Weissman, 1981)

$$\frac{\overline{\phi_r(0)}}{R_0^2} = \frac{10\,\alpha_H}{N_{tot}} \ln 10 \simeq \frac{1}{20\,N_{tot}} . \tag{6.30}$$

This high level of noise could only be achieved if the number of traps was comparable with the number of carriers in the sample. In a metal, where the number of mobile carriers is comparable with the number of atoms, there is no possibility of finding a sufficient number of traps. Therefore, number fluctuations can be eliminated as a potential mechanism for explaining the reported $1/f$ spectra in metal films and whiskers.

The situation is different in an extrinsic, non-degenerate semiconductor, where the number of majority carriers is orders of magnitude less than the number of atoms and may well be comparable with the number of trapping centres. The shallow donors or acceptors are unlikely to contribute significantly towards the noise because they are so close to the band edges that their occupancy hardly fluctuates at all. A distribution of trapping levels around the centre of the band gap could, however, provide a range of lifetimes which may account for the observed $1/f$ fluctuations. A difficulty with this idea is that fluctuations through mid-gap trapping centres are likely to be highly temperature sensitive, which is not usually found in practice.

A certain amount of experimental evidence exists which supports the view that $1/f$ resistance fluctuations are not due to a fluctuation in number. The open-circuit thermo-e.m.f. between two specimens of the same material at different temperatures depends on the carrier concentration, and hence should reflect any number fluctuation that occurs. However, Hooge and Gaal (1971b) observed no $1/f$ noise in an open-circuit thermocell, and Kleinpenning (1974) found no evidence to support the number fluctuation hypothesis in thermo-e.m.f. measurements made on near-intrinsic samples of germanium. This would seem to indicate clearly that number fluctuations are not responsible for $1/f$ resistance fluctuations.

However, the number fluctuation hypothesis cannot be entirely discounted: only recently some measurements of $1/f$ noise in various types of silicon resistor were interpreted in terms of a fluctuation in number (Jones, 1981), and Hall-effect measurements by Brophy and Rostoker (1955) and Kleinpenning (1980) are indicative of a fluctuation in the number of carriers.

6.6.6 Mobility fluctuations

The alternative to a fluctuation in number is a fluctuation in mobility. Mobility fluctuations could explain the $1/f$ fluctuations in the Hall coefficient and also be consistent with the Hooge law, in which the noise power is inversely proportional to the total number of carriers, provided the carriers experienced independent fluctuations in mobility. This idea has been suggested by Kleinpenning and Bell (1976).

It appears, however, that independent mobility fluctuations associated with the individual carriers is a mechanism of doubtful validity. In order to explain the observed $1/f$ spectra, the fluctuations in mobility would have to show characteristic times which are very long, say greater than one second, but no such times exist: the mean-free time of a carrier is on the order of picoseconds and even transit times are usually less than a millisecond. Moreover, as pointed out by Weissman (1981), the shape of the $1/f$ noise spectrum is more or less independent of the transit time. These simple physical arguments would seem to eliminate mobility fluctuations as a potential source of $1/f$ noise, except in connection with temperature fluctuations which affect the mobility. The temperature fluctuation model is discussed later in § 6.7.3.

6.6.7 Volume versus surface effects

Hooge (1969) proposed his hypothetical law in a paper entitled '$1/f$ noise is no surface effect'. Not long afterwards counterclaims appeared, for example by Mirceau et al. (1972) in an article called '$1/f$ noise: still a surface effect', and since then the debate has continued without reaching a categorical conclusion.

If, as originally argued by Hooge, the 'universal' constant α_H in equation (6.29) had indeed been independent of the material of the specimen, this would have constituted strong evidence that $1/f$ noise originates in the volume of the specimen. But α_H is now known to vary between samples, by orders of magnitude in some cases, which weakens the 'volume' argument considerably. Moreover, in semiconductors surface treatments strongly influence the $1/f$ noise, suggesting that in these materials the surface is the origin of the noise, a view which is supported by the absence of $1/f$ noise in JFETs, where surface effects are minimal. In metals the evidence favouring either a surface or a bulk effect is less abundant, although it is certainly true that $1/f$ noise in a metal film often approximately follows the Hooge law, suggesting that a bulk mechanism is at work. On the other hand, the $1/f$ noise observed by Dutta et al. (1977) in copper whiskers was approximately three orders of magnitude greater than that predicted by equation (6.29), which could be taken to indicate that in this case a boundary mechanism is predominant.

The situation is not clarified by the observations of $1/f$ noise in electrolytes (Hooge and Gaal, 1971a). Presumably the noise is not a surface effect, since it is difficult to imagine how surface energy states could exist in an electrolytic solution, yet it does not follow the Hooge law but shows a level which varies with the ionic concentration. Perhaps this indicates merely that $1/f$ noise in aqueous electrolytes is a volume effect and that the Hooge hypothesis may describe bulk $1/f$ fluctuations in solids but not in liquids.

6.6.8 Temperature-dependent spectra

The first systematic measurements of the temperature dependence of $1/f$ noise in metals were reported by Eberhard and Horn (1977). They examined the noise in thin films of silver and copper, 100–1600 Å thick, which were prepared by thermal evaporation onto a sapphire substrate. The dimensions of the metal films were typically $500 \times 10 \ \mu m^2$. In order to ensure that $1/f$ noise predominated over thermal noise throughout the frequency band of interest, which extended from 0.2 Hz to 200 Hz, high current densities of $2 \times 10^6 \ A/cm^2$ or more were used. This caused substantial joule heating of the samples, a problem which was overcome by using the specimen itself to monitor its own temperature: the temperature/resistance characteristic of a sample was measured using low current densities, and the $1/f$ noise associated with a high current density was measured as a function of resistance, which was then converted to temperature via the calibration curve. Incidentally, Eberhard and Horn comment that the ambient atmosphere had no effect on the $1/f$ noise: measurements in air and helium and in vacuo gave identical results.

An example of one of Eberhard and Horn's spectra for Ag at 390 K is shown in Fig. 6.2. The spectrum varies as $1/|f|^\alpha$ over three decades of frequency, where the exponent α equals 1.03 ± 0.06. A small increase in α was observed as the temperature was reduced, amounting to about 20% when the temperature had fallen to 150 K. This 20% increase in α occurred in the Ag and Cu films and appeared to be independent of the sample thickness.

Figure 6.3 shows the temperature-dependence of two of Eberhard and Horn's $1/f$ spectra from 800 Å thick Ag and Cu films. (The hatched area represents room temperature data obtained by Voss and Clarke (1976) on Ag films.) In the figure the magnitude of the noise at 20 Hz is shown, but in view of the insensitivity of α to the temperature, the curves should be more or less

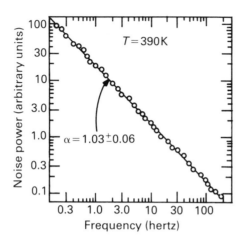

Fig. 6.2 – Noise power spectral density for Ag at 390 K (from Eberhard and Horn (1977), by kind permission of the American Physical Society).

invariant under a change of frequency. Note the rapid drop in the spectra as the temperature is reduced below room temperature, and the peaks which occur at approximately 410 K and 490 K for Ag and Cu, respectively. Similar spectra were observed for samples of different thicknesses. In all cases examined by Eberhard and Horn, the resistance of the samples was characteristic of a bulk metal, and showed the expected linear resistance versus temperature behaviour over the range of temperature investigated.

The effect of the substrate on the temperature–dependence of $1/f$ noise in thin metal films was investigated by Dutta et al. (1978). Their experimental set-up was essentially the same as that used by Eberhard and Horn (1977), except that fused quartz as well as sapphire substrates were used. (The thermal conductivity of quartz is essentially independent of temperature and is substantially lower than that of sapphire, which increases rapidly with decreasing temperature.) Two interesting features of the noise were revealed: (a) above room temperature the substrate has little effect on the $1/f$ spectra of Ag or Cu films and (b) below room temperature the noise in Ag is also insensitive to the substrate but in Cu the quartz produced a significant departure from the behaviour shown in Fig. 6.3 in that the noise flattened off below 300 K instead of falling rapidly with decreasing temperature. The interpretation Dutta and his co-workers put on this is that two types of $1/f$ noise are present in metals, one of which (type A noise) is only weakly temperature dependent and another (type B noise) which is strongly temperature dependent. In Ag the type B noise is so large that it dominates at all the temperatures examined, whereas in Cu on quartz the type A noise is relatively high, causing the cross-over from type A to type B to fall within the temperature range under observation. No explanation for the origin of the two types of noise is known.

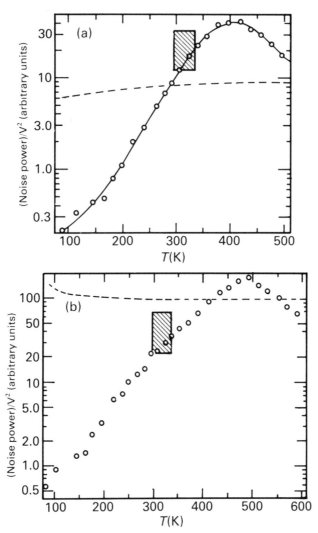

Fig. 6.3 – Temperature-dependence of 1/f noise at 20 Hz for (a) 800 Å thick Ag film and (b) 800 Å thick Cu film (from Eberhard and Horn (1977), by kind permission of the American Physical Society). The broken lines are from Voss and Clarke's temperature fluctuation theory.

6.6.9 Amorphous and polycrystalline materials

Some of the earliest measurements of 1/f noise were made on polycrystalline materials (Bernamont, 1937), and since then there have been numerous reports of the phenomenon in amorphous and polycrystalline materials of various kinds. One of the most recent papers on the subject discusses 1/f noise in cermet materials for thick film resistors, which have application in microelectronic circuits (Prudenziati *et al.,* 1981). The conduction mechanisms in such materials include hopping of carriers between conducting grains and localized states in the glassy structure, and tunnelling of electrons between near-neighbouring grains. Deep energy levels in the glass appear to play an important role in the conduction process.

Generally, the $1/f$ spectra observed in amorphous or polycrystalline materials are accurately proportional to the square of the dc level and closely follow the $|f|^{-1}$ dependence on frequency. In certain glasses the $1/f$ noise is independent of temperature. This has been determined by Sayer and Prasad (1979) for a vanadium phosphate glass over the temperature range 77 K to 300 K. Over the same temperature range they also found that the conductivity increased by six orders of magnitude and the number of carriers, determined from electron-spin-resonance measurements, remained constant. The implication is that the change in conductivity was due to a change in mobility, and since this was not accompanied by a change in the noise this appears to be another argument against the hypothesis that mobility fluctuations (which presumably scale with the mobility) are responsible for $1/f$ noise.

6.7 PHYSICAL MECHANISMS AND THEORETICAL MODELS

The statistical properties of $1/f$ noise, the structure of certain waveforms showing $1/f$ spectra, and some of the experimental evidence on $1/f$ fluctuations have been discussed above. With this background established, it is now appropriate to examine some of the theoretical models of the phenomenon that have been proposed. Obviously, the diversity and in some instances the apparent inconsistency of the experimental observations make the construction of a satisfactory theoretical model extremely difficult. Indeed, no such model is known at present. The models that do exist fall well short of a universal description. We begin with a very specific and rather unusual case of $1/f$ noise.

6.7.1 $1/f$ noise in a lossy capacitor

A lossy capacitor can be represented as a capacitance C in parallel with a resistor R representing the dielectric losses (Fig. 6.4). The resistor shows thermal noise, represented by the parallel noise current generator $i_n(t)$ in the figure, which gives rise to an open circuit voltage fluctuation, $v_n(t)$, at the terminals. Van der Ziel (1975) pointed out that, provided the loss tangent is independent of frequency, the output voltage noise shows a $1/f$ spectrum. The argument is as follows.

If the complex dielectric constant of the lossy dielectric material in the capacitor is $\epsilon = \epsilon' - j\epsilon''$, then the loss tangent is defined as

$$\tan \delta = \epsilon''/\epsilon' \ , \tag{6.31}$$

and the admittance of the parallel RC combination is

$$Y = j\omega \, (\epsilon' - j\epsilon'')\epsilon_0 A/d \ , \tag{6.32}$$

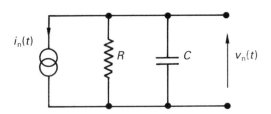

Fig. 6.4 – Equivalent circuit of a lossy capacitor.

where A and d are the cross-sectional area and the thickness of the dielectric, respectively, and ϵ_0 is the permittivity of free space. On comparing equation (6.32) with the expression for the admittance of the equivalent circuit in Fig. 6.4, the capacitance and resistance can be expressed as

$$C = \epsilon' \epsilon_0 \, A/d \, , \qquad R = \frac{d}{\epsilon'' \epsilon_0 \, \omega A} \, , \tag{6.33}$$

and it follows, with the aid of equation (6.31), that

$$R = \frac{1}{\omega C \tan \delta} \, . \tag{6.34}$$

Now, the power spectral density of the thermal noise current generator is

$$\overline{S_i(\omega)} = 4k\theta/R \tag{6.35}$$

and that of the voltage fluctuations is

$$\overline{S_v(\omega)} = \frac{\overline{S_i(\omega)} \, R^2}{(1 + \omega^2 C^2 R^2)} \, . \tag{6.36}$$

On combining the last three equations we find that

$$\overline{S_v(\omega)} = \frac{4k\theta}{|\omega| \, C} \sin \delta \, \cos \delta \, , \tag{6.37}$$

which is a $1/|f|$ law when the loss angle, δ, is independent of frequency.

This condition is not satisfied at low frequencies, where the loss resistance is merely a measure of the conductivity, σ, of the dielectric:

$$R = d/\sigma A \, . \tag{6.38}$$

As this expression is independent of frequency, the loss tangent in equation (6.34) is inversely proportional to frequency, and hence $\overline{S_v(\omega)}$ in equation (6.37) shows a frequency dependence of the form $(1 + \omega^2 R^2 C^2)^{-1}$. Thus, the lossy capacitor does not show a $1/|f|$ spectrum at very low frequencies.

At higher frequencies, however, where dielectric relaxation effects are responsible for the losses, tan δ is observed to be essentially independent of frequency. This is difficult to explain in terms of a single relaxation time, but may be due to a distribution of relaxation times. A possible distribution function, which leads to a frequency-independent loss angle over a limited range of frequency, has been discussed by van der Ziel (1979). Thus, within this limited frequency range, equation (6.37) can be said to represent a $1/|f|$ law. Van der Ziel's argument is not repeated here because it is the same in all but the finest detail as that given below in connection with $1/f$ noise in a semiconductor with a distribution of trapping times.

6.7.2 Surface trapping in semiconductors

Perhaps the most popular model of $1/f$ noise is the trapping model involving a wide spread of time constants. The idea, which is applicable to semiconductors, is that if a free carrier is immobilized by falling into a recombination centre or trap, it is no longer available for conduction and the resistance is modulated accordingly. Assuming that the trapping process obeys Poisson statistics, the modulation associated with a single trap takes the form of a random telegraph signal, which shows a relaxation-type spectrum. If a population of traps with a distribution of trapping times is present, the overall spectrum is the summation of the relaxation spectra from each species of trap, and this sum can show a $1/f$ dependence over a limited range of frequency determined by the spread of the lifetimes. A general argument along these lines has already been introduced in § 6.4.2, and this is developed below.

The power spectral density of a random telegraph signal is derived in § 7.5. According to the result given there, the spectral density of the number fluctuation due to a given species of trap is of the form

$$\frac{4\,\tau_z}{(1 + \omega^2\tau_z^2)}$$

where τ_z is the time constant of the trap. In order to represent the assumed spread of trapping times, a distribution function $p(\tau_z)$ is now introduced, which satisfies the normalization condition

$$\int_0^\infty p(\tau_z)\,d\tau_z = 1 \ . \tag{6.39}$$

The power spectral density of the *total* number fluctuation, $n(t)$, is then

$$\overline{S_n(\omega)} = 4\,\overline{\phi_n(0)} \int_0^\infty \frac{\tau_z p(\tau_z)}{(1 + \omega^2\tau_z^2)}\,d\tau_z \ , \tag{6.40}$$

where $\overline{\phi_n(0)}$ is the mean-square value of $n(t)$.

In order to specify $p(\tau_z)$, McWhorter (1956) assumed that the distribution of trapping times arises from the tunnelling of charge from the semiconductor surface to traps located in the oxide. For a trap at depth w from the interface, the time constant is

$$\tau_z = \tau_0 \exp(\gamma w) , \tag{6.41}$$

where τ_0 and γ are constants and, according to McWhorter, $\gamma \simeq 10^8 \ cm^{-1}$. For a homogeneous distribution of traps between the depths w_1 and w_2, corresponding to time constants τ_1 and τ_2, respectively, it follows from equation (6.41) that

$$p(\tau_z)\,d\tau_z = \begin{matrix} \dfrac{d\tau_z/\tau_z}{\ln(\tau_2/\tau_1)} & \text{for } \tau_1 \leqslant \tau_z \leqslant \tau_2 \\[2mm] 0 & \text{otherwise} \end{matrix} . \tag{6.42}$$

When equation (6.42) is substituted into equation (6.40) the power spectral density of the number fluctuation is found to be

$$\begin{aligned} \overline{S_n(\omega)} &= \frac{4\overline{\phi_n(0)}}{\ln(\tau_2/\tau_1)} \int_{\tau_1}^{\tau_2} \frac{d\tau_z}{(1 + \omega^2 \tau_z^2)} \\[2mm] &= \frac{4\,\overline{\phi_n(0)}}{\ln(\tau_2/\tau_1)} \frac{[\tan^{-1}\omega\tau_2 - \tan^{-1}\omega\tau_1]}{\omega} , \end{aligned} \tag{6.43}$$

which, as discussed in § 6.4.2, closely approximates a $1/|f|$ law over the range of frequency where $\omega\tau_2 \gg 1$ and $0 \leqslant \omega\tau_1 \ll 1$.

As McWhorter's model is based on a surface mechanism, one might expect the resultant spectrum to be different in character from Hooge's hypothetical law (equation (6.29)), which purports to represent volume fluctuations. However, this is not necessarily the case, as pointed out by van der Ziel (1979). If $\overline{\phi_n(0)}$ is put equal to βN_{tot}, where β is a constant and N_{tot} is the total number of carriers in the semiconductor, then the relative number fluctuation over the $1/|f|$ range is

$$\frac{\overline{S_n(\omega)}}{N_{tot}^2} = \frac{\overline{S_r(\omega)}}{R_0^2} = \frac{\{\beta/\ln(\tau_2/\tau_1)\}}{N_{tot}\,|f|} , \tag{6.44}$$

which is identical with Hooge's law if α_H is set equal to $\beta/\ln(\tau_2/\tau_1)$.

Surface trapping via tunnelling to states in the oxide layer is a mechanism for $1/f$ noise in semiconductors which is appealing in its simplicity. It would seem to be supported by the fact that in JFETs, where surface effects are minimal, $1/f$ noise is essentially absent whereas in MOSFETs, with a relatively large semiconductor/oxide interface, the $1/|f|$ law is a prominent feature of the noise spectrum. However, the model, although it is continually being refined (see, for example, van der Ziel, (1978)), is not an unqualified success. Experiment shows that the *level* of the noise is highly sensitive to the condition of the surface, but the spectral *shape* usually remains invariant, following the $1/|f|$ law, irrespective of the surface treatment. This suggests that each successive layer of oxide contributes its own $1/|f|$ spectrum, rather than extending the low frequency end of the existing spectrum by providing relatively deep traps with correspondingly long time constants.

Further evidence against the surface trapping model has been reported by Voss (1978), who made low temperature (4.2 K) measurements of $1/f$ noise in the inversion layer of specially fabricated silicon MOSFETs. The oxide was doped with sodium ions (Na^+), which were mobile at room temperature and could be made to drift either towards or away from the Si/SiO_2 interface by the application of a positive or negative gate voltage. Below about 200 K the Na^+ ions became frozen in position. With a sufficiently high concentration of Na^+ ions, around 10^{12} cm^{-2}, a broad structure associated with the density of states in the Na^+ impurity band appeared in the conductance versus gate voltage characteristic. The noise measurements were made in the vicinity of an impurity band conduction peak, where the transconductance, G_m, was zero. Voss argued that theories based on oxide trapping predict that the intensity of the $1/f$ noise is proportional to G_m^2. In fact, he found experimentally that the noise goes through a maximum when G_m is zero, and that the envelope of the noise-power versus gate-voltage curve follows the conductivity. This led him to the conclusion that the $1/f$ noise is 'an intrinsic transport property of the hopping conduction and is not due to charge trapping'.

It should be borne in mind, of course, that Voss's experiment involved a very specific device structure operating at liquid helium temperature. Nevertheless, it undoubtedly weakens the case for surface trapping as the general mechanism responsible for $1/f$ noise in semiconductors.

6.7.3 Temperature fluctuations in metal films

When a resistor is in thermal equilibrium with its surroundings at mean temperature θ_0, it experiences temperature fluctuations due to thermal energy exchange with the environment. These temperature fluctuations in turn give rise to a resistance fluctuation, $r(t)$, about the mean resistance R_0. If $[\theta(t) - \theta_0] \equiv \Delta\theta(t)$ is the average temperature fluctuation taken over the volume of the specimen, the resistance fluctuation is

$$r(t) = \beta R_0 \Delta\theta(t) , \tag{6.45}$$

where $\beta = (1/R_0)(dr/d\theta)$ is the temperature coefficient of resistance. It follows from equation (6.26) that when a steady current, I, is flowing through the resistor, the power spectral density of the voltage fluctuation across the resistor arising from the temperature fluctuations is

$$\overline{S_v(\omega)} = \beta^2 I^2 R_0^2 \overline{S_{\Delta\theta}(\omega)} , \tag{6.46}$$

where $\overline{S_{\Delta\theta}(\omega)}$ is the power spectral density of $\Delta\theta(t)$.

It is evident from equation (6.46) that temperature fluctuations in the resistor will produce a voltage fluctuation at the terminals when a steady current is flowing through the device. The possibility that this mechanism may be responsible for $1/f$ noise in homogeneous resistors has been investigated by several authors, and notably by Clarke and Voss (1974) and Voss and Clarke (1976) in connection with thin, continuous metal films. (Semiconductors and discontinuous metal films in general show a level of $1/f$ noise which is too high to be accounted for by the temperature fluctuation hypothesis.) The Voss and Clarke theoretical model is based on a solution of the diffusion equation for an infinite medium, which is then averaged over the finite volume of the resistor to give the average temperature fluctuation, $\Delta\theta(t)$. The result is a power spectrum, $\overline{S_{\Delta\theta}(\omega)}$, which decays with increasing frequency, but which does not contain a region varying as $1/|f|$ over many decades. Voss and Clarke argue that a metal film on a glass substrate is a poor approximation to a uniform medium, implying that the presence of the boundaries may modify the spectrum in such a way as to produce an extensive $1/|f|$ region. Accordingly, they explicitly introduce an empirical $1/|f|$ region into the spectrum, between high and low frequency limits determined by the length, l, and width, w, of the film. On employing the normalization condition

$$\overline{\Delta\theta^2} = \frac{1}{2\pi} \int_0^\infty \overline{S_{\Delta\theta}(\omega)} \, d\omega , \tag{6.47}$$

they eventually obtain

$$\overline{S_v(\omega)} = \frac{\beta^2 I^2 R_0^2 \overline{\Delta\theta^2}}{[3 + 2\ln(l/w)] |f|} . \tag{6.48}$$

From thermodynamics, the mean-square value of the temperature fluctuations is

$$\overline{\Delta\theta^2} = k\theta_0^2/C_v , \tag{6.49}$$

where k is Boltzmann's constant and C_v is the heat capacity of the specimen. For a metal at room temperature $C_v \simeq 3Nk$, where N is the total number of atoms in the sample, and hence the spectrum of the voltage fluctuations in equation (6.48) may be written as

$$\overline{S_v(\omega)} = \frac{\beta^2 I^2 R_0^2 \theta_0^2}{3N[3 + 2\ln(l/w)] |f|} . \tag{6.50}$$

The thermal diffusion spectrum in equation (6.50) differs from Hooge's hypothetical law (equation (6.29)) in that it contains the temperature coefficient of resistance in the numerator and involves a geometrical term in the denominator. The latter term, however, hardly represents a significant departure from Hooge's law since it involves only the logarithm of the dimensions, which is a very slowly varying function. In contrast, the presence of β in equation (6.50) is significant, and experimental evidence to support the thermal diffusion theory rather than Hooge's law has been presented by Voss and Clarke: they found no detectable $1/f$ noise in manganin, which has a temperature coefficient of resistance very close to zero, in accord with the theoretical prediction from equation (6.50). Further experimental support for Voss and Clarke's semi-empirical formula was also reported by Clarke and Hsiang (1975), who made measurements of low-frequency noise in tin films at the superconducting transition, where $\beta \simeq 155$ K^{-1} is extremely large compared with the room temperature value of approximately 5×10^{-3} K^{-1}. Below 100 Hz, Clarke and Hsiang found very good agreement between their measurements of the noise spectrum in a tin film on a glass substrate and the predictions of equation (6.50).

According to the theory of thermal diffusion, the temperature fluctuations in metal films should show a degree of spatial coherence, characterized by a frequency-dependent correlation length $\lambda(f) \simeq (D/f)^{\frac{1}{2}}$, where D is the thermal diffusivity of the film. Correlation measurements on a bismuth film at room temperature by Voss and Clarke and on tin at the superconducting transition temperature by Clarke and Hsiang are in good agreement with the theoretical prediction, which provides yet more evidence in favour of the temperature fluctuation theory.

It appears then that in certain cases $1/f$ noise in metal films may be attributable to equilibrium temperature fluctuations, but, despite the initial success of the model, recent investigations have led to the conclusion that in general thermal diffusion is not responsible for the observed spectra. Moreover, the absence of $1/f$ noise in manganin, which provided strong support for the temperature fluctuation theory, has been queried by Hooge (1976), who reports that Vandamme measured contact noise on solid manganin bars and found nothing unusual about the observed spectrum. Hooge concludes that there is 'nothing special in the $1/f$ noise of manganin with its low temperature coefficient of resistivity'. Apart from the uncertainty about the behaviour of manganin, the thermal diffusion model of $1/f$ noise shows serious shortcomings: it does not give rise to an explicit $1/|f|$ region of significant extent, it leads to a low-frequency roll-off, and it predicts a relatively weak, monotonic dependence on temperature. On all of these counts, contrary experimental data have been found. But perhaps the most convincing evidence that $1/f$ noise in metal films supported on a substrate is not due to Voss and Clarke's temperature fluctuation mechanism comes from recent spatial correlation measurements made by Schofield *et al.* (1981).

They took two gold films separated by an electrically insulating layer so thin that the thermal fluctuations in one film were strongly correlated with those in the other film. Thus, if $1/f$ noise were due to thermal diffusion, the noise in the two films should have been strongly correlated. The experiments showed that there was essentially no causal connection between the $1/f$ noise in the two films, the measured coherence coefficient being less than $1/100$ of that predicted from the thermal fluctuation theory. The inevitable conclusion is that, in general, $1/f$ noise in thin metal films is not due to equilibrium energy fluctuations modulating the resistance of the sample. A similar conclusion was reached by Black *et al.* (1981) from correlation experiments performed on chrome films.

6.7.4 Random-service queueing

A number-fluctuation theory of $1/f$ noise in semiconductors was published by Bell (1958), based on the concept of the random-service queue. According to the theory, the charge carriers in the conduction band constitute the members of the queue, which is 'serviced' by traps or recombination centres: carriers join or leave the queue at random times on being emitted from or captured by a trap. Bell argues that the waiting time of a given carrier in the conduction band, before being trapped, can lie anywhere within a very wide range of values, and implies that this wide distribution of times could be responsible for the $1/f$ noise observed in semiconductors.

There are several difficulties with the physics of the queueing theory, an obvious one being that the resistance of a sample follows the instantaneous number of free carriers in the material, and hence in the context of resistance fluctuations and $1/f$ noise, the length of time a given carrier has spent in the conduction band is irrelevant. The important factor is the distribution of the occupancy times of the traps, as discussed in § 6.7.2. As we have seen, a single species of trap gives rise to a relaxation-type spectrum rather than $1/f$ noise, although a variety of trap species may produce a $1/|f|$ spectral law. A more basic objection to the queueing theory is that Bell's analysis fails to produce a $1/|f|$ law, as he acknowledged in the original paper. On account of these shortcomings, Bell's queueing hypothesis is no longer regarded as a possible mechanism for $1/f$ noise in semiconductors.

6.7.5 Quantum theory of $1/f$ noise

The ubiquity of $1/f$ noise suggests that the phenomenon is a fundamental and inevitable feature of charge transportation. This idea has been formalized by Handel (1975), who has developed a quantum theory of $1/f$ noise in which the charge carriers interact with the quantized electromagnetic field. In passing through a circuit element, the carriers are scattered by arbitrary potential barriers and as a result may emit low frequency photons. Although the energy of the photon emission is extremely small, it is sufficient, according to Handel, to modulate the current flowing in the element in such a way as to produce $1/f$ and $1/\Delta f$ noise. The theory is a quantized version of an earlier theory by Handel (1968, 1971), based on turbulence of current carriers in metals and semiconductors.

Handel's theory is complicated, and Tremblay (1978) was unable to verify the detailed calculations. Apparently, an attempt to resolve the differences between the two authors is to appear in a joint publication. Meanwhile, the theory cannot be regarded as substantiated. However, if it is correct, the mechanism sets a lower limit to the level of $1/f$ noise in electronic devices, although if other sources of $1/f$ noise are also present, this lower limit could be substantially exceeded.

6.8 CONCLUDING REMARKS

$1/f$ noise is an enigmatic phenomenon. In electronic devices it is almost invariably present, as has been known for many years, and yet its physical origin remains as obscure today as it has ever done. Numerous experimental and theoretical studies of $1/f$ noise have revealed its many-faceted and intractable character; and after all the effort that has been expended, there are few categorical statements that may be made about it. The experimental data that are available often appear to be contradictory or else they are open to interpretation. Thus, there is a school of thought which firmly believes that $1/f$ noise is a bulk phenomenon, an idea which is strongly opposed by those who believe it originates at the surface. Similarly, mobility and number fluctuations each have their supporters, although there is no incontestable evidence that would eliminate either one as a working hypothesis. Perhaps the one indication of systematic behaviour is Hooge's empirical law, which states that the level of the noise is inversely proportional to the total number of carriers in the sample. But Hooge's law is not universally valid, and even for metal films, in connection with which it was originally formulated, it is sometimes seriously in error.

Theoretically, the situation is just as unsatisfactory. Many theories have been proposed, but few survive careful scrutiny. Some are simply unphysical, whilst others do not fulfil the claim that they produce a $1/|f|$ law. The most widely accepted theory at present is McWhorter's surface trapping model, which deals with a very specific mechanism in semiconductors. Manifestations of $1/f$ noise in metals and other materials remain largely unexplained; and the strong temperature-dependence of $1/f$ noise in metal films that has recently been reported is equally difficult to understand. It is possible that Handel's quantum theory of $1/f$ noise will provide some or all of the answers to the questions raised by the experimental observations, but that is looking well into the future. Meanwhile, one is left with the suspicion that perhaps none of the other theories of $1/f$ noise that are at present available will stand the test of time.

REFERENCES

J. A. Barnes and D. W. Allan (1966), A statistical model of flicker noise, *Proc. IEEE*, **54**, 176–178.

H. G. E. Beck and W. P. Spruit (1978), 1/f noise in the variance of Johnson noise, *J. Appl. Phys.*, **49**, 3384–3385.

D. A. Bell (1955), Distribution function of semiconductor noise, *Proc. Phys. Soc. B* **68**, 690–691.

D. A. Bell (1958), Semiconductor noise as a queueing problem, *Proc. Phys. Soc.*, **72**, 27–32.

D. A. Bell (1960), *Electrical Noise*, Van Nostrand, London, Chapter 10.

D. A. Bell (1980), A survey of 1/f noise in electrical conductors, *J. Phys. C: Solid State Phys.*, **13**, 4425–4437.

D. A. Bell and K. Y. Chong (1954), Current noise in composition resistors, *Wireless Engr.*, **31**, 142–144.

D. A. Bell and S. P. B. Dissanayake (1975), Variance fluctuations of 1/f noise, *Elect. Lett.*, **11**, 274.

T. H. Bell Jr. (1974), Representation of random noise by random pulses, *J. Appl. Phys.*, **45**, 1902–1904.

J. Bernamont (1937), Fluctuations de potentiel aux bornes d'un conducteur metallique de faible volume parcouru par un courant, *A. de Phys.*, 7, 71–140.

R. D. Black, M. B. Weissman and E. M. Fliegle (1981), Lack of spatial cross-correlation in 1/f noise in chrome films, *Proc. 6th International Conf. on Noise in Physical Systems held at the National Bureau of Standards, Gaithersburg, MD, USA, April 6–10, 1981*, p. 152.

J. J. Brophy (1968), Statistics of 1/f noise, *Phys. Rev.*, **166**, 827–831.

J. J. Brophy (1970), Low-frequency variance noise, *J. Appl. Phys.*, **41**, 1697–1701.

J. J. Brophy and N. Rostoker (1955), Hall effect noise, *Phys. Rev.*, **100**, 754–756.

M. A. Caloyanides (1974), Microcycle spectral estimates of 1/f noise in semiconductors, *J. Appl. Phys.*, **45**, 307–316.

J. Clarke and T. Y. Hsiang (1975), Low-frequency noise in tin films at the superconducting transition, *Phys. Rev. Lett.*, **34**, 1217–1220.

J. Clarke and R. F. Voss (1974), 1/f noise from thermal fluctuations in metal films, *Phys. Rev. Lett.*, **33**, 24–27.

P. Dutta, J. W. Eberhard and P. M. Horn (1977), 1/f noise in copper whiskers, *Solid State Communications*, **21**, 679–681.

P. Dutta, J. W. Eberhard and P. M. Horn (1978), 1/f noise in metal films: the role of the substrate, *Solid State Communications*, **27**, 1389–1391.

J. W. Eberhard and P. M. Horn (1977), Temperature dependence of 1/f noise in silver and copper, *Phys. Rev. Lett.*, **39**, 643–646.

I. Flinn (1968), Extent of the 1/f noise spectrum, *Nature*, **219**, 1356–1357.

M. Gardner (1978), White and brown music, fractal curves and one-over-f fluctuations, *Scientific American,* **238** (4), 16–32 (April 1978).

I. S. Gradshteyn and I. M. Ryzhik (1965), *Tables of Integrals, Series and Products,* 4th edn, p. 317. Academic Press, N.Y.

D. Halford (1968), A general mechanical model for $|f|^{\alpha}$ spectral density random noise with special reference to flicker noise $1/|f|$, *Proc. IEEE,* **56**, 251–258.

P. H. Handel (1968), Instabilities and turbulence in semiconductors, *Phys. Stat. Sol.,* **29**, 299–306.

P. H. Handel (1971), Turbulence theory for the current carriers in solids and a theory of $1/f$ noise, *Phys. Rev. A,* **3**, 2066–2073.

P. H. Handel (1975), $1/f$ noise—an 'infrared' phenomenon, *Phys. Rev. Lett.,* **34**, 1492–1495; Nature of $1/f$ phase noise, *Phys. Rev. Lett.,* **34**, 1495–1498; Quantum theory of $1/f$ noise, *Phys. Lett.* **53A**, 438–440.

R. J. Hawkins and G. G. Bloodworth (1971), Measurements of low-frequency noise in thick film resistors, *Thin Solid Films,* **8**, 193–197.

C. F. Hiatt, A. van der Ziel and K. M. van Vliet (1975), Generation–recombination noise produced in the channel of JFETs, *IEEE Trans. Elect. Dev.,* **ED–22**, 614–616.

F. N. Hooge (1969), $1/f$ noise is no surface effect, *Phys. Lett. A,* **29**, 139–140.

F. N. Hooge (1976), $1/f$ noise, *Physica,* **83B** 14–23.

F. N. Hooge and J. L. M. Gaal (1971a), Fluctuations with a $1/f$ spectrum in the conductance of ionic solutions and in the voltage of concentration cells, *Phillips Res. Rep.,* **26**, 77–90.

F. N. Hooge and J. L. M. Gaal (1971b), Experimental study of $1/f$ noise in thermo E.M.F., *Phillips Res. Rep.,* **26**, 345–358.

F. N. Hooge and A. M. H. Hoppenbrouwers (1969a), Amplitude distribution of $1/f$ noise, *Physica,* **42**, 331–339.

F. N. Hooge and A. M. H. Hoppenbrouwers (1969b), $1/f$ noise in continuous thin gold films, *Physica,* **45**, 386–392.

F. N. Hooge, J. Kedzia and L. K. J. Vandamme (1979), Boundary scattering and $1/f$ noise, *J. Appl. Phys.,* **50**, 8087–8089.

F. N. Hooge and L. K. J. Vandamme (1978), Lattice scattering causes $1/f$ noise, *Phys. Lett.,* **66A**, 315–316.

J. B. Johnson (1925), The Schottky effect in low frequency circuits, *Phys. Rev.,* **26**, 71–85.

B. K. Jones (1981), Excess conductance noise in silicon resistors, *Proc. 6th Int. Conf. on Noise in Physical Systems held at the National Bureau of Standards, Gaithersburg, MD, USA, April 6–10, 1981,* pp. 206–209.

B. K. Jones and J. D. Francis (1975), Direct correlation between $1/f$ and other noise sources, *J. Phys.D.,* **8**, 1172–1176.

T. G. M. Kleinpenning (1974), $1/f$ noise in the thermo e.m.f. of intrinsic and extrinsic semiconductors, *Physica,* **77**, 78–98.

T. G. M. Kleinpenning (1980), 1/*f* noise in Hall effect: fluctuations in mobility, *J. Appl. Phys.*, **51**, 3438.

T. G. M. Kleinpenning and D. A. Bell (1976), Hall effect noise: fluctuations in number or mobility?, *Physica*, **81B**, 301–304.

N. N. Lebedev (1965), *Special Functions and their Applications* (translated from the Russian by Richard A. Silverman), Prentice-Hall, p. 36.

C. Maccone (1981), $1/f^x$ noises and Riemann–Liouville fractional integral/ derivative of the Brownian motion, *Proc. 6th Int. Conf. on Noise in Physical Systems held at the National Bureau of Standards, Gaithersburg, MD, USA, April 6–10, 1981*, pp. 192–195.

S. Machlup (1981), Earthquakes, thunderstorms and other 1/*f* noises, *Proc. 6th Int. Conf. on Noise in Physical Systems held at the National Bureau of Standards, Gaithersburg, MD, USA, April 6–10, 1981*, pp. 157–160.

B. Mandelbrot (1967), Some noises with 1/*f* spectrum, a bridge between direct current and white noise, *IEEE Trans. Inf. Theory*, **IT–13**, 289–298.

B. Mandelbrot and J. W. Ness (1968), Fractional Brownian motions, fractional noises and applications, *SIAM Rev.*, **10**, 422–437.

I. R. M. Mansour, R. J. Hawkins and G. G. Bloodworth (1968), Measurement of current noise in M.O.S. transistors from 5×10^{-5} to 1 Hz, *Radio and Elect. Eng.*, **35**, 212–216.

A. L. McWhorter (1956), *Semiconductor Surface Physics* (Ed. R. H. Kingston), University of Pennsylvania Press, Philadelphia.

A. Mirceau, A. Roussel and A. Mitonneau (1972), 1/*f* noise: still a surface effect, *Phys. Lett.*, **41A**, 345–346.

T. Musha (1981), 1/*f* fluctuations in biological systems, *Proc. 6th Int. Conf. on Noise in Physical Systems held at the National Bureau of Standards, Gaithersburg, MD, USA, April 6–10, 1981*, pp. 143–146.

M. Prudenziati, B. Morton and A. Masoero (1981), Temperature dependence of 1/*f* noise in thick film resistors, *Proc. 6th Int. Conf. on Noise in Physical Systems held at the National Bureau of Standards, Gaithersburg, MD, USA, April 6–10, 1981*, pp. 202–205.

V. Radeka (1969), $1/|f|$ noise in physical measurements, *IEEE Trans. Nucl. Sci.*, **NS–16**, 17–35.

M. Sayer and B. Prasad (1979), Electrical noise in semiconducting oxide glasses, *J. Non. Cryst. Solids*, **33**, 345–349.

J. H. Schofield, D. H. Darling and W. W. Webb (1981), 1/*f* noise in continuous metal films is not due to temperature fluctuations, *Proc. 6th Int. Conf. on Noise in Physical Systems held at the National Bureau of Standards, Gaithersburg, MD, USA, April 6–10, 1981*, pp. 147–150.

H. Schönfeld (1955), Beitrag zum 1/*f*-Gesetz beim Rauschen von Halbleitern, *Z. Naturforsch*, Teil **A10**, 291–300.

M. Stoisiek and D. Wolf (1976), Recent investigations on the stationarity of 1/*f* noise, *J. Appl. Phys.*, **47**, 362–364.

U. J. Strasilla and M. J. O. Strutt (1974), Narrow band variance noise, *J. Appl. Phys.*, **45**, 1423–1428.

J. L. Tandon and H. R. Bilger (1976), $1/f$ noise as a non-stationary process: experimental evidence and some analytical conditions, *J. Appl. Phys.*, **47**, 1697–1701.

A. van der Ziel (1950), On the noise spectra of semi-conductor noise and of flicker effect, *Physica*, **16**, 359–372.

A. van der Ziel (1975), Limiting flicker noise in MOSFETs, *Solid State Elect.*, **18**, 1031.

A. van der Ziel (1978), Flicker noise in semi-conductors: not a true bulk effect, *Appl. Phys. Lett.*, **33**, 883–884.

A. van der Ziel (1979), Flicker noise in electronic devices, *Advances in Elect. and Phys.*, **49**, 225–297.

R. F. Voss (1977), *Proc. 1st Symposium on $1/f$ fluctuations held at Saskawa Hall, Tokyo, Japan, 11–13 July 1977*, p. 199.

R. F. Voss (1978), $1/f$ noise and percolation in impurity bands in inversion layers, *J. Phys. C*, **11**, L923–L926.

R. F. Voss and J. Clarke (1976), Flicker $(1/f)$ noise: equilibrium temperature and resistance fluctuations, *Phys. Rev. B*, **13**, 556–573.

R. F. Voss and J. Clarke (1978), '$1/f$ noise' in music: music from $1/f$ noise, *J. Acoust. Soc. Am.*, **63**, 258–263.

M. B. Weissman (1981), Survey of recent $1/f$ noise theories, *Proc. 6th Int. Conf. on Noise in Physical Systems held at the National Bureau of Standards, Gaithersburg, MD, USA, April 6–10, 1981*, pp. 133–142.

7

Burst noise

7.1 INTRODUCTION

Electrical noise in the form of random 'bursts' sometimes occurs in various types of solid state device, including p–n junction diodes and transistors, tunnel diodes and carbon composition resistors. It appears that burst noise is not universally present, but is usually detectable in only a small proportion of devices of a specific type.

In its simplest form, the phenomenon manifests itself as a bistable, step waveform of uniform amplitude, with randomly distributed time intervals between steps, closely resembling a random telegraph signal. On occasion, however, more complicated step waveforms are observed, involving three or more levels. The bistable waveform is said to be symmetrical if the mean times spent in each of the two levels are equal, a condition which is not in general observed in burst-noise waveforms, many of which show strong asymmetry.

It has been suggested that the mechanism responsible for burst noise in reverse-biased p–n junctions is the irregular on–off switching of a surface channel. Burst noise in forward-biased junctions is generally acknowledged nowadays to be due to defects in the vicinity of the junction. The precise nature of these defects is somewhat uncertain, though recent experimental evidence indicates that they are slip lines and dislocations in the crystallographic structure, rather than metallic impurities.

7.2 REVERSE-BIASED p–n JUNCTIONS

One of the earliest references to burst noise in junction devices was made by Montgomery (1952) in a discussion of noise in germanium n–p–n transistors. The phenomenon was present in a small minority of devices, which displayed 'bursts of noise of a very irregular character'. Montgomery observed that devices which had been damaged by excessive biases tended to exhibit the effect.

A typical bistable burst noise waveform is sketched in Fig. 7.1(a). It consists of a random, step waveform, upon which is superimposed white noise. After

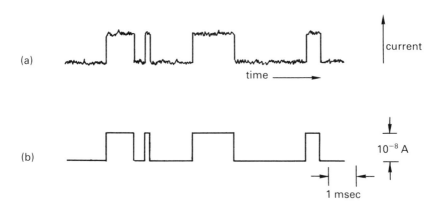

Fig. 7.1 – Typical burst noise current waveform (a) as observed with white noise superimposed and (b) after clipping.

clipping this waveform to remove the white noise component, it appears much like a random telegraph signal, as shown in Fig. 7.1(b). Such a waveform is said to be symmetrical when the mean time spent in the two levels is the same, and asymmetrical when there is a significant departure from this condition. The scales shown in Fig. 7.1 are representative of the burst noise traces often observed in practice.

The first quantitative report of burst noise in a junction device was made by Pay (1956) in connection with measurements on a reverse-biased, point contact germanium diode. An interesting feature of Pay's results, and one which distinguishes his observations from those of subsequent investigators, is that the symmetry of the waveform depends strongly on the reverse current flowing through the junction: there is strong asymmetry at 700 μA, but this becomes less pronounced with increasing current, until at 1000 μA the burst noise is more or less symmetrical. The mean repetition rate measured by Pay also varies with the reverse current, but relatively slowly.

Burst noise in reverse-biased germanium and silicon junctions has been reported more recently by Card and Chaudhari (1965), for reverse biases between 0.7 volts and the breakdown voltage of the junction. They clipped the burst noise waveforms, in order to obtain clean traces like that in Fig. 7.1(b), and measured the probability density functions for the durations of the upper and lower levels in the random square wave. They found that these probability functions are Poisson exponentials (see § 7.5). The height of the current steps they observed is approximately 10^{-8} A and the mean duration of a burst is typically of the order of one millisecond.

Card and Chaudhari also reported the presence of burst noise in GaSb tunnel diodes operating in the injection region with a bias of 0.2 v. They also observed it occasionally in tin oxide films, carbon films and carbon composition resistors, although in these devices the phenomenon was evanescent, seldom persisting long enough for statistical measurements to be made.

The statistical behaviour of bistable burst noise in reverse-biased germanium p–n junctions has been carefully studied by Wolf and Holler (1967). They made their measurements on the base-emitter diodes of diffused p–n–p germanium transistors biased in the reverse direction, well below breakdown. The collector contact was left floating. Of the 20 samples they examined, four exhibited burst noise. They varied the temperature between −10°C and +20°C, the upper and lower limits being imposed by thermal noise at the high end, which distorted the random square wave so that it could no longer be safely extracted, and by the switching rate at the low end, which became too small for measurements to be made. Figure 7.2 shows typical examples of the reverse-current, burst noise waveforms they observed.

The experimental results reported by Wolf and Holler are all for a single, selected transistor. The range of reverse bias over which the burst noise was clearly observed is indicated by the horizontal bar in Fig. 7.3, showing the reverse characteristic of the junction at a fixed temperature. As can be seen from the figure, soft breakdown occurs in the junction at a reverse bias of 51 V.

Several interesting results emerged from Wolf and Holler's study. They found that, at a constant reverse bias, the average number of current pulses per

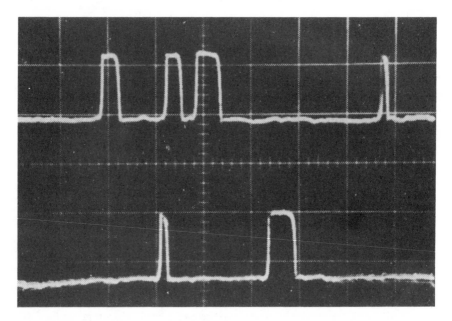

Fig. 7.2 – Waveform of burst noise in a reverse biased (7.5 V), germanium junction. The horizontal scale is 20 nA/div. and the vertical scale is 20 msec/div. (After Wolf and Holler (1967), by kind permission of the American Institute of Physics.)

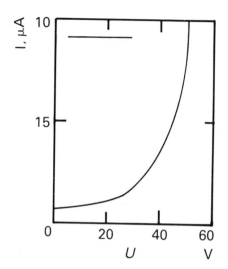

Fig. 7.3 – Reverse current/voltage characteristic of a germanium junction, with the bar indicating the range over which burst noise was observed. (After Wolf and Holler (1967), by kind permission of the American Institute of Physics.)

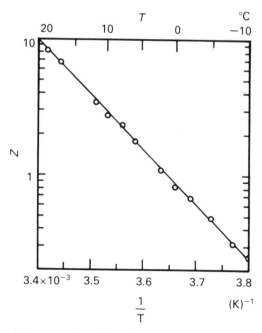

Fig. 7.4 – Average number of bursts per second, z, as a function of inverse temperature, for a reverse bias of 7.5 V. The circles are experimental measurements. (After Wolf and Holler (1967), by kind permission of the American Institute of Physics.)

second, z, varied exponentially with the reciprocal of the absolute temperature, as shown in Fig. 7.4. The solid line in the figure is the curve

$$z = \text{const.} \times \exp\left(-q\phi/k\theta\right) , \tag{7.1}$$

where k is Boltzmann's constant and θ is the absolute temperature. According to Wolf and Holler, $q\phi$ may be an activation energy, whose value, found from the slope of the curve, is $q\phi = 0.79 \pm 0.04$ eV. This is larger than the energy gap of germanium. z was found to be independent of reverse bias over the range 5 V to 30 V, which contrasts with the observations made by Pay on a point contact diode.

The probability density distributions of the positive and negative current pulses were measured at temperatures $0°$, $10°$ and $19.5°$C. All the results were found to be of the form

$$p(t_\pm) = t_\pm^{-1} \exp\left(-t_\pm/\tau_\pm\right) , \tag{7.2}$$

where $t_\pm$ is the duration of a positive or negative pulse and $\tau_\pm$ is the average length of t_+ or t_-. The experimental results are shown in Fig. 7.5, where the exponential dependence is illustrated. It can be seen from these graphs, by comparing the values of the ordinates when $t_+ = t_- = 0$, that the ratio τ_+/τ_- varies with temperature; it takes the values 40 at $0°$C, 35 at $10°$C and 32 at $19.5°$C, indicating that as the temperature rises the waveform becomes less asymmetrical.

The probability densities were found to be independent of reverse bias, but the pulse height, ΔI, increased with increasing reverse bias, eventually approaching a limiting value, as shown in Fig. 7.6.

Finally, the power spectrum of the noise was measured by Wolf and Holler, and found to vary as f^{-2} for frequencies greater than 150 Hz (Fig. 7.7).

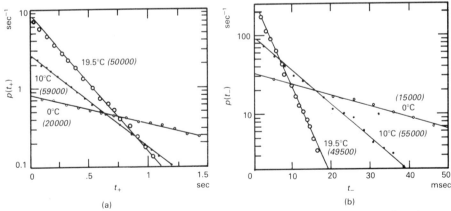

Fig. 7.5 — Distribution densities of (a) positive and (b) negative current pulses at temperatures $0°$C, $10°$C and $19.5°$C. The circles are experimental points and the numbers in parentheses indicate the number of events analysed. (After Wolf and Holler (1967), by kind permission of the American Institute of Physics.)

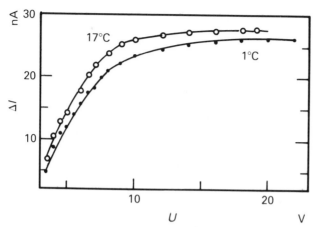

Fig. 7.6 — Pulse height, ΔI, as a function of reverse bias U. The experimental points are the open and closed circles, for temperatures of 17°C and 1°C, respectively. (After Wolf and Holler (1967), by kind permission of the American Institute of Physics.)

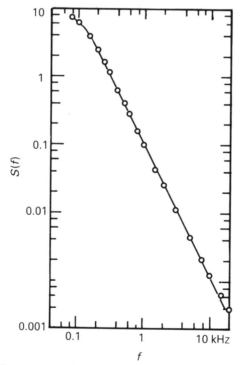

Fig. 7.7 — Power spectral density of burst noise. The open circles are experimental points and the straight portion of the solid line varies as f^{-2}. (After Wolf and Holler (1967), by kind permission of the American Institute of Physics.)

The origin of burst noise in reverse-biased junctions is still not entirely certain. Several mechanisms have been suggested, one of the first being random thermal fluctuations causing the on–off switching of a surface conduction path. This was proposed by Card and Chaudhari, and received support later from Wolf and Holler. All these authors rejected the idea that the phenomenon is associated with breakdown in the body of the junction, on the grounds that it can occur at reverse biases well below the breakdown voltage.

In view of this conclusion, it is perhaps surprising that a second mechanism for burst noise which has received considerable attention in the literature concerns microplasmas. Microplasma effects are known to be associated with breakdown in p–n junctions, as discussed by Shockley (1961) and studied by McKay and McAfee (1953), McKay (1954) and Chynoweth and McKay (1956, 1959).

Microplasma noise (as distinct from burst noise) is observed as a step waveform having an amplitude of approximately 10^{-5} A (Champlin, 1959). The microplasmas themselves are localized high-field regions within the junction, on the order of several hundred angstroms across, in which crytallographic flaws or imperfections containing traps immobilize high densities of charge. This trapped charge supports the process of avalanching when it occurs. The formation and subsequent collapse of a microplasma occur randomly, giving rise to the observed random step changes in the junction current. Rose (1957) has proposed a model, based on an analogy with gas discharge phenomena, to explain microplasma effects; and Haitz (1964) has also discussed a model for the electrical behaviour of a microplasma.

The step waveform which characterizes microplasma noise shows obvious similarities to burst noise waveforms, but there are important differences between them. In particular, the amplitude of burst noise is about three orders of magnitude lower than that of microplasma noise, and the switching rate of burst noise is essentially voltage independent, whereas the duty cycle of microplasma noise varies with the applied voltage.

Schenck (1967, 1968), who has considered burst noise in connection with device degradation and failure, recognizes the difficulties in interpreting burst noise as a microplasma phenomenon. At the same time he suggests that these difficulties are not insuperable, and that they do not absolutely preclude the possibility that burst noise originates in the formation and collapse of microplasmas. He argues, for instance, that just because burst noise occurs well below the body breakdown voltage, it does not necessarily follow that the phenomenon is not a breakdown effect: it could be associated with surface breakdown, which is known to occur at a voltage which is lower than that for body breakdown (Garrett and Brattain, 1956).

Schenck proposes a qualitative microplasma model for burst noise, based on the premise that the internal series resistance of the path leading to the high-field region in the junction is $10^8 - 10^{11}$ Ω. As he points out, this is much larger than the $10^3 - 10^5 \ \Omega$ associated with conventional microplasmas. He postulates the following sequence of events occurring at a microplasma site initially biased with a voltage above the breakdown voltage.

A carrier either generated within or diffusing into the high-field region, triggers an avalanche. As the current builds up, the voltage drop along the high internal series resistance increases until the voltage drop across the high-field region falls below the turn-on value, at which point the discharge ceases. Some of the carriers released in the discharge will probably be trapped in the immediate vicinity of the microplasma. Of those carriers that are re-emitted from the traps after the discharge has ended, some will trigger it into action again, and in this way a series of short avalanche-current bursts could be sustained, with the process eventually terminating when, by chance, there is no re-emitted carrier available to trigger a further discharge. The site will then remain quiescent until the whole process is initiated again by a carrier diffusing into or generated within the high-field region.

Schenck suggests that each of the observed burst-noise pulses, having an amplitude of approximately 10^{-9} A and lasting of the order of a millisecond, is in fact an 'averaged form' of one of the short series of discharges. The main difficulty with this model is that it does not provide an explanation of the high internal series resistance in the current path to the high-field region, on which it relies so heavily. By postulating such a resistance, the correct amplitude for the current pulses follows automatically; but this would seem to be begging the question. Conventional microplasmas do not show such high resistance current paths, i.e. microplasma pulses show amplitudes which are several orders of magnitude greater than the amplitudes of burst-noise pulses, and it is not sufficient merely to *assume* otherwise and then on the basis of this assumption draw the conclusion that burst noise is a microplasma effect. Schenck does intimate that, in silicon devices, the high internal resistance could be associated with the silicon/silicon dioxide interface, but does not present any quantitative evidence to support this view.

The phenomenon of burst noise and its origin have been discussed by Leonard and Jaskolski (1969), in connection with wideband integrated circuit amplifiers. They concluded from their study that 'popcorn noise'[†] (burst noise) occurs when the reverse current-voltage characteristic of the base-collector junction shows negative resistance regions around the 'knee' in the curve appearing at the onset of breakdown. When the reverse bias was such that the operating point coincided with a negative resistance region, Leonard and Jaskolski claim that localized light emission was observed. Now, microplasmas are known to emit light (Chynoweth and McKay, 1956), and this led Leonard and Jaskolski to infer that burst noise and microplasma noise are equivalent. They did not, however, attempt to reconcile the disparities alluded to above between microplasma noise and burst noise.

† So-called by analogy with the sound of corn cooking in a pot.

Leonard and Jaskolski's interpretation has been called into question by Knott (1970), who made measurements of burst noise in discrete silicon planar transistors. He was unable to corroborate the hypothesis that burst noise in transistors is due to microplasma effects in the reverse-biased collector-base junction; instead, his experimental results indicate that, in these devices, burst noise originates at or near the surface of the forward-biased emitter-base junction. This conflict between Leonard and Jaskolski's findings and those of Knott could mean that two different types of burst noise exist. Indeed, Oren (1971) has argued that burst noise is a complicated phenomenon which can arise from several, sometimes competing, factors. In his view, there is no reason to suppose that a single mechanism is responsible for all the burst-noise waveforms that have been observed. It is undoubtedly true that, in certain respects, burst noise in forward-biased junctions shows different characteristic behaviour from that in reverse-biased junctions, suggesting that in these two cases different mechanisms are at work. The attributes of burst noise in forward-biased junctions are discussed below.

7.3 FORWARD-BIASED p–n JUNCTIONS

Burst noise in silicon planar transistors (n–p–n and p–n–p) biased in the active mode, was reported by Giralt et al. (1965, 1966). The structure of the noise resembled that of burst noise in reverse-biased junctions in that it appeared as a step waveform, sometimes with two levels and sometimes with more than two. The magnitude of the steps remained constant in time, but showed a dependence on temperature and the biasing level of the transistor. The duration of the pulses varied randomly, with a probability distribution given by Poisson statistics.

These early experimental results were confirmed by Martin et al. (1968), who reported observing current pulses of between 10 μs and several minutes duration, statistically distributed according to the Poisson law. They also found that the character of the burst noise could be strongly modified, to the extent that occasionally the phenomenon either appeared or disappeared, by storing the (unbiased) transistor at 200°C for a few hours. The modification was usually accompanied by a drift in the gain, h_{fe}, of the transistor.

Martin et al. examined the burst noise in the base current of various transistor types (2N 2484, 2N 3707, 2N 2222, BC 183) and found that the magnitude of the pulses, ΔI_B, varied with temperature and emitter-base voltage, V_{EB}, according to the relation

$$\Delta I_B = \Delta I_{B0} \exp -(\theta_0/\theta) \exp (q V_{EB}/nk\theta) , \qquad (7.3)$$

where ΔI_{B0}, θ_0 and n are parameters whose values depend on the sample under test. The latter two parameters fall in the ranges $3500K \leqslant \theta_0 \leqslant 7000K$, $1.7 \leqslant n \leqslant 2.6$, and typically take the values $\theta_0 \simeq 7000K$, $n \simeq 2$. These values are not inconsistent with an interpretation proposed by Martin et al., that the burst pulses are surface recombination—current instabilities due to abrupt changes of surface potential across a zone showing a high rate of recombination. The fluctuations in the potential across the zone could trigger random changes in the occupancy of the traps, thus accounting for the observed burst noise. The same authors recognize that such a mechanism would involve a complicated distribution of traps in the forbidden gap of the semiconductor, and suggest as an alternative explanation the random 'on—off' switching of an inversion channel close to the surface.

The question of whether the burst noise in forward-biased junctions is a surface effect or a bulk effect, has been investigated by Hsu and Whittier (1969), who performed measurements on gate-controlled devices. The gate electrodes were fabricated over the junctions of diodes and transistors, allowing the experimenters to control the surface conditions. Two types of burst noise were observed, one gate-voltage dependent and the other gate-voltage independent. In the case of the former, it was possible to switch the burst noise on and off by varying the gate voltage, which had the advantage that the same device could be examined in the presence or absence of burst noise. For this reason, Hsu and Whittier concentrated their attention on gate-voltage dependent burst noise, and excluded from their study those devices showing gate-voltage independent noise.

Figure 7.8 shows a typical burst-noise waveform associated with the forward

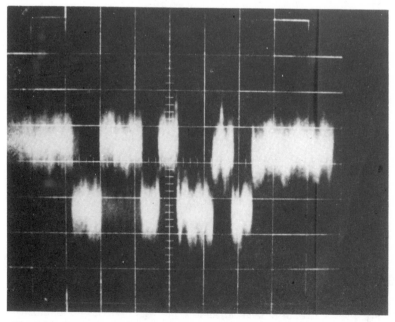

Fig. 7.8 — Typical burst noise waveform in the forward current of a p—n junction diode. (After Hsu and Whittier (1969), by kind permission of Pergamon Press.)

current of a p—n junction diode, as observed by Hsu and Whittier. The amplitudes of the pulses they observed were always less than a few tenths of a microampere, and the pulse widths varied from a few microseconds upwards, with no detectable upper limit.

Figure 7.9, also taken from Hsu and Whittier, shows the effect of the gate voltage on the forward current—voltage characteristics of a p—n junction. The device structure used in the measurements is shown inset in the figure. The solid lines are the characteristics of one of the p—n junctions investigated. The peaks in these characteristics, falling at gate voltages of approximately −10 volts and +10 volts, correspond to depletion of the surface of the n-region (substrate) and the surface of the p^+ region, respectively. In certain junctions, a departure from the behaviour indicated by the solid lines occurred, at gate voltages greater than about +10 volts: an excess current was observed to flow in the field-induced junction, as indicated by the broken lines in the figure. Local concentrations of defects in the inversion layer could have been responsible for this excess current, or it may have been due to a tunnelling mechanism (Reddi, 1967). According to Hsu and Whittier, gate-voltage dependent burst noise is exclusively associated with the excess current: those junctions which exhibited no excess current never showed burst noise, whereas burst noise was observed in those devices which did

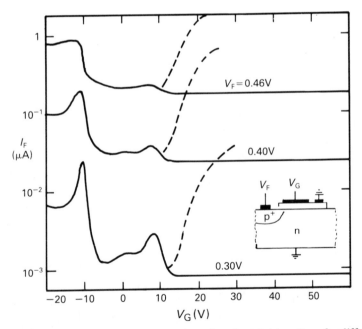

Fig. 7.9 — Curves of forward current as a function of gate voltage for different forward biases. (After Hsu and Whittier (1969), by kind permission of Pergamon Press.)

exhibit an excess current, but only when the gate bias was such that surface inversion occurred in a manner allowing the excess current to flow. Moreover, the gate-voltage threshold for the onset of burst noise was found to be the same as that of the excess current in the forward current-gate voltage characteristic. Similar behaviour was observed in the junction transistors examined by Hsu and Whittier.

Several investigations of burst noise in bipolar transistors followed the early observations: Jaeger and Broderson (1970) proposed a phenomenological model for use in low-frequency noise calculations; Luque *et al.* (1970) found that the amplitude of the burst-noise pulses depends on the forward emitter-base voltage but not on the reverse collector-base voltage; a spectral analysis and an experimental study of the statistics of burst noise pulses was reported by Martin and Blasquez (1971); and the degradation in the performance of integrated circuits due to burst noise was discussed by Conti and Corda (1974). The essential conclusions that emerge from these investigations are: (1) the amplitude of the burst noise pulses in the base current obeys a relation of the form given by equation (7.3); (2) the value of n in this expression is approximately equal to 2, but is greater than m in the expression

$$I_B = I_{B0} \exp (q V_{EB}/mK\theta) \tag{7.4}$$

for the base current; (3) the burst repetition rate depends linearly on the emitter current, I_E, but does not depend on V_{CB}; (4) the length of the less probable state decreases with increasing I_E; (5) mechanical stress, caused by pressing the transistor die with a steel point, produces in almost every case a change in the average rate at which the bursts occur: (6) burst noise occurs much more frequently in n—p—n transistors than in p—n—p devices; and (7) the spectrum of the noise is of the form $\{1 + (f/f_0)^2\}^{-1}$, where f_0 depends on the biasing condition.

The origin of burst noise in forward-biased junctions has been considered by several authors, all of whom invoke some form of imperfection in the crystal structure as the underlying cause of the phenomenon. Luque *et al.* propose a mechanism involving the appearance and disappearance of a large-scale recombination centre. They suggest that the centre could be a 60° edge dislocation, since these are known to be efficient recombination centres because of a 'dangling bond' (Shockley 1953, Read 1954) which acts as a deep acceptor. Such a dislocation in an n-type semiconductor causes bending of the energy bands, and a depleted cylindrical zone around the dislocation line is created. In p-type material, the Fermi level falls below the energy level of the centre, and no depletion region exists. Luque *et al.* argue that, owing to mechanical stresses, a very high density of dislocations can exist in a layer immediately below the emitter contact, that momentum transfer from the emitter-current electrons produces motion of these dislocations through the transistor, and the different recombination currents associated with a given centre in different regions of an n–p–n device, account for the observed burst-noise pulses. The whole process can continue indefinitely, producing a statistically stationary burst-noise waveform, if a new dislocation appears under the emitter contact to relieve the stress created by the departure of one of its fellows on its journey through the transistor.

A model involving surface imperfections does seem to account for several of the features characterising burst noise. This view finds support in the work of Martin *et al.* (1972), who show that, on a single wafer, the percentage of transistors exhibiting burst noise is proportional to the density of surface dislocations on the wafer. Other authors, including Blasquez (1973) and Conti and Corda (1974), have also been able to identify a connection between crystallographic-defect density and the proportion of transistors showing burst noise. On the other hand, Broderson *et al.* (1971) suggest that metallic precipitates are responsible for burst noise, and Hsu *et al.* (1970) propose a model based on this type of imperfection.

Hsu *et al.* reasoned as follows: the number of charge carriers in a typical burst-noise pulse having an amplitude of 10^{-8} A and a duration of 1m sec, is of the order of 10^8. It is extremely unlikely that the mechanism responsible for producing the current pulse should activate all of these carriers independently. Much more probable is an action in which a single event triggers the flow of all the carriers in the pulse. This would occur if a single recombination–generation centre were located adjacent to a defect with a high rate of recombination. Such a defect could be a metal precipitate.

The proposed model relies on an action in which the current through the precipitate is modulated by a change in the occupancy of the neighbouring recombination–generation centre. Imagine the defect located in the metallurgical p–n junction, either at the surface or in the bulk, as illustrated schematically in Fig. 7.10(a). The defect extends throughout the junction, connecting the n and p regions. Hsu *et al.* proceed by assuming that a high potential barrier exists between the metal precipitate and the n-type semiconductor, which acts as a rectifying contact, whilst the potential barrier between the metal and the p-type semiconductor is low, constituting an ohmic contact (Fig. 7.10(b)). The recombination– generation centre trigger is located in the space-charge region of the rectifying barrier on the n-side of the junction.

When a forward bias is applied to the junction, a large proportion of the potential drop takes place across the rectifying metal–semiconductor contact, thereby reducing the height of the barrier (Fig. 7.11(a)), and a current flows through the junction. Now, if the occupancy of the recombination–generation centre changes, through the capture of an electron say, the barrier height is raised and the junction current is reduced (Fig. 7.11(b)). Thus, according to the model, a single capture or generation event could modulate the flow of a large number of carriers in a way which is qualitatively consistent with the observations of burst noise. In particular, the amplitude of the burst noise pulses obtained

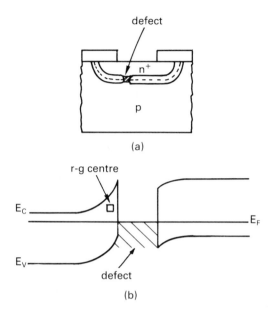

Fig. 7.10 – Model for burst noise. (a) Defect located in the metallurgical junction, and (b) energy band diagram around the defect.

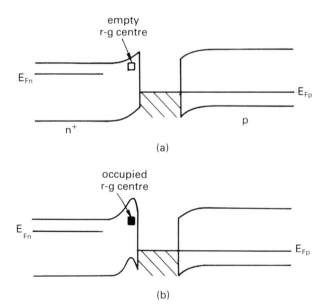

Fig. 7.11 – Effect of change of occupancy of the r–g centre on the rectifying barrier. (a) Centre empty, barrier low and large current flow. (b) Centre occupied by an electron, barrier raised and current reduced.

from the model shows an exponential dependence on applied bias, of the form given by equation (7.3).

The models described above, of Luque *et al.* and Hsu *et al.*, attribute the phenomenon of burst noise to two fundamentally different types of crystallographic imperfection: in the former model, the noise pulses are associated with the motion of dislocations in the crystal structure, whereas in the latter, metal precipitates are held to be responsible for the noise. Roedel and Viswanathan (1975) attempted to distinguish between these two possibilities by introducing modifications into the manufacturing process of linear integrated circuit operational amplifiers. First, they prevented the precipitation of metal atoms at the emitter-base junction, without affecting the dislocation density, by the use of an HCl annealing technique. This treatment produced a noticeable but unspectacular increase in the number of specimens showing little or no burst noise. However, a much more dramatic improvement was obtained when the effect of thermal shock was minimized, by slowing the rate at which the devices were pulled from the 1000°C emitter furnace from 20 cm/min to 2 cm/min. At the faster pull rate, virtually all the devices had emitter-base junction dislocations, but at the slower rate 80 per cent of the devices were entirely free of dislocations, presumably because the wafers had time to heal themselves in the hot zone of the furnace. The results of the two types of treatment are summarized in Table 7.1. Apparently, dislocations are the primary source of burst noise in forward-biased junctions.

This conclusion is strongly supported by the work of Blasquez (1978), on wafers of planar, epitaxial n–p–n transistors. He performed statistical experiments on several dozen wafers, with the aim of identifying the nature of the crystallographic defects primarily responsible for burst noise in forward-biased junctions. The effects of defects due to thermal shock (slip lines), of diffusion-induced dislocations within and outside the emitters of highly doped transistors, and of metal (gold and copper) precipitates, were investigated. The results of the study show that the metallic impurities/precipitates did not act as sources of burst noise, whereas the slip lines due to thermal shock and the diffusion-induced dislocations due to a high emitter doping level both produced burst noise.

This evidence, together with the earlier work of Roedel and Viswanathan, seems incontrovertible: the main source of burst noise in forward-biased p–n junctions is crystallographic dislocations and not metallic precipitates.

7.4 BURST NOISE IN RESISTORS

Burst noise has been reported in various types of resistor by Pay (1956), Bell (1960) and Card and Mauretic (1963). According to Bell, the phenomenon appears in the form of irregularly spaced impulses superimposed on the statistically regular current noise. This is rather different in character from the step waveforms found in junction devices and described in the previous sections.

Pay found that the rate at which the bursts occur in carbon composition and carbon film resistors tends to increase with increasing current but to decrease slowly if a moderate current is left on for a long time. There is probably some thermal mechanism underlying this behaviour since there was a tendency for the original rate of bursts to return after the devices had been left off load for some time. Chong (1953) and Bell and Chong (1954) also found that burst noise in carbon composition resistors could be much modified, either by leaving the devices on load for a long time or by briefly passing a heavy current through them. The latter type of treatment led to irreversible changes in the noise waveform which Bell attributes to the burning out of doubtful contacts.

7.5 THE STATISTICS OF BISTABLE BURST NOISE

We have already mentioned that a bistable burst-noise waveform can be represented as a random telegraph signal. The statistical properties of such a signal have been discussed by Rice (1944, 1945) on the assumption that the probability of a transition from one level to the other is governed by the Poisson law. Rice's treatment is based on an earlier analysis of the problem by Kenrick (1929), the latter being of some historical interest since it appears to be one of the first applications of the correlation function method to the problem of determining the power spectrum of a random waveform. The argument developed below essentially follows the analysis of Rice.

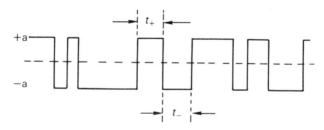

Fig. 7.12 – Sketch of a random telegraph signal.

The function $x(t)$ representing the bistable waveform in Fig. 7.12 can take either of two values, which we designate $+a$ and $-a$. If we assume that the probability of a transition from one level to the other in the time interval $(t, t + dt)$ is $\nu\, dt$, and that this probability is independent of events occurring outside the interval, then the probability of *exactly* m transitions in the interval $(0, T)$ is given by the Poisson distribution (see Appendix 1):

$$p(m, T) = \frac{(\nu T)^m}{m!}\, e^{-\nu T} .$$

(7.5)

It is easily shown, by calculating the first moment about the origin of this distribution, that ν is the mean number of transitions per second.

Table 7.1

Effect of metal precipitates and dislocations on burst noise (bn). These results were obtained by Roedal and Viswanathan (1975) from measurements on linear integrated circuit operational amplifiers (by kind permission, © 1975 IEEE).

Treatment	Number of samples	High bn > 400 pA	Moderate bn	Low bn < 100 pA
Normal	233	70%	24%	6%
Metal precipitates removed	20	4%	44%	16%
Dislocations removed	61	2%	16%	82%

If τ_+ and τ_- are the average times spent in the upper and lower levels, respectively, then the probability density of the times t_+ and t_- spent in the two levels is

$$p(t_\pm) = \tau_\pm^{-1} \exp - (t_\pm/\tau_\pm) \quad . \tag{7.6}$$

This result follows immediately from the Poisson distribution, by forming the product of the probability of no transitions in the interval $t_\pm$ with the probability of one transition between $t_\pm$ and $t_\pm + \mathrm{d}t_\pm$. Equation (7.6) states that the times spent in the two levels are exponentially distributed, in agreement with the statistical measurements of burst noise in Fig. 7.5 by Wolf and Holler (1967).

In order to derive the power spectrum of burst noise, the product $x(t)x(t+\tau)$ is formed and then averaged to give the autocorrelation function, which in turn gives the power spectrum through the Wiener–Khintchine theorem. The product $x(t)x(t + \tau)$ is equal to $+a^2$ if an even number of transitions occur in the interval $(t, t + \tau)$, or to $-a^2$ if an odd number of transitions occur in the same interval. Therefore, the average value of $x(t)x(t + \tau)$ is

$$\begin{aligned}\overline{x(t)x(t+\tau)} = {} & a^2 \times \text{probability of an even number of transistions in } (t, t+\tau) \\ & - a^2 \times \text{probability of an odd number of transitions in } (t, t+\tau) \ .\end{aligned} \tag{7.7}$$

The two probabilities in this expression depend only on the length of the interval $|\tau|$ and not on when the interval starts. Hence, by setting $T = |\tau|$ in equation (7.5), we have from equation (7.7)

$$\overline{x(t)x(t+\tau)} = a^2 [p(0, |\tau|) + p(2, |\tau|) + \ldots]$$
$$- a^2 [p(1, |\tau|) + p(3, |\tau|) + \ldots]$$
$$= a^2 e^{-\nu|\tau|} [1 - \nu|\tau| + \frac{\nu^2|\tau|^2}{2!} - \frac{\nu^3|\tau|^3}{3!} + \ldots]$$
$$= a^2 e^{-2\nu|\tau|} . \tag{7.8}$$

But the average on the left of this expression is equal to the autocorrelation function, $\overline{\phi_x(\tau)}$, of the waveform; that is

$$\overline{\phi_x(\tau)} = a^2 e^{-2\nu|\tau|} , \tag{7.9}$$

and the corresponding power spectrum is

$$\overline{S_x(\omega)} = 4 \int_0^\infty \overline{\phi_x(\tau)} \cos \omega\tau \, d\tau$$

$$= 4a^2 \int_0^\infty \exp - (2\nu\tau) \cos (\omega\tau) \, d\tau$$

$$= \frac{2a^2/\nu}{(1 + \omega^2/4\nu^2)} . \tag{7.10}$$

The power spectrum of burst noise in equation (7.10) has the same form as that of a relaxation process; it is essentially flat at low frequencies, below the half-power frequency $f_0 = \nu/\pi$, and decays as ω^{-2} at higher frequencies. This spectral shape has been observed in different types of device by several investigators, including Wolf and Holler, Hsu and Whittier, and Martin and Blasquez.

It is interesting that, apart from a scaling factor, the only parameter appearing in the power spectrum of burst noise is the mean rate of transition from one level to the other. Jaeger and Broderson measured a mean transition rate (which equals twice the burst rate) of $\nu = 738$ sec^{-1}, corresponding to a half-power frequency of $\nu/\pi = 235$ Hz. Their measured power spectrum showed the form predicted by equation (7.10), with a half-power frequency of 255 Hz. This value agrees with the previous value to within the accuracy of the experimental measurement. Other authors have also measured the power spectrum of burst noise, and in all cases the results have conformed to the spectral shape given by equation (7.10).

However, an alternative expression to equation (7.10) has been derived by Hsu and Whittier, on the basis of a model in which bistable burst noise is represented as a random linear superposition of square pulses, each of duration τ_d. By invoking Carson's theorem, they arrive at a power spectrum having a frequency dependence of the form

$$\frac{\sin^2(\omega\tau_d/2)}{(\omega\tau_d/2)^2},$$

from which they obtain the overall power spectrum of the burst noise by assuming that the pulse widths are governed by an unspecified distribution function. There are two difficulties with this approach. One is that the result given above shows zeros and sharply peaked maxima, which is not in agreement with what is observed by Hsu and Whittier themselves or by others, though it could perhaps be argued that the unknown distribution function would lead to an improvement by introducing a certain amount of smoothing. But secondly, and more fundamentally, the model itself appears to be unsound because bistable burst noise is not a random superposition of square pulses, and Carson's theorem is not the appropriate means for handling the statistics of the noise waveform. This may be appreciated from the fact that, in the random superposition representation, the pulses can overlap, resulting in a synthesized waveform showing more than two levels. It is difficult to see how such a waveform, which may show a multiplicity of levels, can be equated with a bistable step waveform; and it is equally difficult to identify its sharply peaked power spectrum with that of bistable burst noise. We conclude that the appropriate model of burst noise is that leading to the spectral form given in equation (7.10).

REFERENCES

Bell D. A. (1960), *Electrical Noise*, Van Nostrand, London, p. 262.

Bell D. A. and Chong K. Y. (1954), Current noise in composition resistors, *Wireless Eng.,* **31**, 142–144.

Blasquez G. (1973), Ph.D. Thesis, University of Toulouse.

Blasquez G. (1978), Excess noise sources due to defects in forward biased junctions, *Solid State Elect.,* **21**, 1425–1430.

Broderson A. J., Cook K. B. and Chenette E. R. (1971), Conference sur le bruit de fond des composants actifs semi-conducteurs, *Colloque C.N.R.S. No. 204,* C.N.R.S., Paris.

Card W. H. and Chaudhari P. K. (1965), Characteristics of burst noise, *Proc. IEEE (letters),* **53**, 652–653.

Card W. H. and Mauretic A. (1963), Burst noise in semiconductor devices, *Second symposium on the Physics of Failure, Chicago,* pp. 268–283.

Champlin K. S. (1959), Microplasma fluctuations in silicon, *J. Appl. Phys.*, **30**, 1039–1050.

Chong K. Y. (1953), M.Sc. Thesis, University of Birmingham.

Chynoweth A. G. and McKay K. G. (1956), Photon emission from avalanche breakdown in silicon, *Phys. Rev.*, **102**, 369–376.

Chynoweth A. G. and McKay, K. G. (1959), Light emission and noise studies of individual microplasms in silicon p–n junctions, *J. Appl. Phys.*, **30**, 1811–1813.

Conti M. and Corda G. (1974), Noise sources identification in integrated circuits through correlation analysis, *IEEE J. Solid State Circuits*, **SC–9**, 124–133.

Garrett C. G. B. and Brattain W. H. (1956), Some experiments on, and a theory of, surface breakdown, *J. Appl. Phys.*, **27**, 299–306.

Giralt G., Martin J. C. and Mateu-Perez F. X. (1965), Sur un phenomene de bruit dans les transistors, caracterise par des creneaux de courant d'amplitude constante, *CR Acad. Sci.*, **261**, 5350–5353.

Giralt G., Martin J. C. and Mateu-Perez F. X. (1966), Burst noise of silicon planar transistors, *Elect. Lett.*, **2**, 228–229 (in French).

Haitz R. H. (1964), Model for the electrical behaviour of a microplasma, *J. Appl. Phys.*, **35**, 1370–1376.

Hsu S. T. and Whittier R. J. (1969), Characterization of burst noise in silicon devices, *Solid State Elect.*, **12**, 867–878.

Hsu S. T., Whittier R. J. and Mead C. A. (1970), Physical model for burst noise in semiconductor devices, *Solid State Elect.*, **13**, 1055–1071.

Jaeger R. C. and Broderson A. J. (1970), Low frequency noise sources in bipolar junction transistors, *IEEE Trans. Elect. Dev.*, **ED–17**, 128–134.

Kenrick G. W. (1929), The analysis of irregular motions with applications to the energy frequency spectrum of static and of telegraph signals, *Phil. Mag.*, *Ser. 7*, **7**, 176–196.

Knott K. F. (1970), Burst noise and microplasma noise in silicon planar transistors, *Proc. IEEE (letters)*, **58**, 1368–1369.

Leonard P. L. and Jaskolski S. V. (1969), An investigation into the origin and nature of 'popcorn noise', *Proc. IEEE (letters)*, **57**, 1786–1788.

Luque A., Mulet J., Rodriguez T. and Segovia R. (1970), Proposed dislocation theory of burst noise in planar transistors, *Elect. Lett.*, **6**, 176–178.

Martin, J. C. and Blasquez, G., (1971), Sur le spectre de bruit en creneaux, *Solid State Elect.*, **14**, 89–93.

Martin J. C. and Blasquez G., de Cacqueray A., de Brebisson M. and Schiller C. (1972), L'effet des dislocations cristallines sur le bruit en creneaux des transistors bipolaires au silicium, *Solid State Elect.*, **15**, 739–744.

Martin J. C., Esteve D. and Blasquez G. (1968), Burst noise in silicon planar transistors, *Conference on Physical Aspects of Noise in Electronic Devices, University of Nottingham, UK*.

McKay K. G. (1954), Avalanche breakdown in silicon, *Phys. Rev.*, **94**, 877–884.

McKày K. G. and McAfee K. B. (1953), Electron multiplication in silicon and germanium, *Phys. Rev.,* **91**, 1079–1084.

Montgomery H. C. (1952), Transistor noise in circuit applications, *Proc. IRE,* **40**, 1461–1471.

Oren R. (1971), Discussion of various views on popcorn noise, *IEEE Trans. Elect. Dev.,* **ED–18**, 1194–1195.

Pay R. G. (1956), M.Sc. Thesis, University of Birmingham.

Read, W. T. Jr. (1954), Theory of dislocations in germanium, *Phil. Mag.,* **45**, 775–796.

Reddi V. G. K. (1967), Influence of surface conditions on silicon planar transistor current gain, *Solid State Elect.,* **10**, 305–334.

Rice S. O. (1944; 1945), Mathematical analysis of random noise, *Bell Syst. Tech. J.,* **23**, 282–332; **24**, 46–156.

Roedel R. and Viswanathan C. R. (1975), Reduction of popcorn noise in integrated circuits, *IEEE Trans. Elect. Dev.,* **ED–10**, 962–964.

Rose D. J. (1957), Microplasmas in silicon, *Phys. Rev.,* **105**, 413–418.

Schenck J. F. (1967), Progressive failure mechanisms of a commercial silicon diode, in *Physics of Failure in Electronics,* vol. 5, pp. 18–35 (edited by T. S. Shilliday and J. Vacarro, RADC).

Schenck J. F. (1968), Burst noise and walkout in degraded silicon devices, *Proc. of the 1967 IEEE Sixth Ann. Reliability Physics Symp., Los Angeles, California,* pp. 31–39.

Shockley W. (1953), Dislocation and edge states in the diamond crystal structure, *Phys. Rev.,* **91**, 228.

Shockley W. (1961), Problems related to p–n junctions in silicon, *Solid State Elect.,* **2**, 35–67.

Wolf D. and Holler E. (1967), Bistable current fluctuations in reverse-biased p–n junctions of germanium, *J. Appl. Phys.,* **38**, 189–192.

8

Noise in oscillators

8.1 INTRODUCTION

Solid-state devices such as tunnel diodes, Impatts and Gunn diodes are employed nowadays as microwave oscillators in a wide range of applications. These devices are inherently noisy, which in some applications is of little significance, but in others it may be an important consideration. This would be the case, for example, with a microwave receiver in which the mixer, driven by a local oscillator, precedes the amplifier stage. Naturally, the physical mechanisms responsible for the noise depend on the internal structure of the oscillator. We are not concerned with the details of these noise-producing mechanisms in this chapter, but assume merely that the oscillator contains an internal, white noise generator which gives rise to a randomly fluctuationg signal at the output. The spectral characteristics of the output noise in solid state oscillators have been reported in the literature, for example by Josenhans (1966), Scherer (1968), Ashley et al. (1968) and Ashley and Palka (1970).

The essential components of a stable oscillator are a frequency selective circuit, a device showing a negative differential conductance (or a feedback loop with power gain), and a device with a non-linear characteristic which acts as a limiter on the amplitude of oscillation. Usually, the negative conductance and the non-linear characteristic are provided by the same device.

Non-linear elements may be classified as 'fast' or 'slow', depending on how quickly they respond to a signal applied to the terminals. The time taken by a slow non-linear element to respond is several cycles, whereas a fast device responds essentially instantaneously. Only in the latter case is noise significant, and hence we shall confine our attention to fast non-linear, negative conductance oscillators.

This type of oscillator was analysed originally by van der Pol (1927), who investigated the free and forced oscillations of the system. It can be represented very simply as a parallel LCR circuit shunted by the negative, non-linear conductance (or, of course, by the dual circuit consisting of a series LCR circuit with the negative conductance connected in series). Under the appropriate conditions, such a circuit is unstable and will be triggered into oscillation by the noise inherent in the system. The amplitude of this free oscillation stabilizes at a level which is governed by the degree of non-linearity in the negative conductance element. As with most problems involving non-linear effects, a general analysis of oscillator circuits is difficult to achieve, although it is possible to treat certain types of non-linear characteristic on an individual basis.

The particular form of non-linear conductance discusssed by van der Pol is a smoothly varying function which can be represented as a rapidly converging power series in the output voltage. Oscillators based on a non-linear element of this form are known as van der Pol oscillators. Most of the solid-state devices commonly employed as the non-linear element in negative conductance oscillators show a non-linear characteristic which approximates to the van der Pol form. Hence, we shall concentrate our attention on this type of non-linearity, beginning with a discussion of the essential operating characteristics of a van der Pol oscillator in preparation for the subsequent discussions of oscillator noise.

8.2 FREE OSCILLATIONS IN THE VAN DER POL OSCILLATOR

Figure 8.1 shows the equivalent circuit of a negative conductance oscillator, consisting of a parallel LC circuit with a loss conductance G_L shunted by the non-linear, negative conductance $-G$ (G is positive). When $G > G_L$ the circuit is unstable and noise inherent in the system will trigger the circuit into free oscillation. An oscillatory voltage, $v(t)$, then appears at the output.

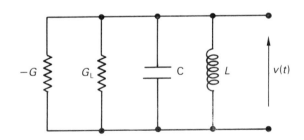

Fig. 8.1 — Equivalent circuit of a negative conductance oscillator.

The detailed form of the non-linearity determines the characteristics of the oscillator. Assuming that G can react instantaneously to rapid changes in $v(t)$, the non-linear conductance can be expressed as a power series in $v(t)$ as follows:

$$G = G_0 \left[1 + \alpha v(t) + \beta v^2(t) + \ldots\right] , \qquad (8.1)$$

where $\alpha, \beta \ldots$ are constants and G_0 is the (positive) value of G when $v(t) = 0$.

When the series in equation (8.1) converges so rapidly that terms beyond the quadratic term are negligible, we have

$$G = G_0 \left[1 + \alpha v(t) + \beta v^2(t) \right] , \tag{8.2}$$

which is the type of non-linearity investigated by van der Pol (1927). As he demonstrated, the essential properties of the oscillator are determined, not by the linear term, but by the quadratic term in equation (8.2). An alternative form of non-linearity, which forms the basis of the Robinson (1959) oscillator, shows a sharp knee in the characteristic at some value of $|v(t)|$, beyond which saturation occurs. Although the Robinson and van der Pol characteristics are quite different, the characteristics of the respective oscillators are similar, although the latter tends to be rather more noisy because $1/f$ noise in the non-linear conductance essentially modulates G_0, which in turn produces amplitude and phase fluctuations in the output. Faulkner and Meade (1968) have found evidence for this in a Gunn-effect oscillator (which approximates to a van der Pol oscillator) in the form of correlation between $1/f$ noise in the bias level and FM noise in the output.

The differential equation for the output voltage in Fig. 8.1 is

$$C\frac{\mathrm{d}v(t)}{\mathrm{d}t} + (G_L - G)v(t) + \frac{1}{L} \int v(t)\mathrm{d}t = 0 , \tag{8.3}$$

which, when combined with equation (8.2) and differentiated with respect to time, becomes

$$C\frac{\mathrm{d}^2 v(t)}{\mathrm{d}t^2} + \{G_L - G_0 - 2\alpha G_0 v(t) - 3\beta G_0 v^2(t)\} \frac{\mathrm{d}v(t)}{\mathrm{d}t} + \frac{v(t)}{L} = 0 . \tag{8.4}$$

This is a non-linear, homogeneous (i.e. there is no source term on the right-hand side) differential equation, and its solution represents the free oscillation at the output. Van der Pol (1927, 1934) has shown formally that the oscillation frequency is the resonance frequency of the LC combination, and that the amplitude of the output builds up quickly to a limiting value determined by the coefficient of the quadratic term in equation (8.2). This limiting behaviour may be understood from the following physical argument.

When $v(t)$ is small, the non-linear conductance approximates to the linear, negative conductance $-G_0$, and given that $G_0 > G_L$ the net conductance of the circuit is negative. The circuit is then unstable, it goes into free oscillation, and the amplitude of the oscillations increases exponentially with time (instead of being damped exponentially, as would be the case in a circuit with positive conductance). When the amplitude of $v(t)$ becomes sufficiently large for the term $\beta v^2(t)$ in equation (8.2) to be comparable with -1 (assuming β is negative), a change of sign occurs in G, and the non-linear element acts as a positive conductance. At this point the oscillation ceases to increase in amplitude and instead maintains a steady stable level.

It is not necessary to solve equation (8.4) in order to determine the steady amplitude of the oscillation and to appreciate the significance of the quadratic term in equation (8.2). Instead, we employ the condition that in the steady state the energy absorbed by the loss conductance G_L is equal to the energy supplied by the negative conductance $-G$ (Robinson 1962, 1974), to obtain the relationship

$$\int_0^{T_0} G_L v^2(t)\, dt = \int_0^{T_0} G v^2(t)\, dt \ , \tag{8.5}$$

where T_0 is the period of oscillation. On setting

$$v(t) = v_0 \cos \omega_0 t \ , \tag{8.6}$$

where $\omega_0 = 2\pi/T_0$ is the angular frequency, we have from equations (8.2) and (8.5),

$$\frac{G_L v_0^2}{2} = \frac{G_0 v_0^2}{T_0} \int_0^{T_0} \{\cos^2 \omega_0 t + \alpha v_0 \cos^3 \omega_0 t + \beta v_0^2 \cos^4 \omega_0 t\}\, dt \ . \tag{8.7}$$

Now, the function $\cos \omega_0 t$ raised to an odd power integrates to zero over one period, leaving only integrals of the form

$$\int_0^{T_0} \cos^{2n} \omega_0 t\, dt = \frac{2\pi}{\omega_0} \frac{(2n-1)!!}{(2n)!!} \ , \quad n = 1, 2, \ldots \tag{8.8}$$

in equation (8.7). It follows that

$$\frac{G_L v_0^2}{2} = \frac{G_0 v_0^2}{2} + \frac{3}{8} \beta G_0 v_0^4 \ , \tag{8.9}$$

which has the solution

$$v_0^2 = - \left(\frac{4}{3}\right) \frac{(G_0 - G_L)}{\beta G_0} \ . \tag{8.10}$$

Bearing in mind that $G_0 > G_L$, the right-hand side of this equation is positive when $\beta < 0$, which is consistent with the qualitative argument given above.

Note that α, the coefficient of the linear term in equation (8.2), does not feature in the expression for the amplitude of the stable oscillations: the level is determined by β, the coefficient of the quadratic term. As the quadratic non-linearity in the conductance characteristic is reduced, the free oscillation amplitude of the oscillator increases in accord with the inverse relationship in equation (8.10).

The angular frequency of the free oscillation, ω_0, is equal to the angular resonance frequency of the reactive elements in the oscillator circuit of Fig. 8.1, i.e. $\omega_0 = 1/\sqrt{LC}$. This result emerges from the solution of equation (8.4), as discussed by van der Pol.

8.3 FORCED OSCILLATIONS

Figure 8.2 shows the equivalent circuit of an oscillator driven by an external current generator. The external source can have a profound effect on the output of the oscillator. The differential equation for the output voltage is now the inhomogeneous version of equation (8.4), in which the source term is included on the right-hand side:

$$C\frac{d^2v(t)}{dt^2} + \{G_L - G_0 - 2\alpha G_0 v(t) - 3\beta G_0 v^2(t)\}\frac{dv(t)}{dt} + \frac{v(t)}{L} = \frac{di_s(t)}{dt} \qquad (8.11)$$

and, as before, it is the quadratic term in the non-linear conductance (i.e. the cubic term in the current) which is responsible for the interesting behaviour of the oscillator.

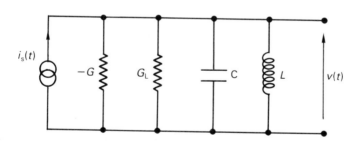

Fig. 8.2 – Equivalent circuit of an oscillator driven by an external current source.

The solution of equation (8.11) has been derived in detail by van der Pol (1934) for the case when the external source drives the oscillator with a harmonic signal. As we might expect, the analysis is lengthy but the essential conclusions may be summarised briefly. The two important factors governing the nature of the output voltage are the amplitude of the driving term and the frequency of the external current generator in relation to the resonance frequency of the circuit.

When the frequency of the source is close to that of the free oscillations, the output locks in phase and frequency to the source and the free oscillations are completely suppressed, provided the level of the source is sufficiently high. Outside the band of frequencies where frequency locking occurs, the external source has little effect on the output, which then consists almost entirely of the free oscillation.

It must be emphasized that the spectral width of the frequency locking region depends on the level of the external generator. If that generator were to represent inherent noise in the oscillator, instead of a harmonic signal, the phenomenon of frequency locking could be safely ignored since, in any practical circuit, the energy of the noise source over the coherence time of the fluctuations is many orders of magnitude less than that of the free oscillation. Thus, the free oscillation will persist when the noise generator is present, but now its amplitude and phase will show random fluctuations. These fluctuations are the subject of the remaining discussions in this chapter.

8.4 THE OUTPUT NOISE

As suggested above, the inherent noise in a van der Pol oscillator can usually be represented by a single generator as shown in Fig. 8.3. This is not the most general representation, but it is appropriate for most solid-state, negative conductance oscillators. In a Gunn-effect oscillator, for example, the noise generator $i_n(t)$ in Fig. 8.3 would represent the non-equilibrium Johnson noise of the hot electron population outside the propagating domain (see Chapter 10).

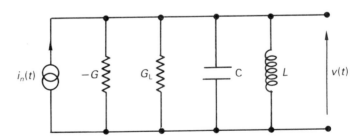

Fig. 8.3 – Equivalent circuit of a noisy oscillator, showing a parallel noise current generator.

Several authors have investigated the effects of the noise on the oscillator output, most of whom assume that $i_n(t)$ is a white noise source. They include Garstens (1957), Edson (1960), Mullen (1960) and Kurokawa (1966, 1969). Kurokawa's analysis has been extended by Ohtomo (1972) to include $1/f$ noise. In order to elucidate the arguments, we shall also assume that $i_n(t)$ has a uniform power spectrum, since this leads to a certain simplification of the analyses. It is also appropriate to a number of practical, solid-state oscillators.

The output of the noisy oscillator illustrated in Fig. 8.3 is the solution of the inhomogeneous equation

$$C \frac{dv(t)}{dt} + (G_L - G)v(t) + \frac{1}{L} \int v(t) \, dt = i_n(t) , \tag{8.12}$$

where the term on the right represents the white noise source and G is the non-linear conductance defined in equation (8.2). Because of the presence of the non-linear terms, a general solution of this equation is difficult to achieve. As with many non-linear problems, the usual approach is to seek an approximate solution by employing a linearization technique; but even then the analysis is complicated.

The noise has two effects on the free oscillations at the output: it modulates the amplitude and it also introduces a randomly fluctuating phase shift. Thus, the output voltage can be expressed in the form

$$v(t) = v_0 [1 + a(t)] \cos \{\omega_0 t - \psi(t)\} , \tag{8.13}$$

where $a(t)$ is the amplitude modulation, $\psi(t)$ is the phase modulation and v_0 is the amplitude of the free oscillations in the absence of noise, whose value is given in equation (8.10). The functions $a(t)$ and $\psi(t)$ represent stationary, stochastic processes and, since the oscillator circuit is extremely narrow band, they vary very slowly with time; or in other words, their spectral content lies well below the frequency of the free oscillation. The problem now is to find a solution of equation (8.12) having the general form of equation (8.13).

Since $a(t)$ and $\psi(t)$ are stochastic processes, they can be specified only in terms of their statistical properties. In particular, we shall be concerned with their power spectral densities and the cross-spectral density between them. These quantities are obtained from the solution of equation (8.12), in an analytical procedure which clearly demonstrates the influence of the quadratic non-linearity on the spectral character of the output noise. Before discussing the details of this solution, it is instructive to consider the general form of the power spectrum of the output function specified in equation (8.13).

8.5 THE OUTPUT SPECTRUM

Perhaps the most straightforward way of deriving an expression for $\overline{S_v(\omega)}$, the power spectral density of the output voltage in equation (8.13), is first to construct the autocorrelation function, $\phi_v(\tau)$, which may then be used in conjunction with the Wiener–Khintchine theorem to give $\overline{S_v(\omega)}$.

By definition, the autocorrelation function of $v(t)$ is

$$\overline{\phi_v(\tau)} = \lim_{T \to \infty} \frac{1}{T} \int_{-T/2}^{T/2} \overline{v(t) \, v(t + \tau)} \, dt , \tag{8.14}$$

where T is the observation time, τ is the delay, and the overbar indicates an ensemble average. The integrand here is

$$\overline{v(t)v(t+\tau)} = \overline{R(t)R(t+\tau)\cos\{\omega_0 t - \psi(t)\}\cos\{\omega_0(t+\tau) - \psi(t+\tau)\}}\ ,$$

(8.15)

where the expression in equation (8.13) has been substituted for $v(t)$ and for brevity we have set $R(t) = v_0[1 + a(t)]$. Now, the product of the trigonometric terms in equation (8.15) can be expressed in the form

$$\cos\{\omega_0 t - \psi_1\}\cos\{\omega_0(t+\tau) - \psi_2\} =$$

$$\tfrac{1}{2}[\cos\omega_0\tau\cos(\psi_2 - \psi_1) + \sin\omega_0\tau\sin(\psi_2 - \psi_1)] +$$

$$+ \tfrac{1}{2}[\cos 2\omega_0 t\cos\{\omega_0\tau - (\psi_1 + \psi_2)\} - \sin 2\omega_0 t\sin\{\omega_0\tau - (\psi_1 + \psi_2)\}]\ ,$$

(8.16)

where $\psi_1 \equiv \psi(t)$ and $\psi_2 \equiv \psi(t+\tau)$. When this expression is substituted into equation (8.15), which is then integrated according to equation (8.14), the second term in square brackets gives a zero result. This is due to the presence of the deterministic terms $\cos(2\omega_0 t)$ and $\sin(2\omega_0 t)$, which alone show a time dependence after the ensemble averaging, since $a(t)$ and $\psi(t)$ are stationary processes. The remaining part of the integrand, since the ensemble averaging makes it independent of t, can be taken outside the integral to give

$$\overline{\phi_v(\tau)} = \tfrac{1}{2}\overline{R(t)R(t+\tau)\,[\cos\omega_0\tau\cos\{\psi(t+\tau) - \psi(t)\} +}$$

$$\overline{\sin\omega_0\tau\sin\{\psi(t+\tau) - \psi(t)\}]}\ .$$

(8.17)

A simplification can be achieved in equation (8.17) by recognizing that the phase difference $|\psi(t+\tau) - \psi(t)|$ is very much less than unity. Then, on expanding the trigonometric terms containing the phase difference in equation (8.1) to second order in the argument, we find that

$$\overline{\phi_v(\tau)} \simeq \tfrac{1}{2}\overline{R(t)R(t+\tau)\,[\{1 - \tfrac{1}{2}[\psi^2(t+\tau) + \psi^2(t)] + \psi(t+\tau)\,\psi(t)\}\cos\omega_0\tau}$$

$$\overline{+ \{\psi(t+\tau) - \psi(t)\}\sin\omega_0\tau]}$$

$$= (v_0^2/2)\{[1 - \overline{\phi_\psi(0)} + \overline{\phi_\psi(\tau)} + \overline{\phi_a(\tau)}]\cos\omega_0\tau$$

$$+ [\overline{\phi_{\psi a}(\tau)} - \overline{\phi_{a\psi}(\tau)}]\sin\omega_0\tau\},$$

(8.18)

where $\overline{\phi_a(\tau)}$ and $\overline{\phi_\psi(\tau)}$ are the autocorrelation functions of the amplitude and phase fluctuations, respectively, and

$$\overline{\phi_{\psi a}(\tau)} = \overline{\phi_{a\psi}(-\tau)} = \lim_{T \to \infty} \frac{1}{T} \int_{-T/2}^{T/2} \overline{\psi(t + \tau)\, a(t)}\, dt \qquad (8.19)$$

defines the cross correlation between $a(t)$ and $\psi(t)$. Note that in arriving at the result in equation (8.18), use was made of the fact that the mean values of the amplitude fluctuation, $\overline{a(t)}$, and the phase difference, $[\overline{\psi(t + \tau)} - \psi(t)]$, are zero.

The power spectral density of $v(t)$ can now be obtained from equation (8.18) with the aid of the Wiener–Khintchine inversion integral

$$\overline{S_v(\omega)} = 2 \int_{-\infty}^{\infty} \overline{\phi_v(\tau)} \exp(-j\omega\tau)\, d\tau \ , \qquad (8.20)$$

where the bilateral form for the integral has been used for convenience in dealing with the cross correlation functions $\overline{\phi_{\psi a}(\tau)}$ and $\overline{\phi_{a\psi}(\tau)}$. On making the substitution for $\overline{\phi_v(\tau)}$, the power spectral density is found to be

$$\begin{aligned}
\overline{S_v(\omega)} = (v_0^2/4) \{ & 4\pi\, [1 - \overline{\phi_\psi(0)}]\, [\delta(\omega - \omega_0) + \delta(\omega + \omega_0)] \\
& + [\overline{S_a(\omega - \omega_0)} + \overline{S_a(\omega + \omega_0)}] \\
& + [\overline{S_\psi(\omega - \omega_0)} + \overline{S_\psi(\omega + \omega_0)}] \\
& + [\overline{C_{\psi a}(\omega - \omega_0)} - \overline{C_{\psi a}(\omega + \omega_0)}] \} \ . \qquad (8.21)
\end{aligned}$$

The new functions in this expression are defined as follows: $\overline{S_a(\omega)}$ and $\overline{S_\psi(\omega)}$ are the power spectral densities of $a(t)$ and $\psi(t)$, respectively, and $\overline{C_{\psi a}(\omega)} = 2\mathrm{Im}\, \overline{S_{\psi a}(\omega)}$, where $\overline{S_{\psi a}(\omega)}$ is the cross spectral density between $a(t)$ and $\psi(t)$ and Im denotes 'imaginary part of'. The symmetry of the expression on the right of equation (8.21) is such that only positive values of ω need be considered, in which case a factor of 2 must be introduced to account for the negative range of frequencies. Assuming then that $\omega > 0$, and bearing in mind that $a(t)$ and $\psi(t)$ are low frequency functions whose spectra take significant values only when $\omega \ll \omega_0$, it is evident that the terms $\overline{S_a(\omega + \omega_0)}$, $\overline{S_\psi(\omega + \omega_0)}$ and $\overline{C_{\psi a}(\omega + \omega_0)}$ in equation (8.21) are negligible when ω is in the vicinity of ω_0 (which is the region of interest). It follows that the output power spectra density of the oscillator can be expressed as

$$\overline{S_v(\omega)} = \frac{v_0^2}{2} \{ 4\pi [1 - \overline{\phi_\psi(0)}]\, \delta(\omega - \omega_0) + \overline{S_a(\omega - \omega_0)} + \overline{S_\psi(\omega - \omega_0)} \\
+ \overline{C_{\psi a}(\omega - \omega_0)} \} \qquad (8.22)$$

for frequencies around the free oscillation frequency.

The term containing the delta function in equation (8.22) represents the effect of the noise on the free oscillation: a reduction in the strength of the delta function is produced by the phase fluctuation. Of the remaining terms, representing noise sidebands on either side of the carrier frequency, the first is the AM noise, the second is the PM noise and the third expresses the AM–PM coherence in the sidebands.

In the absence of any coherence, the noise is symmetrical about the free oscillation frequency since $\overline{S_a(\omega - \omega_0)}$ and $\overline{S_\psi(\omega - \omega_0)}$ are even functions of $(\omega - \omega_0)$. On the other hand, the function $\overline{C_{\psi a}} \, (\omega - \omega_0)$ is an odd function of $(\omega - \omega_0)$ and hence if the amplitude and phase fluctuations are not independent the noise sidebands show a degree of asymmetry. As shown in Appendix 5, when $i_n(t)$ is a white noise generator, the amplitude and phase fluctuations are independent (i.e. $C_{\psi a} \, (\omega - \omega_0) = 0$) and the noise sidebands are symmetrical about ω_0.

8.6 THE LINEARIZATION TECHNIQUE

The result in equation (8.22) gives a useful insight into the general character of the output spectrum of the noisy oscillator. In order to go beyond this, the various terms appearing in the general expression must be specified, which brings us to the solution of the non-linear, inhomogeneous equation (8.12). The solution is achieved through a linearization procedure in which the only frequency-mixing retained is that between the noise and the free oscillation: frequency shifting arising from the noise beating against itself is a very small effect and is ignored.

We begin the discussion by writing the output, $v(t)$, of the oscillator as the sum of the free oscillation, $v_f(t) = v_0 \cos \omega_0 t$, and a term, $v_n(t)$, representing the effect of the noise:

$$v(t) = v_f(t) + v_n(t) \; . \tag{8.23}$$

On comparing this expression with that in equation (8.13), the noise term is seen to be

$$v_n(t) = v_1(t) \cos \omega_0 t + v_2(t) \sin \omega_0 t \; , \tag{8.24}$$

where $v_1(t)$ and $v_2(t)$ are slowly varying functions given by the expressions

$$v_1(t) = v_0 \left[\{1 + a(t)\} \cos \psi(t) - 1 \right] \tag{8.25a}$$

and

$$v_2(t) = v_0 \sin \psi(t) \; . \tag{8.25b}$$

If equations (8.25) are expanded to first order in the variables $a(t)$ and $\psi(t)$, we find that $v_1(t) \simeq v_0 a(t)$ and $v_2(t) \simeq v_0 \psi(t)$, and thus to this order of approximation the output noise in equation (8.24) is

$$v_n(t) \simeq v_0 \left[a(t) \cos \omega_0 t + \psi(t) \sin \omega_0 t \right] \; . \tag{8.26}$$

It is implicit here of course that $|a(t)|$ and $|\psi(t)|$ are both very much less than unity.

We now recall equation (8.12) and substitute for $v(t)$ from equation (8.23) to obtain

$$\left\{ C\frac{dv_f}{dt} + (G_L - G_0)\,v_f + \frac{1}{L}\int v_f\,dt - \beta G_0 v_f^3 \right\} +$$

$$+ \left\{ C\frac{dv_n}{dt} + (G_L - G_0)\,v_n + \frac{1}{L}\int v_n\,dt - \beta G_0 v_n(3v_f^2 + 3v_n v_f + v_n^2) \right\} = i_n(t) ,$$

$$(8.27)$$

where the functional dependence of v_f and v_n on t has been omitted and for convenience the coefficient α in the expansion of the non-linear conductance (equation (8.2)) has been set equal to zero. (There is no loss of generality here since α plays no significant part in determining the output of the oscillator.) Now, the first term in parenthesis in equation (8.27) is identically zero (cf. equation (8.3) for the free oscillation), and hence the equation we have to solve for the output noise is

$$C\frac{dv_n}{dt} + (G_L - G_0)v_n + \frac{1}{L}\int v_n\,dt - \beta G_0 v_n (3v_f^2 + 3v_n v_f + v_n^2) = i_n(t) . \quad (8.28)$$

Note that intermodulation products will arise from the non-linear term in this equation, associated with the noise beating against itself and against the free oscillation. We shall return to this point shortly.

In the absence of the non-linearity (i.e. if β were zero), the spectral density of the noise voltage obtained from equation (8.28) would be

$$\overline{S_v(\omega)} = \frac{\overline{S_i}}{(G_L - G_0)^2 \left[1 + Q_0^2 \left(\dfrac{\omega}{\omega_0} - \dfrac{\omega_0}{\omega} \right)^2 \right]}, \quad (8.29)$$

where $\overline{S_i}$ is the spectral density of the white noise current generator $i_n(t)$, $\omega_0 = 1/\sqrt{LC}$ and $Q_0 = \omega_0 C/|G_L - G_0|$ is the external Q-factor of the oscillator. Evidently, according to this crude approximation, the noise is simply shaped by the narrow-band filtering action of the oscillator circuit. This conclusion, however, is over-simplistic and will be modified when the effects of the non-linear term in equation (8.28) are taken into account.

Equation (8.28) is made tractable by linearizing the non-linear term. Three terms, placed in parenthesis, contribute to the non-linearity, but two of these, the second and the third containing the noise fluctuation $v_n(t)$, are negligibly small compared with the first, which represents the free oscillation. Therefore, the only non-linear term we need consider in equation (8.28) is that involving the product $v_f^2 \, v_n$, which is equivalent to retaining the intermodulation products between the free oscillation and the noise whilst neglecting all noise/noise interactions.

From equation (8.26) and the definition of the free oscillation as $v_f(t) = v_0 \cos \omega_0 t$, and by expressing all the trigonometric functions in terms of exponentials, this product is

$$v_f^2 v_n = \frac{v_0^3}{8} \left[(2f + f^*) \exp (j\omega_0 t) + (2f^* + f) \exp (-j\omega_0 t) + f \exp (3j\omega_0 t) + f^* \exp (-3j\omega_0 t) \right] \, ,$$

(8.30)

where

$$f \equiv f(t) = a(t) - j\psi(t)$$

(8.31)

is a slowly varying, complex function of time. Now, bearing in mind that we are interested only in frequencies around the frequency of free oscillation (since the oscillator circuit is highly frequency selective), it is evident that the third harmonic terms in equation (8.30) make an insignificant contribution to the output noise and hence may be safely neglected, allowing us to write

$$v_f^2 v_n \simeq \frac{v_0^3}{8} \left[(2f + f^*) \exp (j\omega_0 t) + (2f^* + f) \exp (-j\omega_0 t) \right]$$

$$= \frac{v_0^2}{4} \left[2 v_n(t) + v_q(t) \right] \, ,$$

(8.32)

where

$$v_q(t) = v_0 \left[a(t) \cos \omega_0 t - \psi(t) \sin \omega_0 t \right] \, .$$

(8.33)

When the result in equation (8.32) is substituted back into equation (8.28), we obtain

$$C \frac{dv_n}{dt} - (G_L - G_0)(v_n + v_q) + \frac{1}{L} \int v_n dt = i_n(t) \, ,$$

(8.34)

where the expression in equation (8.10) has been substituted for v_0^2. Equation (8.34) is a linear differential equation in the fluctuating processes $v_n(t)$ and $v_q(t)$, and hence the first step in solving the non-linear differential equation (8.28) has been achieved. On examining these two equations, it is clear that the effect of the non-linear term in equation (8.28) is to contribute to the (linear) resistive term in equation (8.34). This occurs through the beating of the free oscillation with the noise, as expressed through equation (8.32). It is interesting to note that this frequency shifting of the spectral components of the noise is a phenomenon which is peculiarly non-linear in origin: linear systems never give rise to such an effect.

Although the differential equation describing the output noise has now been linearized, certain difficulties still remain. This is readily appreciated from inspection of equation (8.34), which contains *two* fluctuating processes, $v_n(t)$ and $v_q(t)$ (which between them fully specify the amplitude and phase fluctuations). The formal approach to the problem is to construct two simultaneous, linear differential equations from equation (8.34) by exploiting the orthogonality of the functions $\sin \omega_0 t$ and $\cos \omega_0 t$ over one cycle. The procedure, which is rather lengthy, eventually leads to the required solutions for the AM and PM noise spectra and also the AM–PM coherence, as detailed in Appendix 5. An alternative approach is given below which is more straightforward, but which is lacking in mathematical rigour. However, it leads to the correct spectral shapes, and illustrates well the important physical attributes of the AM and PM noise spectra. It is based on the supposition that equation (8.34) may be solved for $a(t)$ by setting $\psi(t)$ equal to zero and vice versa. An obvious objection to this procedure is that *all* the energy in $i_n(t)$ drives, say, the amplitude fluctuation when $\psi(t)$ is forced to be zero. In order to rectify this, we assume that the average energy in $i_n(t)$ is equally shared between $a(t)$ and $\psi(t)$, a situation which is accommodated by introducing a factor of $1/\sqrt{2}$ into the source term when equation (8.34) is split into AM and PM components. The simplified analysis then leads to the correct power spectral densities for the AM and PM noise, as may be confirmed by comparison with the results obtained more formally in Appendix 5.

8.7 THE AM NOISE SPECTRUM

In order to investigate the AM noise, we now neglect the effects of the phase fluctuation in equation (8.34) by setting $\psi(t) = 0$. From the definitions in equations (8.26) and (8.33) we then have $v_n(t) = v_q(t) = v_a(t)$, where $v_a(t) = v_0 a(t) \cos \omega_0 t$, and thus equation (8.34) reduces to the form

$$C\frac{dv_a}{dt} - 2(G_L - G_0) v_a + \frac{1}{L} \int v_a \, dt = \frac{i_n(t)}{\sqrt{2}} \, , \qquad (8.35)$$

where the factor of $1/\sqrt{2}$ on the right-hand side represents the proportion of the total noise energy which is 'driving' $a(t)$. Equation (8.35) is a linear differential equation in the variable $v_a(t)$, which can be solved for the power spectral density of the noise using familiar Fourier transform techniques. On transforming both sides, the power spectral density of $v_a(t)$ is found almost immediately to be

$$\overline{S_{v_a}(\omega)} = \frac{\overline{S_i}}{8(G_L - G_0)^2 \left[1 + \frac{Q_0^2}{4}\left(\frac{\omega}{\omega_0} - \frac{\omega_0}{\omega}\right)^2\right]} \, , \qquad (8.36)$$

where $\overline{S_i}$ is again the power spectral density of $i_n(t)$ and Q_0 is still the Q-factor of the oscillator circuit when $\beta = 0$.

A comparison of the result in equation (8.36) with that in equation (8.29), where the effect of the non-linear term in the conductance has been neglected, shows that the non-linearity has two effects on the power spectrum of the AM noise: it *reduces* the peak level by a factor of 8 and it *broadens* the bandwidth of the noise by a factor of 2. The latter effect could have been anticipated from inspection of equation (8.35), since the resistive term there contains a factor of 2 which would have been absent had the non-linearity been ignored.

The level of the noise spectrum in relation to the amplitude of the free oscillation can be examined by setting $\omega = \omega_0$ in equation (8.36), in which case we find

$$\overline{S_{v_a}(\omega_0)} = \frac{\overline{S_i}}{8(G_L - G_0)^2}$$

$$= \frac{2\overline{S_i}}{9\,\beta^2 G_0^2 v_0^4} . \tag{8.37}$$

This indicates that the AM noise is reduced as the oscillator is driven harder, an effect which is indeed observed in practice. It does not continue to indefinitely high levels of v_0, however, because terms beyond second order eventually become significant in the expansion of the non-linear conductance.

The expression in equation (8.36) is the power spectral density of the AM noise voltage $v_a(t) = a(t)v_0 \cos \omega_0 t$. If $\overline{S_a(\omega)}$ is the power spectral density of $a(t)$, it is easily shown that

$$\overline{S_{v_a}(\omega)} = \frac{v_0^2}{4} \left[\overline{S_a(\omega - \omega_0)} + \overline{S_a(\omega + \omega_0)} \right]$$

$$\simeq \frac{v_0^2}{4} \overline{S_a(\omega - \omega_0)} , \tag{8.38}$$

where the approximation holds because $\overline{S_a(\omega)}$ is a low-frequency function and we are interested only in frequencies around the carrier frequency. When this result is substituted into equation (8.36), we find that

$$\overline{S_a(\omega - \omega_0)} = \frac{8\,\overline{S_i}}{9\beta^2\,G_0^2\,v_0^6 \left[1 + Q_0^2 \left(\dfrac{\omega - \omega_0}{\omega_0} \right)^2 \right]} , \tag{8.39}$$

where the high-Q approximation, namely $(\omega^2 - \omega_0^2)^2/\omega^2\omega_0^2 \simeq 4(\omega - \omega_0)^2/\omega_0^2$, has been employed. Equation (8.39) specifies the AM noise power spectral density appearing in the general expression for the output spectrum in equation (8.22).

8.8 THE PM NOISE SPECTRUM

We now derive the PM noise spectrum from equation (8.34) by setting the amplitude fluctuation to zero. It then follows that $v_n(t) = -v_q(t) = v_\psi(t)$, where $v_\psi(t) = \psi(t) v_0 \sin \omega_0 t$, and thus in this case the resistive term in equation (8.34) disappears altogether, leaving the equation

$$C \frac{dv_\psi}{dt} + \frac{1}{L} \int v_\psi \, dt = \frac{i_n(t)}{\sqrt{2}} \tag{8.40}$$

to be solved. The factor of $1/\sqrt{2}$ here represents the proportion of the noise energy in $i_n(t)$ which is 'driving' $\psi(t)$. The same techniques as those described above lead to the following expression for the power spectral density of $v_\psi(t)$:

$$\overline{S_{v_\psi}(\omega)} = \frac{\overline{S_i}}{2Q_0^2(G_L - G_0)^2 \left(\dfrac{\omega - \omega_0}{\omega_0} \right)^2} \cdot \tag{8.41}$$

Thus, the PM noise shows a *very narrow band* spectrum centred on the free oscillation frequency and, as with the AM spectrum, its level is reduced when the oscillator is driven harder.

Since $v_\psi(t) = \psi(t) v_0 \sin \omega_0 t$, the spectral density of $\psi(t)$ is related to $\overline{S_{v_\psi}}(\omega)$ in equation (8.41) through the expression

$$\overline{S_{v_\psi}(\omega)} = \frac{v_0^2}{4} \overline{S_\psi(\omega - \omega_0)} \tag{8.42}$$

and hence

$$\overline{S_\psi(\omega - \omega_0)} = \frac{8 \, \overline{S_i}}{9 \beta^2 G_0^2 v_0^6 Q_0^2 \left(\dfrac{\omega - \omega_0}{\omega_0} \right)^2}, \tag{8.43}$$

which specifies the PM noise power spectral density appearing in equation (8.22). Note that in the immediate vicinity of ω_0, the PM noise spectrum is very much larger than that of the AM noise, but the tails of the two spectra are equal in level. These features are illustrated in Fig. 8.4.

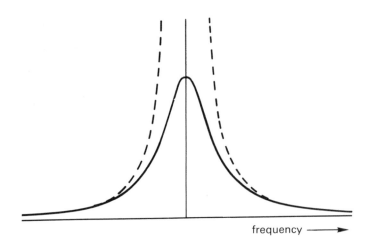

Fig. 8.4 — Sketch of the spectral shapes of the AM (solid line) and PM (broken line) noise spectra from equations (8.39) and (8.43), respectively. The curves are symmetrical about the frequency of free oscillation.

8.9 CONCLUDING REMARKS

Noise in negative conductance oscillators is a complex phenomenon. A complete analytical solution of the problem would take into account beating between different noise frequencies as well as between noise and the carrier frequency, and would include all the harmonic terms generated by these non-linear interactions. Such a solution is not possible and, even if it were, would probably be of little utility. Presumably, if ever it could be accomplished, it would be lengthy and complicated, and would differ from the solution discussed above only in insignificant details.

The analysis described above and in Appendix 5 provides a clear insight into the major physical effects of the non-linearity on the noise. The important mechanism in this context is the beating of the noise with the free oscillation, which shifts certain noise components into the spectral band of interest. In effect, these newly introduced components double the size of the circuit conductance (i.e. halve the Q-factor of the circuit) as far as the AM noise is concerned, and eliminate the circuit conductance entirely in connection with the PM noise. Thus, the AM noise spectrum is broadened by the non-linearity whereas the PM spectrum is very narrow band. The tails of the two spectra are equal in level, since at frequencies out of the vicinity of the carrier frequency the spectra are both determined by the reactive components in the circuit, which are unaffected by the van der Pol non-linearity.

The theory of oscillator noise discussed above is a comparatively simple account, based on a single-resonance LCR circuit containing a non-linear conductance. In practice, oscillator circuits may be more complicated than this, possibly containing a non-linear reactance, for example, or perhaps several non-linear elements. Such circuits are beyond the scope of the present discussion, which is intended to establish the basic essentials of oscillator noise. Moreover, the more complicated circuits are not usually amenable to an analytical treatment and must be investigated using a computer modelling approach. An example of this is contained in a recent paper by Sautereau et al. (1981).

REFERENCES

J. R. Ashley and F. M. Palka (1970), Noise properties and stabilization of Gunn and avalanche oscillators and amplifiers, G—MTT Int. Microwave Symp. Dig., pp. 161–164.

J. R. Ashley, C. B. Searles and F. M. Palka (1968), The measurement of oscillator noise at microwave frequencies, IEEE Trans. Microwave Theory Tech. (special issue on noise), MTT—16, 753–760.

W. A. Edson (1960), Noise in oscillators, Proc. IRE, 48, 1454–1466.

E. A. Faulkner and M. L. Meade (1968), Flicker noise in Gunn diodes, Elect. Lett., 4, 226–227.

M. Garstens (1957), Noise in non-linear oscillators, J. Appl. Phys., 28, 352–356.

J. Josenhans (1966), Noise spectra of Read diode and Gunn oscillators, Proc. IEEE (Lett.), 54, 1478–1479.

K. Kurokawa (1966), Noise in synchronized oscillators, IEEE Trans. Microwave Theory Tech., MTT—16, 234–240.

K. Kurokawa (1969), Some basic characteristics of broadband negative resistance oscillator circuits, Bell Syst. Tech. J., 48, 1937–1956.

J. A. Mullen (1960), Background noise in non-linear oscillators, Proc. IRE, 48, 1467–1473.

M. Ohtomo (1972), Experimental evaluation of noise parameters in Gunn and avalanche oscillators, IEEE Trans. Microwave Theory Tech., MTT—20, 425–437.

F. N. H. Robinson (1959), Nuclear resonance absorption circuit, J. Sci. Instrum., 36, 481–487.

F. N. H. Robinson (1962), Noise in Electrical Circuits, Oxford University Press, Chapter 8.

F. N. H. Robinson (1974), Noise and Fluctuations in Electronic Devices and Circuits, Oxford University Press, Chapter 18.

J. F. Sautereau, J. Graffeuil and J. C. Martin (1981), Time domain large signal noise modelling in microwave oscillators, Proc. Sixth International Conference on Noise in Physical Systems, National Bureau of Standards, Gaithersburg, Maryland, USA, April 6–10, 1981, pp. 47–50.

E. F. Scherer (1968), Investigation of the noise spectra of avalanche oscillators, IEEE Trans. Microwave Theory Tech., MTT—16, 781–788.

B. van der Pol (1927), Forced oscillations in a circuit with a non-linear resistance, *Phil. Mag. Series 7*, **3**, 65–80.

B. van der Pol (1934), The non-linear theory of electric oscillators, *Proc. IRE*, **22**, 1051–1086.

9

Tunnel diodes and parametric amplifiers

9.1 INTRODUCTION

The tunnel diode is a p–n junction in which the bulk regions are degenerate, owing to very high doping levels. Two tunnel currents flow across the junction, one in each direction, and both show full shot noise. The current-voltage characteristic of a tunnel diode shows a region of negative resistance at small forward biases. This may be utilized to achieve amplification. The noise figure of the amplifier depends on the series resistance of the bulk regions of the diode, the magnitude of the negative resistance, and the equivalent thermal noise conductance of the shot noise in the tunnel current.

A negative resistance is also the basis of parametric, or variable parameter, amplification. In this case, the negative resistance is achieved through a non-linear interaction between the signal and a high-frequency pump. The mixing element is usually the non-linear capacitance of a reverse-biased p–n junction. Parametric amplifiers are narrow bandwidth, high-frequency amplifiers, usually operating in the microwave region. Generally, they are somewhat noisier but considerably cheaper than masers, and quieter than travelling wave amplifiers, although the latter show a broader bandwidth and higher gain. An important contribution to the noise in a parametric amplifier is the thermal noise in the idler circuit, at the difference frequency between the pump and the signal: this gives rise to a noise current generator in the signal circuit at the signal frequency, whose power is proportional to the magnitude of the negative conductance.

9.2 THE TUNNEL DIODE

The tunnel diode is a p–n junction in which both bulk regions are so heavily doped that they are degenerate. Figure 9.1 shows the energy band structure of such a junction in thermal equilibrium: the Fermi level lies above the bottom of the conduction band on the n-side of the junction and below the top of the valence band on the p-side. The majority carrier densities in a device with the band structure shown in the figure are on the order of 10^{19} cm^{-3}.

As a consequence of the heavy doping, the width of the junction in a tunnel diode is extremely narrow, on the order of 100 Å. This is comparable with the mean separation of the impurity atoms in the host lattice. A quantum mechanical argument shows that, under these conditions, it is possible for electrons to tunnel through the junction from the n-region to the p-region and also in the reverse direction, from the p-region to the n-region. In other words, the wave functions of the electrons can penetrate the junction, giving a finite probability of transfer from one side to the other.

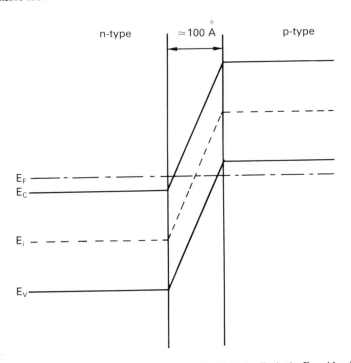

Fig. 9.1 — Equilibrium energy levels in the Esaki diode. E_F is the Fermi level, E_C and E_V are the conduction and valence bands, and E_i is the intrinsic Fermi level.

In thermal equilibrium, there can be no net current flow across the junction and hence the flows in either direction must be equal. When a reverse bias is applied, that is, one which increases the height of the potential barrier, there is a rapid increase in current owing to the enhanced flow of electrons tunnelling across the junction from the valence band on the p-side to the conduction band on the n-side; that is, under reverse bias, the tunnel diode appears to exhibit a Zener breakdown voltage of zero. Under forward bias, when the potential barrier is reduced below its equilibrium height, the current at first increases steadily with increasing applied voltage. This is due to electron tunnelling from the occupied levels in the conduction band on the n-side to the vacant levels in the valence band on the p-side. As the applied voltage is further increased, the current passes through a maximum, after which it declines rapidly; that is, the current-voltage characteristic in this region shows a negative resistance. This comes about because the probability of an electron tunnelling through the junction is high only between levels at the same energy (i.e. energy is conserved). When the forward bias depresses the valence band on the p-side below the conduction band on the n-side, the tunnelling probability falls dramatically and the tunnelling current falls along with it. On increasing the applied voltage still further, the current passes through a minimum and then begins to rise again, owing to the same mechanisms of minority carrier injection and depletion-layer recombination as found in non-degenerate p–n junctions. A sketch of the current-voltage characteristic of a tunnel diode is shown in Fig. 9.2.

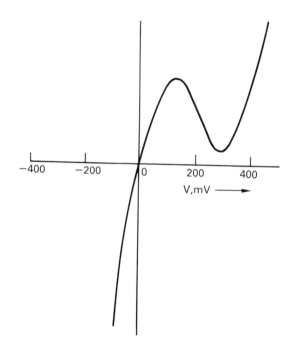

Fig. 9.2 — Typical current-voltage characteristic of a tunnel diode.

The maximum and minimum currents in the forward part of the current-voltage characteristic are known as the peak and valley currents, respectively. Typically, the ratio between them is about 10:1. The minimum occurs at an applied voltage in the region of one hundred to three hundred millivolts, depending on the material of the junction, the doping level and the temperature. In an ideal junction, the current at the minimum would be zero, corresponding to a tunnelling probability of zero, but in practice an excess current is observed to flow. This is attributed by Yajima and Esaki (1958) to electrons tunnelling to and from energy states in the band gap, associated with impurities and vacancies in the lattice structure. The excess current has been studied by Meyerhofer *et al.* (1962) in degenerate germanium junctions, and by Chynoweth *et al.* (1961) in silicon tunnel diodes; and Kane (1961) has discussed a theory of the excess current in tunnel junctions.

The feature of outstanding interest in the current-voltage characteristic of the tunnel diode is the region of negative resistance between the peak and valley currents. The slope of this part of the characteristic can be very steep, corresponding to a resistance as small as 1 ohm or less. Esaki (1958), whose name is now associated with the tunnel diode, was the first to observe the 'anomalous' behaviour of the current-voltage characteristic, during the course of an investigation of internal field emission in very narrow germanium p–n junctions. Not long after the appearance of his pioneering paper, the same type of current-voltage characteristic, exhibiting a region of negative resistance under forward bias, was reported in degenerate silicon junctions (Esaki and Miyahara, 1960).

Compared with minority carrier injection, electron tunnelling is extremely fast, with a time constant of about 10^{-12} seconds. This means that the Esaki diode can be used as a very fast switch or for high frequency (microwave) amplification. The factor limiting the response time of the device is not the tunnelling time constant but the capacitance of the diode, which is high per unit area because of the narrow junction width ($\simeq 1 \ \mu F/cm^2$). In order to achieve a fast switching time ($\simeq 10^{-9}$ seconds), the area of the junction must be kept as small as possible. Fortunately, suitably small junctions can be fabricated in which the working current is not reduced to an unacceptably low level, because the tunnelling current density is very high.

In addition to silicon and germanium, a number of III–V compound semiconductors have been used to produce tunnel diodes. These include gallium arsenide (Hołonyak and Lesk, 1960), indium antimonide (Batdorf *et al.,* 1960. Hulme 1961), and indium arsenide (Kleinknecht, 1961). The reasons for investigating these materials included a quest for reduced response times, improved noise performance, and lower series resistance in the bulk regions. Certain difficulties quickly become apparent, however. For example, in InSb the junction must be cooled to liquid nitrogen temperature in order to reduce the current due to carrier injection to a negligible level: the small energy gap in this material (0.18 eV at 300 K) implies a relatively low potential barrier which is easily crossed by thermally excited carriers at room temperature.

9.3 NOISE IN TUNNEL DIODES

9.3.1 Shot noise in the tunnel currents

The power spectrum of the noise in the current associated with electrons tunnelling through the potential barrier can be expressed in terms of an equivalent saturated diode current, I_{eq}, as follows:

$$\overline{S_{iT}(\omega)} = 2q\,I_{eq}\ . \tag{9.1}$$

Assuming that the electrons cross the barrier independently of each other, the noise is full shot noise in each of the tunnel currents, and thus I_{eq} is the *sum* of the oppositely directed tunnel currents:

$$I_{eq} = I_{c \to v} + I_{v \to c}\ , \tag{9.2}$$

where $I_{c \to v}$ and $I_{v \to c}$ are, respectively, the currents associated with electrons tunnelling from the conduction band in the n-region to the valence band in the p-region and from the valence band in the p-region to the conduction band in the n-region.

Given that current mechanisms other than tunnelling are negligible, the diode current flowing in the external circuit is

$$I = I_{c \to v} - I_{v \to c}\ , \tag{9.3}$$

which is the *difference* between the two tunnel currents crossing the potential barrier. The equivalent saturated diode current in equations (9.1) and (9.2) can be expressed in terms of the diode current, I, according to the following argument (Pucel, 1961).

The two tunnel currents can each be represented as an integral over the electronic energy states available for tunnelling. These integrals, which were originally formulated by Esaki (1958), take the forms

$$I_{c \to v} = A \int_{E_c}^{E_v} f(E)\,Z(E,V)\,[1 - f(E + qV)]\ \mathrm{d}E \tag{9.4a}$$

and

$$I_{v \to c} = A \int_{E_c}^{E_v} f(E + qV)\,Z(E,V)\,[1 - f(E)]\ \mathrm{d}E\ , \tag{9.4b}$$

where E is the energy, V is the applied voltage, A is a factor which depends on the material of the semiconductor, and

$$f(E) = [1 + \exp\{(E - E_F)/k\theta\}]^{-1} \tag{9.5}$$

is the Fermi–Dirac distribution function, in which E_F is the Fermi level. The function $Z(E, V)$ in equations (9.4) is a measure of the probability that an electron tunnels through the potential barrier, involving the density of states in the conduction and valence bands.

It is a matter of straightforward algebra to show from these equations that the integrands in equation (9.4) are related as follows:

$$f(E + qV)[1 - f(E)] = f(E)[1 - f(E + qV)] \exp(-qV/k\theta) .$$
(9.6)

Hence, the tunnel currents in equations (9.4) are also related in a simple way, through the expression

$$I_{c \to v} = I_{v \to c} \exp(qV/k\theta) ,$$
(9.7)

and therefore from equations (9.2) and (9.3) we find that

$$I_{eq} = I \coth(qV/2k\theta) .$$
(9.8)

An alternative derivation of this result, in which I_{eq} is expressed as an overlap integral, has been given by Bates (1961).

Equation (9.8) allows us to write the power spectrum of the tunnel current noise in equation (9.1) in terms of the diode current I:

$$\overline{S_{iT}(\omega)} = 2qI \coth(qV/2k\theta) .$$
(9.9)

For small applied bias, when $V \ll k\theta/q$, the equivalent saturated diode current in equation (9.8) can be written approximately as

$$I_{eq} \simeq \frac{2k\theta}{q} g ,$$
(9.10)

where $g = I/V$ is the incremental conductance of the diode. Therefore, in the limiting case of zero bias, the power spectrum in equation (9.9) reduces to

$$\overline{S_{iT}(\omega)} = 4k\theta g ,$$
(9.11)

which is the expression for thermal noise in a conductance g, which is as we would expect for the equilibrium condition.

One advantage of equation (9.9), in which the power spectral density of the tunnelling current noise is formulated in terms of the diode current and applied voltage, is that it can be used to check the assumption that electrons tunnel through the potential barrier independently of each other. Agouridis and van Vliet (1962) reported measurements of I_{eq} in a ZJ56A tunnel diode biased into the lower region of positive slope in the forward current-voltage characteristic.

They found a flat noise spectrum over the range 100 kHz–30 MHz, with good agreement between the measured value of I_{eq} and that calculated from equation (9.8). Their conclusion was that the tunnel currents show full shot noise, implying that the tunnelling of individual carriers is indeed independent.

9.3.2 Noise in the excess current

The excess current region of the current-voltage characteristic of a tunnel diode extends around the valley voltage. An 'excess' noise component is associated with this region of the characteristic, which shows a $1/f$ spectral density, and which predominates over the noise associated with the tunnelling currents and the current due to minority carrier injection.

The excess noise was reported first by Yajima and Esaki (1958) and subsequently by Montgomery (1961). The latter author made excess noise measurements on germanium and gallium arsenide tunnel junctions and found that the $1/f$ spectral behaviour extends from well above 1 kHz down to the lowest frequency he measured, which was 30 Hz. One possible explanation for this spectrum is that the density of occupied energy levels in the band gap shows a time dependence which is commensurate with a $1/f$ spectral shape. Since tunnelling to and from these levels is the mechanism responsible for the excess current, such a time dependence would explain the observed $1/f$ spectrum of the excess noise.

In germanium tunnel diodes, the spectral density of the excess noise is proportional to some power of the excess current. The index in this power law relationship is close to 2. Montgomery (1961), for example, made measurements on three germanium specimens and found power indices of 2.15, 1.95 and 1.7 at a measurement frequency of 1 kHz. This near square-law dependence of the spectral density of the excess noise on the excess current was originally reported by Yajima and Esaki (1958). Montgomery was unable to find a similar power law relationship in GaAs tunnel diodes because, in the samples he studied, the minority carrier injection current was significant in the excess current region around the valley voltage. This made it difficult to determine the relationship between the excess noise and the excess current.

9.3.3 Noise in the injection current

When a tunnel diode is biased into the upper region of positive conductance in its current-voltage characteristic, it shows shot noise at frequencies above about 10 MHz, and at lower frequencies a noise level in excess of shot noise. The low frequency noise has been attributed by Agouridis and van Vliet (1962) to a combination of shot noise in the minority carrier injection current and excess noise associated with the excess current. The shot noise observed at higher frequencies is the normal p–n junction noise associated with minority carrier injection.

9.4 TUNNEL DIODE AMPLIFIERS

The fact that a tunnel diode exhibits a region of negative resistance in its current-voltage characteristic means that the device can be used for amplification. Tunnel diode amplifiers received a good deal of attention in the literature in the late 1950s and early 1960s, mainly because of their high-frequency, low-noise potential. It was also recognized that they show certain advantages over non-linear susceptance amplifiers; for example, in the latter type of amplifier the negative conductance is achieved by a non-linear interaction involving an RF pump, whereas in the tunnel diode amplifier no pump is necessary since the negative conductance is an inherent feature of the diode's current-voltage characteristic.

Sommers (1959) developed a tunnel diode which was used by Chang (1959) as the active element in a negative conductance amplifier. Subsequently, Chang (1960) discussed the optimum noise performance of tunnel diode amplifiers. This was followed by the appearance in the literature of a number of related papers, by van der Ziel and Tamiya (1960), Hines and Anderson (1960), Penfield (1960), Tiemann (1960), Nielson (1960) and van der Ziel (1961a, 1961b). A general account of the tunnel diode as a circuit element was given by Pucel (1960), much of which concerns the noise performance of tunnel diode amplifiers.

9.4.1 The equivalent circuit

The equivalent circuit of a tunnel diode when it is biased into the negative resistance region of the characteristic is shown in Fig. 9.3. The capacitance C is the capacitance of the junction, $-R = -1/G$ is the negative resistance of the junction at the operating point (i.e. R is positive), and R_b represents the series resistance of the bulk regions of the diode. The inductance of the leads and other parasitic elements have not been included in this equivalent circuit.

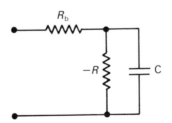

Fig. 9.3 – Equivalent circuit of a tunnel diode biased into the negative resistance region of the characteristic.

Although R_b is small, because of the high doping levels in the bulk regions, it is nevertheless non-zero. This means that, beyond a cut-off frequency, f_c, the diode no longer shows a negative resistance, and it does not then show any gain.

The cut-off frequency is readily calculated by setting the real part of the impedance of the equivalent circuit equal to zero. This gives

$$f_c = \frac{1}{2\pi RC} \left(\frac{R}{R_b} - 1 \right)^{\frac{1}{2}} \simeq \frac{1}{2\pi C\sqrt{RR_b}} , \qquad (9.12)$$

where the approximation holds because, in a good diode, $R \gg R_b$. It would
appear from equation (9.12) that the cut-off frequency cannot be increased by
reducing the junction area, since the product $C\sqrt{RR_b}$ is independent of area.
However, Pucel (1960) argues that this is not so, and that f_c can in fact be
raised by reducing the junction area. He suggests that there are variations in the
barrier thickness due to fluctuations in the impurity concentration over the area
of the junction, and that the tunnel currents are concentrated at those points
where the barrier is thinnest. Thus, R and R_b are not inversely proportional to
the junction area, whereas the capacitance C is essentially unaffected by the
doping level fluctuations and still scales with the area. According to this reasoning, a reduction in the area of the junction leads to a reduction in C and a
corresponding increase in f_c.

9.4.2 The gain of a tunnel diode amplifier

The equivalent circuit of a tunnel diode amplifier is shown in Fig. 9.4. In the
figure, $i_s(t)$ is the (sinusoidal) signal current source with conductance G_s, G_L is
the load conductance, and L is a tuning inductance.

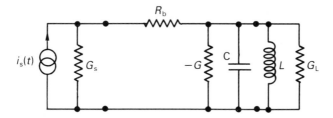

Fig. 9.4 – Equivalent circuit of a tunnel diode amplifier.

The gain of the amplifier can be expressed in terms of the transducer gain,
η, (Haus and Adler, 1957), defined as the *actual* power delivered to the load to
the *available* power from the source. By examining Fig. 9.4, we see that

$$\eta = \frac{|i_L|^2 / G_L}{|i_s|^2 / 4 G_s} , \qquad (9.13)$$

where i_L is the current flowing in the load conductance. Now, if $\omega_0 = 1/\sqrt{LC}$ is the resonance angular frequency of the parallel L–C combination in Fig. 9.4, a straightforward circuit analysis shows that

$$\frac{|i_L|^2}{|i_s|^2} = \frac{G_L^2}{(G_L + G_s - G)^2 + \omega^2 C^2 \left(1 - \dfrac{\omega_0^2}{\omega^2}\right)^2} , \tag{9.14}$$

and hence

$$\eta = \frac{4\, G_s G_L}{(G_L + G_s - G)^2 + \omega^2 C^2 \left(1 - \dfrac{\omega_0^2}{\omega^2}\right)^2} . \tag{9.15}$$

At resonance, when $\omega = \omega_0$, this expression reduces to

$$\eta_0 = \frac{4\, G_s G_L}{(G_L + G_s - G)^2} . \tag{9.16}$$

It is implicit here that $(G_L + G_s) \neq G$, otherwise $\eta_0 \to \infty$ and the amplifier becomes unstable.

The bandwidth of the amplifier at the half-power points is found from equation (9.15). If the angular frequencies at the upper and lower half-power points are ω_2 and ω_1, respectively, then

$$\omega_{2,1} = \sqrt{\frac{.1}{4C'^2} + \omega_0^2} \pm \frac{1}{2C'} , \tag{9.17}$$

where

$$C' = C/(G_L + G_s - G) . \tag{9.18}$$

Thus, the bandwidth is

$$B = (\omega_2 - \omega_1)/2\pi = 1/2\pi C' = (G_L + G_s - G)/2\pi C , \tag{9.19}$$

and the product of the square root of the gain at resonance and the bandwidth is

$$B\sqrt{\eta_0} = \frac{\sqrt{G_s G_L}}{\pi C} . \tag{9.20}$$

The gain–bandwidth product in equation (9.20) is a useful measure of the performance of the amplifier. It reinforces our earlier assertion that the capacitance, C, should be as small as possible. Obviously, the larger the values taken by G_s and G_L, the larger the gain–bandwidth product. This implies that the amplifier should have a high gain, and that $(G_s + G_L)$ should be approximately (but not precisely) equal to G. From this condition we find that the maximum gain–bandwidth product is

$$(B\sqrt{\eta_0})_{\max} \simeq G/2\pi C , \qquad (9.21)$$

occurring when $G_s = G_L = G/2$. This symmetry between the source and load conductances, necessary to achieve the optimum condition represented by equation (9.21), is not, however, compatible with other aspects of the tunnel diode's behaviour. In particular, it does not correspond to the condition for achieving the minimum noise figure, as we show below.

9.4.3 The noise figure of a tunnel diode amplifier

There are two noise generators to be considered when calculating the noise figure of a tunnel diode: one is the thermal noise voltage generator, $v_n(t)$, associated with the series resistance of the bulk regions and the other is the current generator, $i_n(t)$, representing the shot noise in the tunnel currents. These generators are shown in the equivalent circuit of the tunnel diode amplifier in Fig. 9.5(a). The current generator $i_{ns}(t)$ represents the noise associated with the source.

For the purpose of calculating the noise figure, it is convenient to make a Norton transformation of the equivalent circuit of the tunnel diode, as shown in Fig. 9.5(b). Here the series voltage generator, $v_n(t)$, and the parallel current

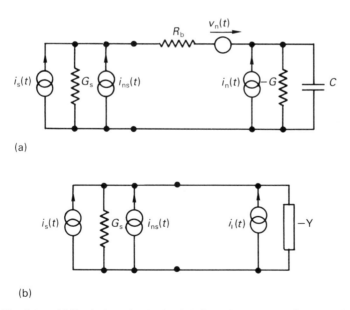

(a)

(b)

Fig. 9.5 – (a) Equivalent circuit showing the noise generators in a tunnel diode amplifier, and (b) the Norton equivalent circuit.

generator, $i_n(t)$, have been transformed into a parallel current generator $i_1(t)$, and the admittance $-Y$ is

$$-Y = \frac{-\left(1 - \dfrac{\omega^2}{\omega_c^2}\right) + j\dfrac{\omega}{\omega_c}\dfrac{R}{\{R_b(R - R_b)\}^{\frac{1}{2}}}}{(R - R_b)\left[1 + \dfrac{\omega^2}{\omega_c^2}\left(\dfrac{R_b}{R - R_b}\right)\right]}, \tag{9.22}$$

where $\omega_c/2\pi = f_c$ is the cut-off frequency of the tunnel diode given in equation (9.12).

The spectral densities of the current and voltage noise generators $i_n(t)$ and $v_n(t)$ are

$$\overline{S_i(\omega)} \simeq 2qI = 4k\theta G_e \tag{9.23a}$$

and

$$\overline{S_v(\omega)} = 4k\theta R_b , \tag{9.23b}$$

where we have approximated the noise associated with the *net* tunnelling current as full shot noise. For an operating point in the negative resistance region of the characteristic, this is justified since the applied voltage is somewhat greater than $2k\theta/q$ and the coth function in equations (9.8) and (9.9) is very close to unity. The conductance G_e in equation (9.23a) is the equivalent thermal noise conductance of the shot noise in the tunnel current. It follows from equations (9.23) and Fig. 9.5 that, as $v_n(t)$ and $i_n(t)$ are uncorrelated, the spectral density of the current generator $i_1(t)$ in Fig. 9.5(b) is

$$\overline{S_1(\omega)} = \frac{\left\{1 + \dfrac{\omega^2}{\omega_c^2}\left(\dfrac{R_b}{R - R_b}\right)\right\}\overline{S_v(\omega)} + R^2\overline{S_i(\omega)}}{(R - R_b)^2\left\{1 + \dfrac{\omega^2}{\omega_c^2}\left(\dfrac{R_b}{R - R_b}\right)\right\}}. \tag{9.24}$$

Assuming that the source is at the same temperature, θ, as the tunnel diode, the spectral density of the source current noise generator is

$$\overline{S_{i_{ns}}(\omega)} = 4k\theta G_s . \tag{9.25}$$

Now, from Fig. 9.5(b), the noise figure of the amplifier can be written simply as

$$F = 1 + \frac{\overline{S_1(\omega)}}{\overline{S_{i_{ns}}(\omega)}}$$

$$= 1 + \frac{\left[R_b \left\{ 1 + \dfrac{\omega^2}{\omega_c^2} \left(\dfrac{R_b}{R - R_b} \right) \right\} + R^2 G_e \right]}{G_s (R - R_b)^2 \left[1 + \dfrac{\omega^2}{\omega_c^2} \left(\dfrac{R_b}{R - R_b} \right) \right]} .$$ (9.26a)

For low frequencies, below the cut-off frequency, this expression reduces to

$$F \simeq 1 + \frac{R_b}{G_s (R - R_b)^2} + \frac{R^2 G_e}{G_s (R - R_b)^2} , \quad \omega < \omega_c .$$ (9.26b)

It is apparent from equation (9.26b) that, in order to minimize the noise figure, the source conductance should be as large as possible. In a high-gain tunnel diode amplifier, we have seen that

$$(G_s + G_L) \simeq G ,$$ (9.27)

though equality between the two sides of this expression is precluded if the amplifier is to remain stable. According to equation (9.27), the maximum value of G_s is

$$G_{s\,max} \simeq G ,$$ (9.28)

giving a low-frequency noise figure

$$F_{min} \simeq 1 + \frac{R_b}{R} + \frac{G_e}{G} .$$ (9.29)

Here, we have used the fact that in a good tunnel diode, $R \gg R_b$. With $G_e = qI/2k\theta = 0.1 \ \Omega^{-1}$, $G = 1 \ \Omega^{-1}$ and $(R_b/R) = 0.1$, the noise figure calculated from equation (9.29) is 1.2, or 0.8 dB. This is somewhat better than the noise figures measured by Pucel (1960), the lowest of which was 7 dB. However, in the device he used for the measurements, the region of negative resistance was not very steep, corresponding to a conductance $G \simeq 0.02 \ \Omega^{-1}$. When this value is substituted into equation (9.29), with the other parameter values remaining as above, a noise figure in accord with the measured value is obtained.

The condition in equation (9.28) giving the minimum noise figure in equation (9.29), implies asymmetry between the source and load conductances. As we have seen, this is not compatible with the condition for maximum gain, which requires symmetrical loading. If the source and load are symmetrical, then

$$G_s \simeq G/2$$ (9.30)

and the noise figure is

$$F \simeq 1 + 2\frac{R_b}{R} + 2\frac{G_e}{G} .$$ (9.31)

In this case, we see that the excess noise figure, defined as $(F - 1)$, is *twice* the minimum excess noise figure. Pucel (1960) observed that the noise figure did indeed fall as the source and load conductances were made progressively more asymmetrical, in accord with the theory described above.

9.5 THE PARAMETRIC AMPLIFIER

As we have seen in connection with the tunnel diode, devices exhibiting a negative resistance can be utilised to achieve amplification. It is also possible to introduce a negative resistance into the signal circuit, and hence achieve amplification, by the use of non-linear susceptances. The technique, known as parametric (or variable parameter) amplification, has application at microwave frequencies.

Although parametric processes in mechanical systems were known to Lord Rayleigh, their potential for achieving amplification was not recognised until relatively recently, when van der Ziel (1948) discussed the mixing properties of non-linear condensers. Subsequently, Suhl (1957) investigated parametric behaviour associated with the phenomenon of ferromagnetic resonance. Certain non-linear dielectrics, for example, barium titanate, have also received attention in connection with parametric amplification, although nowadays the parametric element most commonly employed is a reverse-biased p—n junction, exhibiting a non-linear charge-voltage characteristic.

Van der Ziel's (1948) analysis of parametric amplification was followed by the classic work of Manley and Rowe (1956), who derived relationships between the average powers at different frequencies in non-linear capacitors and inductors. These relationships are quite general and, in particular, they are independent of the shape of the non-linear characteristic, the only proviso being that the latter should be single valued. (An extension of the analysis to include those characteristics showing a hysteresis loop which is no more than double valued also appears in Manley and Rowe's original discussion.) Previously, Hartley (1936) derived similar results, under much less general conditions, for a particular form of capacitance modulator.

9.5.1 Principle of operation

Parametric amplification relies on the fact that when a high-frequency source (the pump) and a low-frequency source (the signal) are applied to a non-linear susceptance, the flow of power at the difference frequency introduces a negative conductance into the signal circuit. The magnitude of the negative conductance increases as the level of the high-frequency source rises, and also as the bandwidth of the amplifier is reduced. A narrow bandwidth has the further advantage of reducing the power that is dissipated in sidebands other than the one of interest.

An equivalent circuit for a parametric amplifier is shown in Fig. 9.6. It consists of three parallel LCR circuits coupled together through a non-linear capacitor (parametric diode). The resonance (angular) frequencies of the three circuits are ω_1 (signal circuit), ω_3 (pump circuit) and $\omega_2 = \omega_3 - \omega_1$ (idler circuit). The amplifier is said to be degenerate if $\omega_1 = \omega_2$, and non-degenerate when ω_1 and ω_2 are widely separated. In the non-degenerate case, three frequency components, namely ω_1, ω_2 and ω_3, appear in the voltage across the non-linear capacitor, whereas only two frequency components appear when $\omega_1 = \omega_2$. The two cases require separate analyses, as discussed by Chang (1964).

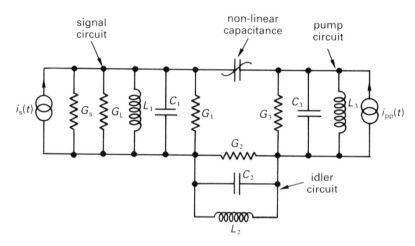

Fig. 9.6 – Equivalent circuit of a parametric amplifier.

Most practical parametric amplifiers are non-degenerate. For this reason, we shall ignore the degenerate case here. The non-degenerate amplifier we discuss below is one in which the non-linear element is a parametric diode whose charge varies quadratically with the voltage across its terminals. The quadratic law is of practical importance because it gives rise to distortionless amplification, whereas higher-power laws do not.

By considering the current and voltage associated with the parametric diode at the three frequencies ω_1, ω_2 and ω_3, the current-voltage relations for the three circuits in Fig. 9.6 can be derived. From these relations, the admittance of the *signal* circuit is found to be (see Appendix 6)

$$Y_s = (G_T - G) + j\left(B_1 - \frac{GB_2}{G_2}\right) , \qquad (9.32)$$

where B_1 and B_2 are the circuit susceptances of the signal and idler branches, excluding the non-linear component of the parametric diode, and

$$G_T = (G_1 + G_s + G_L) . \qquad (9.33)$$

The negative conductance in equation (9.32), $-G$, (G is positive) for a signal frequency ω (which is not necessarily equal to ω_1), is given by the expression

$$G = \beta(\omega) \frac{\omega(\omega_3 - \omega)}{\omega_1 \omega_2} G_T \; ,$$

(9.34a)

where

$$\beta(\omega) = \frac{a_2^2 |I_{pp}|^2 \omega_1 \omega_2}{G_2 G_3^2 G_T \left\{ \left(1 + \dfrac{V_1^2}{v_0^2}\right) + \left(\dfrac{B_2}{G_2}\right)^2 \right\}}$$

(9.34b)

and

$$v_0^2 = \frac{G_2 G_3}{a_2^2 \, \omega_3(\omega_3 - \omega)} \; .$$

(9.34c)

In these expressions, a_2 is the coefficient of non-linearity, as defined in Appendix 6, and $|I_{pp}|$ is the amplitude of the pump. It is implicit in equations (9.34) that there is no overlap between the passbands of the three resonant circuits in Fig. 9.6, so that for all frequencies within the passband of any one of the circuits the remaining two circuits show admittances which are essentially infinite.

9.5.2 The power gain of the amplifier

The power gain, η, of a parametric amplifier is defined here as the ratio of the *actual* power delivered to the load to the *available* power from the source. Thus,

$$\eta = \frac{G_L V_1^2}{\{|I_s|^2 / 4G_s\}} = 4 G_s G_L |Y_s|^{-2} \; ,$$

(9.35)

where V_1 and $|I_s|$ are the amplitudes of the signal current and voltage. On substituting for Y_s from equation (9.32), the power gain becomes

$$\eta = \frac{4 \, G_s G_L}{(G_T - G)^2 + \left(B_1 - \dfrac{G B_2}{G_2}\right)^2} \; .$$

(9.36)

At resonance, when $\omega = \omega_1$, the susceptances B_1 and B_2 are zero, and the gain takes its maximum value

$$\eta_0 = \frac{4 (G_s G_L / G_T^2)}{(1 - \beta_1)^2} \; ,$$

(9.37)

where $\beta_1 = \beta(\omega_1)$.

When $\beta_1 = 0$, the negative conductance in the signal circuit is zero and the amplifier shows no gain. When $\beta_1 = 1$, the gain goes to infinity and the system becomes unstable; that is, the amplifier becomes an oscillator. Between these two extreme conditions, linear amplification can be achieved provided the signal level is sufficiently low.

The bandwidth of the amplifier is determined from equation (9.36). If we ignore the frequency dependence of G, which is negligible throughout the band, we see that the half-power points fall at frequencies which satisfy the condition

$$\left(B_1 - \frac{G B_2}{G_2}\right)^2 = G_T^2 (1 - \beta_1)^2 . \tag{9.38}$$

Now, the susceptances in this expression are as follows:

$$B_1 = G_1 Q_1 \left(\frac{\omega}{\omega_1} - \frac{\omega_1}{\omega}\right) \simeq 2 G_1 Q_1 \frac{(\omega - \omega_1)}{\omega_1} \tag{9.39a}$$

and

$$B_2 = G_2 Q_2 \left(\frac{\omega_3 - \omega}{\omega_2} - \frac{\omega_2}{\omega_3 - \omega}\right) \simeq 2 G_2 Q_2 \frac{(\omega_1 - \omega)}{\omega_2} , \tag{9.39b}$$

where Q_1 and Q_2 are the quality factors of the unloaded signal and idler circuits, and the approximations hold over the band of interest because, in this frequency range, $\omega \simeq \omega_1$. On substituting the approximate expressions in equations (9.39) back into equation (9.38) and solving for the roots of the resultant quadratic in ω, the upper and lower half-power points are found to occur at frequencies

$$\omega_{\pm} = \omega_1 \pm \frac{G_T(1 - \beta_1)}{2\left(\dfrac{G_1 Q_1}{\omega_1} + \dfrac{G Q_2}{\omega_2}\right)} , \tag{9.40}$$

from which it can be seen that, to the order of approximation given here, the passband is symmetrical about the centre frequency ω_1.

From equation (9.40), the bandwidth of the amplifier, normalized to the centre frequency, is

$$B = \frac{\omega_+ - \omega_-}{\omega_1} = \frac{G_T (1 - \beta_1) \omega_2}{(G_1 Q_1 \omega_2 + G Q_2 \omega_1)} . \tag{9.41}$$

On recalling the expression for the gain in equation (9.37), the gain–bandwidth product is therefore

$$\eta_0^{\frac{1}{2}} B = \frac{2 \omega_2 \sqrt{G_s G_L}}{(G_1 Q_1 \omega_2 + G Q_2 \omega_1)} . \tag{9.42}$$

When the amplifier shows a high gain, the condition $G \simeq G_T$ obtains and it then follows that the maximum gain–bandwidth product occurs when

$$G_s = G_L = (G - G_1)/2 \quad . \tag{9.43}$$

Thus,

$$(\eta_0^{\frac{1}{2}} B)_{\max} = \frac{(G - G_1)\omega_2}{(G_1 Q_1 \omega_2 + G Q_2 \omega_1)} \tag{9.44a}$$

$$\simeq \frac{\omega_2}{\omega_1 Q_2} \;, \tag{9.44b}$$

where we have assumed that the second term in the denominator of equation (9.44a) is very much greater than the first term, because of the loading in the signal circuit, and also that $G \gg G_1$. Evidently, a high gain–bandwidth product can be achieved by increasing the ratio ω_2/ω_1. From equation (9.44b), the fractional bandwidth when the power gain is 100, $\omega_2/\omega_1 = 10$ and $Q_2 = 1000$, is 0.1%, thus illustrating the narrow band nature of the parametric amplifier.

9.5.3 The noise figure

The noise in a parametric amplifier originates in the circuit conductances G_s, G_1 and G_2. Noise from G_3, the conductance of the pump circuit, is negligible because it is swamped by the pump current; and for the purpose of calculating the noise figure, the noise from the load conductance can be ignored since it is normally taken into account in the following stage of amplification.

The significant noise generators in a parametric amplifier are shown in the equivalent circuit in Fig. 9.7. All but one of these generators appear in the signal circuit, the exception being the current generator $i_{n2}(t)$ representing thermal

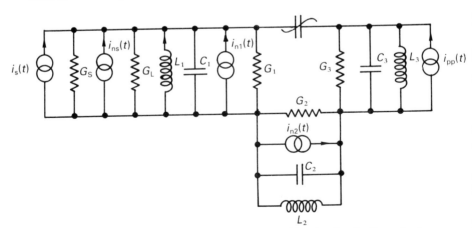

Fig. 9.7 – Equivalent circuit of a parametric amplifier showing the thermal noise generators associated with the circuit conductances.

noise in the idler circuit. In order to calculate the noise figure of the amplifier, it is necessary to establish the contribution $i_{n2}(t)$ makes to the noise in the signal circuit.

On examining the first of the current-voltage relations in equation (A6.12) of Appendix 6, it is apparent that a *voltage* across the idler circuit at frequency ω_2 gives rise to a *current* in the signal circuit at frequency ω_1. Now, at a frequency ω_2, the spectral density of the *voltage* noise generator, $v_{n2}(t)$, representing the voltage fluctuations across G_2 is

$$\overline{S_{v_{n2}}(\omega_2)} = \frac{4k\theta}{G_2} , \tag{9.45}$$

and hence the spectral density at frequency ω_1 of the equivalent current noise generator in the signal circuit is

$$\overline{S_{i_{n2}}(\omega_1)} = \overline{S_{v_{n2}}(\omega_2)} \, \omega_1^2 \, |C'|^2$$

$$= \frac{4k\theta}{G_2} \omega_1^2 \, |C'|^2 . \tag{9.46}$$

In this expression, C' is defined from equations (A6.12) as

$$|C'| = a_2 V_3 = a_2 |I_{pp}|/G_3 , \tag{9.47}$$

where $|I_{pp}|$ is the amplitude of the pump current generator. The effect of the term responsible for the non-linear distortion in the third of equations (A6.12) has been neglected here.

The spectral densities of the current generators representing thermal noise in the conductances G_s and G_1 are

$$\overline{S_{i_{ns}}(\omega_1)} = 4k\theta \, G_s \tag{9.48}$$

and

$$\overline{S_{i_{n1}}(\omega_1)} = 4k\theta \, G_1 . \tag{9.49}$$

Since there is no correlation between any of the noise sources in the circuit, the noise figure of the amplifier can now be written as

$$F(\omega_1) = \frac{\overline{S_{i_{ns}}(\omega_1)} + \overline{S_{i_{n1}}(\omega_1)} + \overline{S_{i_{n2}}(\omega_1)}}{\overline{S_{i_{ns}}(\omega_1)}}$$

$$= 1 + \frac{G_1}{G_s} + \frac{\omega_1^2 \, |C'|^2}{G_2 G_s} . \tag{9.50}$$

Now, through equation (A6.15) in Appendix 6 and equation (9.47), we can relate C' to the negative conductance G:

$$|C'|^2 = \frac{G_2 G}{\omega_1 \omega_2},$$
(9.51)

and it follows that the noise figure can be expressed as

$$F(\omega_1) = 1 + \left\{ \frac{G_1}{G_s} + \frac{\omega_1}{\omega_2} \frac{G}{G_s} \right\}.$$
(9.52)

This result was originally derived by Heffner and Wade (1958), and was discussed subsequently by van der Ziel (1959). It can be generalized to an input frequency ω which may differ from ω_1 but which still lies in the passband of the signal circuit, by replacing ω_1 with ω and ω_2 with $(\omega_3 - \omega)$:

$$F(\omega) = 1 + \left\{ \frac{G_1}{G_s} + \frac{\omega}{(\omega_3 - \omega)} \frac{G}{G_s} \right\}.$$
(9.53)

It is apparent from equations (9.52) and (9.53) that the contribution to the noise figure from the thermal noise in the idler circuit depends on the negative conductance G. For a fixed value of G, the noise figure can be reduced by increasing the ratio ω_2/ω_1. As we have seen, for a fixed value of Q_2, the quality factor of the idler circuit, this will also increase the gain–bandwidth product of the amplifier. Refrigeration can also be used to reduce the noise figure: if the amplifier is cooled to a temperature θ_0 and the signal source is at room temperature, θ, then the term in parenthesis in equation (9.53) is reduced by the factor θ_0/θ.

9.5.4 Concluding remarks

The parametric amplifier described above is one of several variants. These include parametric down-converters, in which the output current is at the idler frequency instead of the signal frequency, parametric up-converters, and parametric amplifiers with pump frequencies lower than the signal frequency. Chang (1964) has given a comprehensive account of these and other aspects of parametric amplification in his book on parametric and tunnel diodes.

REFERENCES

D. C. Agouridis and K. M. van Vliet (1962), Noise measurements on tunnel diodes, *Proc. IRE (Correspondence)*, **50**, 2121.

R. L. Batdorf, G. C. Dacey, R. L. Wallace and D. J. Walsh (1960), Esaki diode in InSb, *J. Appl. Phys.*, **31**, 613–614.

C. W. Bates (1961), Tunnelling currents in Esaki diodes, *Phys. Rev.*, **121**, 1070–1071.

K. K. N. Chang (1959), Low-noise tunnel diode amplifier, *Proc. IRE (Correspondence)*, **47**, 1268–1269.

K. K. N. Chang (1960), The optimum noise performance of tunnel diode amplifier, *Proc. IRE (Correspondence)*, **48**, 107–108.

K. K. N. Chang (1964), *Parametric and Tunnel Diodes*, (Ed. W. L. Everitt), Prentice-Hall, Chapter 6.

A. G. Chynoweth, W. L. Feldmann and R. A. Logan (1961), Excess tunnel currents in silicon Esaki junctions, *Phys. Rev.*, **121**, 684–694.

L. Esaki (1958), New phenomenon in narrow germanium p–n junctions, *Phys. Rev.*, **109**, 603–604.

L. Esaki and Y. Miyahara (1960), New device using the tunnelling process in narrow p–n junctions, *Solid State Elect.*, **1**, 13–21.

R. V. L. Hartley (1936), Oscillations in systems with non-linear reactance, *Bell Syst. Tech. J.*, **15**, 424–440.

H. A. Haus and R. B. Adler (1957), An extension of the noise figure definition, *Proc. IRE (Correspondence)*, **45**, 690–691.

H. Heffner and G. Wade (1958), Gain, bandwidth and noise characteristics of the variable parameter amplifier, *J. Appl. Phys.*, **29**, 1321–1331.

M. E. Hines and W. W. Anderson (1960), Noise performance theory of Esaki (tunnel) diode amplifiers, *Proc. IRE (Correspondence)*, **48**, 789.

N. Holonyak Jr. and I. A. Lesk (1960), Gallium arsenide tunnel diodes, *Proc. IRE*, **48**, 1405–1409.

K. F. Hulme (1961), Indium antimonide tunnel diodes, *Brit. J. Appl. Phys.*, **12**, 651–653.

E. O. Kane (1961), Theory of tunnelling, *J. Appl. Phys.*, **32**, 83–91.

H. P. Kleinknecht (1961), Indium arsenide tunnel diodes, *Solid State Elect.*, **2**, 133–142.

J. M. Manley and H. E. Rowe (1956), Some general properties of non-linear elements – part I. General energy relations, *Proc. IRE*, **44**, 904–913.

D. Meyerhofer, G. A. Brown and H. S. Sommers Jr. (1962), Degenerate germanium I. Tunnel, excess and thermal current in tunnel diodes, *Phys. Rev.*, **126**, 1329–1341.

M. D. Montgomery (1961), Excess noise in germanium and gallium arsenide Esaki diodes in the negative resistance region, *J. Appl. Phys.*, **32**, 2408–2411.

E. G. Nielson (1960), Noise performance of tunnel diodes, *Proc. IRE (Correspondence)*, **48**, 1903–1904.

P. Penfield Jr. (1960), Noise performance of tunnel-diode amplifiers, *Proc. IRE (Correspondence)*, **48**, 1478–1479.

R. A. Pucel (1960), Physical principles of the Esaki diode and some of its properties as a circuit element, *Solid State Elect.*, **1**, 22–33.

R. A. Pucel (1961), The equivalent noise current of Esaki diodes, *Proc. IRE (Correspondence)*, **49**, 1080–1081.

H. S. Sommers (1959), Tunnel diodes as high frequency devices, *Proc. IRE*, **47**, 1201–1206.

H. Suhl (1957), Proposal for a ferromagnetic amplifier in the microwave range, *Phys. Rev.*, **106**, 384–385.

J. J. Tiemann (1960), Shot noise in tunnel diode amplifiers, *Proc. IRE*, **48**, 1418–1423.

A. van der Ziel (1948), On the mixing properties of non-linear condensers, *J. Appl. Phys.*, **19**, 999–1006.

A. van der Ziel (1959), Noise figure of reactance converters and parametric amplifiers, *J. Appl. Phys. (letters)*, **30**, 1449.

A. van der Ziel (1961a), Noise measure of lossy tunnel-diode amplifier stages, *Proc. IRE (Correspondence)*, **49**, 1211–1212.

A. van der Ziel (1961b), Noise measure of distributed negative-conductance amplifiers, *Proc. IRE (Correspondence)*, **49**, 1212–1213.

A. van der Ziel and J. Tamiya (1960), Note on the noise figure of negative conductance amplifiers, *Proc. IRE (Correspondence)* **48**, 796.

T. Yajima and L. Esaki (1958), Excess noise in narrow germanium p–n junctions, *J. Phys. Soc. Japan*, **13**, 1281–1287.

10

Hot electron devices

10.1 INTRODUCTION

The physical principles underlying the operation of hot electron devices are quite different from those governing the behaviour of p–n junction devices such as the bipolar transistor. To a large extent, this is reflected in the nature of the noise in these microwave and millimetre wave devices. The primary noise mechanism in the Impatt, for example, is avalanching, which is an intrinsic feature of the device, associated with the very high electric field. An important, and perhaps more familiar, form of noise is found in the Gunn diode, namely Johnson noise; but even this differs from our usual concept of thermal noise, since it relates to the thermal velocity fluctuations of the hot electron population, which is not in thermal equilibrium with its surroundings.

The devices discussed in this chapter are the Impatt, the Trapatt and the Baritt, which are *transit-time devices,* and the Gunn diode, which is a *transferred-electron device.* The Impatt and the Trapatt are avalanche devices, and consequently are inherently noisy, whereas nowadays, owing to continuing improvements in fabrication technology, the Baritt and the Gunn diode are relatively quiet.

Hot electron devices, their operating principles and characteristics, are topics which are somewhat obscure. Before discussing the noise characteristics of such devices, the physics of their operation is described. But we begin with an account of the basic physics of hot electrons.

10.2 HOT ELECTRONS

When an electric field is applied to a semiconductor or a conductor, the charge carriers gain kinetic energy and an electric current flows. The kinetic energy of the carriers is randomly distributed amongst the carrier population through collisions with the crystal lattice, a process which also increases the vibrational energy of the lattice itself. Thus, the carrier population and the atoms of the lattice have a higher mean thermal energy than in the equilibrium state, or in other words, they have become hotter.

There are two mechanisms responsible for the conduction of heat in a solid, one being the transport of thermal vibrational energy by the lattice waves, or phonons, and the other the transport of thermal kinetic energy by the mobile carriers. In a metal, the latter mechanism usually predominates because of the high density of electrons, and, as a result, the thermal conductivity, K, scales (at a fixed temperature) with the electrical conductivity, σ. The relationship between K and σ is known as the Wiedemann–Franz law (Kittel, 1970). In a metal, the temperatures of the lattice and the electron population (sometimes referred to as an electron gas) are essentially the same.

The situation is different in a semiconductor, where the density of charge carriers can be substantially less than in a metal. In this case, heat conduction occurs principally via phonons. It follows that, if the crystal is mounted on a heat sink, the lattice can be kept relatively cool even though the charge carriers may gain considerable kinetic energy from the electric field. When the field is less than about 10^5 volt/metre, the mean kinetic energy of the charge carriers does not differ significantly from the equilibrium value, but at higher fields the energy of the carriers increases substantially. When this condition prevails, the charge carriers are quite reasonably referred to as "hot".

Hot carriers show different properties from carrier populations which are in thermal equilibrium with the lattice. In particular, the drift-velocity/field characteristic departs from the linear law which applies at lower fields. This is illustrated in Fig. 10.1, where the velocity is sketched as a function of the applied field for electrons in silicon. Holes in silicon and both types of carrier in germanium show a similar sort of behaviour. It can be seen from the figure that, as the field increases, the drift velocity rises progressively less rapidly, until eventually it saturates. Saturation occurs because of electron–phonon collisions which, from an energy balance argument, can be shown to give rise to a velocity which is

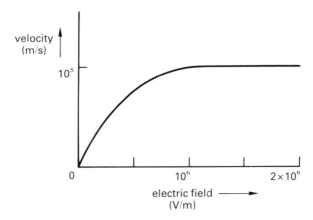

Fig. 10.1 – Sketch of the velocity/field characteristic of electrons in silicon.

independent of the applied field. In silicon, the scattering-limited velocity is approximately 10^5 m/sec.

Hot electrons are energetic charge carriers which are capable, through collisions with the valence electrons of the atoms in the host lattice, of expelling those electrons from the valence band and transferring them to the conduction band. These newly liberated electrons then become available to help expel still more valence electrons, which in turn release yet more, and so on. This process is the mechanism responsible for avalanche breakdown. Avalanche breakdown is observed in certain reverse-biased p—n junctions, which show a reverse current which increases sharply, by orders of magnitude, for a very small change in voltage when the breakdown voltage is reached. Diodes showing this type of reverse characteristic are used as voltage stabilizers, and are often referred to as Zener diodes.

Reverse-biased p—n junctions operating in the avalanche breakdown region can oscillate at very high frequencies, around 1 GHz or higher. This suggests that the phenomenon of avalanche breakdown might be useful for producing power at microwave frequencies. Indeed, this is the case, and semiconductor oscillators of this type are known as avalanche oscillators. There are two basic types of avalanche oscillator, each described by an acronym: one is the IMPATT (IMPact Avalanche and Transit Time) diode and the other is the TRAPATT (TRApped Plasma Avalanche and Triggered Transit). Impatts and Trapatts are similar in structure but they operate in different modes of oscillation. The power produced by the Impatt is typically in the region of a few hundred milliwatts in the frequency range 3 GHz to 50 GHz, whereas a Trapatt may produce a peak power output of several hundred watts in the range 0.5 GHz to 5 GHz. A third type of microwave oscillator, which is not an avalanche device, is the BARITT (BARrier Injection Transit Time). The physics of these devices is discussed in the following section.

The form of the velocity/field characteristic exhibited by silicon and germanium is not universally observed in all semiconductors. Certain of the III—V compound semiconductors, notably GaAs, show a maximum in the velocity as the field is increased, as illustrated in Fig. 10.2 (Fawcett et al., 1970). The maximum is attributed to a process of electron transfer from a high mobility state to a low mobility state as the field is increased. As a result of the transfer, a region of negative differential mobility appears in the velocity/field characteristic. At sufficiently high fields, essentially all the electrons are in the low mobility state, and the velocity saturates in much the same way as in silicon or germanium.

Semiconductor materials which show a region of negative differential mobility in their velocity/field characteristic can exhibit very high frequency (microwave) oscillations in the current when biased with sufficiently high fields. The phenomenon was discovered by J. B. Gunn in 1963 (Gunn, 1964), although it had been predicted in earlier theoretical work. In his experiments, Gunn

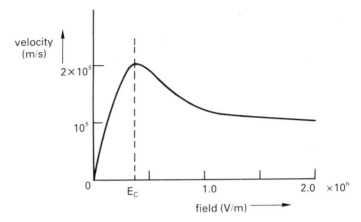

Fig. 10.2 – Sketch of the velocity/field characteristic of electrons in GaAs.

observed oscillations in the current at a frequency of about 1 GHz in specimens of n-type GaAs with ohmic contacts. The mechanism responsible for the oscillations, which appeared when the applied voltage exceeded a critical value, is the transfer of electrons from the high to the low mobility state alluded to above. The Gunn diode is the basis of the transferred-electron oscillator discussed in § 10.6.

10.3 PHYSICS OF TRANSIT-TIME DEVICES

Avalanche and barrier injection devices are two-terminal structures which have been fabricated from Ge, GaAs and, perhaps most commonly, from Si. The original Impatt structure, proposed by Read (1958), is p^+–n–i–n^+ (or n^+–p–i–p^+). It was several years after Read's proposal that his ideas were confirmed, when Johnson *et al.* (1965) were able to demonstrate a working device. Most modern Impatts differ from Read's conception of the device in that they have a p^+–n–n^+ (or p^+–p–n^+) structure, although the principles of operation are essentially the same as those expounded by Read. Trapatts have a similar structure to Impatts but operate in a different mode of oscillation. They were first demonstrated by Prager *et al.* (1967), in advance of the theory of their operation which was not developed until about two years later by Bartelink and Scharfetter (1969) and Clorfeine (1969). The third type of device discussed below is the Baritt diode, whose basic mechanism is punch-through injection rather than avalanche breakdown. Compared with the Impatt, the Baritt is a low-noise device. This is because the avalanche process in the Impatt gives rise to shot noise enhanced by a multiplicative factor, whereas the Baritt shows shot noise which has undergone a certain reduction due to space charge smoothing. On the other hand, the Baritt is inherently less efficient than the Impatt.

10.3.1 The Impatt

Figure 10.3(a) shows the structure of a p^+-n-n^+ Impatt under reverse bias conditions. We shall assume that, in the absence of avalanching, the reverse bias is sufficiently high for the n-region to be fully depleted of carriers, or in other words, that the diode is operating in the punch-through condition. Thus, the charge density in the n-region is uniform and equal to the donor concentration, N_D, and hence the electric field, E, throughout the region depends linearly on distance, with a slope

$$\frac{dE}{dx} = qN_D/\epsilon_0\epsilon_r \; . \tag{10.1}$$

The field profile is sketched in Fig. 10.3(b). Now, the avalanche multiplication process is strongly dependent on the field strength. This dependence may be appreciated by considering the ionization coefficient, α, which shows a variation with field of the form

$$\alpha \propto \exp\left\{-(E_0/E)^m\right\} , \tag{10.2}$$

where E_0 is a constant and the index m lies in the range 1 to 2. As a consequence of the strong dependence of the ionization rate on the field, avalanche multiplication occurs in the Impatt in a localized zone close to the field maximum. Typically, this zone is 1 micron thick (Gilden and Hines, 1966). Throughout the remainder of the n-region, designated the drift region, the field is too low for ionization to occur, although it is still sufficiently high for carriers in this region to be moving at their scattering-limited velocity.

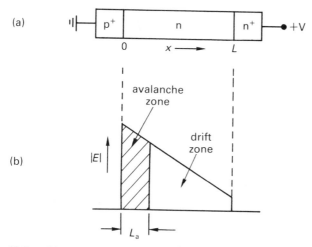

Fig. 10.3 – (a) The structure of a p^+-n-n^+ Impatt, and (b) the electric field profile across the n-region.

The detailed theories of the current-voltage behaviour of the Impatt are complicated and beyond the scope of the present discussion. However, the following qualitative description gives a reasonable account of the physics of the device. It also outlines the mechanism of power generation, which is the basis of the Impatt's utility as a microwave source in such applications as the pump in a parametric amplifier and the local oscillator in a radar.

For simplicity, suppose that the scattering-limited velocities of holes and electrons are the same and equal to v_s, that the space-charge fields due to the free charge carriers are negligible, and that the ionization rates of holes and electrons are equal. Furthermore, assume that the width of the avalanche zone, L_a, is independent of the voltage across the device. The transit time of a carrier across the avalanche zone is then $\tau_a = L_a/v_s$.

In the steady state, breakdown takes place in the avalanche zone, a constant current flows, and the voltage across the avalanche zone is V_b. Now, suppose that the diode voltage varies with time, so that the voltage V_a across the avalanche zone oscillates about V_b, as shown in Fig. 10.4(a). When V_a is greater than V_b the avalanche current rises, and when V_a is below V_b the current falls, as shown in Fig. 10.4(b). The sharp peaks in the avalanche current are a consequence of the highly non-linear nature of the current-generating mechanism. Note that the peaks in the avalanche current occur at the cross-over points in the voltage waveform where the voltage is passing through V_b from above to below. Thus, the avalanche-current peaks lag the peaks in the voltage, which is characteristic of an inductance. The avalanche current is often said to be inductive, although the equivalent inductance is clearly very non-linear.

The total *terminal* current in an Impatt differs significantly from the avalanche current. It consists of two components, an *induced* current and a *capacitive* current, the latter being the average displacement current across the depleted n-region. Now, the capacitive current is purely reactive (i.e. in quadrature with the voltage) and hence does not contribute to the power flow from the device. This is not the case with the induced current, which shows a fundamental Fourier component in antiphase with the voltage, as illustrated in Fig. 10.4(c). Thus, at frequencies around the avalanche frequency, the diode behaves as a negative resistance and, instead of absorbing power like a positive resistance, gives out power. If the device is placed in a suitable resonant circuit, spontaneous fluctuations of voltage and current develop, triggered by the thermal energy in the diode. When this occurs, d.c. power is converted into power at microwave frequencies, and the Impatt acts as an r.f. oscillator. The process is not very efficient, however, a typical efficiency being 10%, and a good deal of heat is generated which has to be conducted away through a substantial heat sink.

The fundamentally important factor in the power generation process is the induced current. The significant features characterizing the waveform of this current component can be understood from the following argument.

The avalanche-current waveform in Fig. 10.4(b) consists of a series of

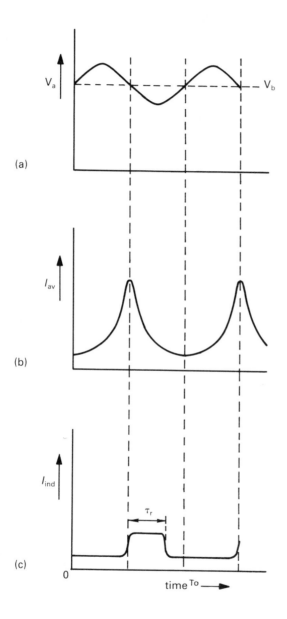

Fig. 10.4 – (a) Oscillations of the voltage across the avalanche zone, V_a, about the breakdown voltage, V_b. (b) The avalanche current, and (c) the induced current showing an approximately rectangular waveform with peaks of width equal to the transit time τ_r.

pulses, each of which is sufficiently narrow for its shape to be unimportant. Thus, each pulse can be safely approximated as a delta function, in which case the avalanche current is

$$I_{av} = Q \sum_k \delta (t - kT_0) , \qquad (10.3)$$

where Q is the charge per current pulse. If we consider just one of these pulses, $Q\delta(t)$, we see that it is associated with a pulse of holes of charge Q leaving the avalanche region via the contact on the extreme left of the n-region, and a pulse of electrons of charge $-Q$ leaving the avalanche region through its boundary on the right. Now, the charge packet of holes immediately passes through the p$^+$ contact and recombination occurs, giving rise to an induced current in the external circuit which is negligible; but the charge packet of electrons does not immediately reach the n$^+$ contact, having first to drift across the depleted n-region, or drift zone (Fig. 10.3(b)). An induced current then flows in the external circuit for the time τ_r during which the electrons are crossing the drift zone. τ_r is usually known as the transit time. It can be shown that, if the electrons are drifting at their scattering-limited velocity, and provided there is negligible recombination in the n-region, the magnitude of the induced current is Qv_s/L. Thus, the waveform of the induced current is more or less rectangular, as shown in Fig. 10.4(c), with the peaks and troughs in antiphase with the peaks and troughs of the voltage. This is the phenomenon referred to earlier when the Impatt was said to behave like a negative resistance.

The Impatt has received widespread attention in the literature. In the mid-1960s a great deal of work was done on the device, and a special issue of the *IEEE Trans. Elect. Dev.* (**ED–13**, No. 1, 1966) was devoted to the bulk-effect and transit-time devices. For a more recent account of the physics of the Impatt, the reader is referred to Misawa (1971) and Carroll (1974).

10.3.2 The Trapatt

The Trapatt oscillator is usually a p$^+$–n–n$^+$ (or p$^+$–p–n$^+$) structure, similar to that of the Impatt, which is biased into the reverse breakdown region. The punch-through condition obtains, so that, in the absence of an avalanche current, the n-region is depleted of carriers. The Trapatt is distinguished from the Impatt in that it is a large-signal device, it operates in a different mode of oscillation, it is more efficient (typically showing an efficiency of 50%) and it generates power at lower frequencies. Usually the Trapatt is operated in a pulsed mode, whereas the Impatt is continuous.

In the discussion of the Impatt, the space-charge fields associated with the charges in the avalanche were ignored. These space-charge fields play an important part in the functioning of the Trapatt. They are responsible for the behaviour that characterizes the Trapatt mode of oscillation, in which the device switches extremely rapidly (in typically 10^{-10} seconds) from a high-impedance, high-voltage state, to a low-impedance, low-voltage state.

The role of the space-charge fields can be understood from the following argument. Suppose that at time $t = 0$, the p^+–n–n^+ Trapatt in Fig. 10.5(a) is reverse-biased to punch-through, but that the peak field in the n-region is well below the field E_a necessary for avalanche breakdown. Under these conditions, the n-region is depleted of mobile charge carriers, and Poisson's equation gives the slope of the field as

$$\frac{dE}{dx} = qN_D/\epsilon_0\epsilon_r \ , \tag{10.4}$$

where N_D is the donor density in the n-region. On integrating equation (10.4), and assuming that at the boundary where $x = L$ the field is zero, we have that

$$E(x) = \frac{qN_D}{\epsilon_0\epsilon_r}(x - L) \ . \tag{10.5}$$

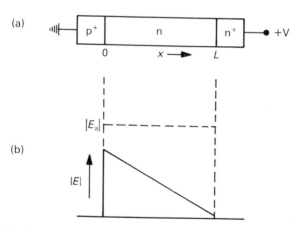

Fig. 10.5 – (a) The structure of a p^+–n–n^+ Trapatt, and (b) the electric field in the n-region at $t = 0$, when the peak field is well below the critical field E_a at which avalanching occurs.

Now suppose that at time $t = 0$ a constant current source, I_0, is switched on to the terminals of the Trapatt. Provided the field is below E_a, the n-region remains depleted of carriers, and the only current that can flow through the region is a displacement current given by

$$I_0 = -\epsilon_0\epsilon_r A \frac{\partial E}{\partial t} \ , \tag{10.6}$$

where A is the cross-sectional area of the device. When equation (10.6) is integrated with respect to t and the initial condition in equation (10.5) is employed to determine the constant of integration, the field is found to increase linearly with time:

$$E(x, t) = \frac{qN_D}{\epsilon_0 \epsilon_r}(x - L) - \frac{I_0 t}{\epsilon_0 \epsilon_r A} \ . \tag{10.7}$$

Equation (10.7) allows us to establish how the *field profile* moves through the n-region. For if we set $E(x, t)$ equal to a fixed value, E', and then differentiate equation (10.7) with respect to time, the constant field point is found to move through the device with a velocity

$$v' = \frac{dx}{dt} = \frac{I_0}{AqN_D} \ . \tag{10.8}$$

Thus the field profile also moves through the n-region with the velocity v' given in equation (10.8). By increasing I_0 and decreasing N_D, this velocity can be made to exceed the scattering-limited velocity of the hot carriers.

Let us now examine how the moving field profile can give rise to oscillatory behaviour in the Trapatt. As we have seen, when the magnitude of the field is below the threshold field required for avalanching, the field profile is linear in time and position. But when the electric field exceeds E_a, the carrier density increases rapidly as a result of avalanche multiplication and the field no longer increases linearly with time. Figure 10.6, showing the field profile at successive

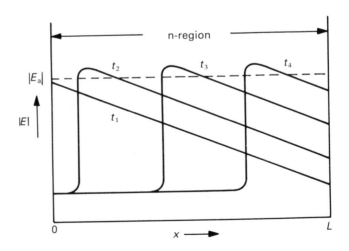

Fig. 10.6 – Electric field profiles in the n-region at successive times $t_1 < t_2 < t_3 < t_4$.

times, illustrates what happens when the critical field is exceeded, assuming that the velocity v' is greater than the scattering-limited velocity.

As soon as the field reaches the critical level, avalanching begins. Carriers are generated by the multiplication process but, because $v' > v_s$, the field keeps increasing. i.e. the distribution of carriers cannot change sufficiently quickly to immediately modify the field. Consequently, the field rises briefly to a level substantially above the critical field, and in so doing creates an extremely large population of carriers through avalanching. These carriers then reduce the field behind the avalanche region (bearing in mind that the whole field profile is moving extremely rapidly through the n-region) to a low level where the carriers are no longer hot, but obey a linear velocity/field law. Electrical neutrality prevails in this region since the carriers arrange themselves, according to the usual conception of dielectric relaxation, to achieve this condition. The holes and electrons generated by the avalanche process constitute a *plasma* which is 'trapped' in the low-field region behind the advancing avalanche region. The region of avalanche multiplication propagates through the device in 100 pico-seconds for a typical Trapatt, and was referred to by Bartelink and Scharfetter (1969) as an *avalanche shock front*.

The voltage across the terminals of the Trapatt is the area under the field profile. It is clear from Fig. 10.6 that this area changes very rapidly (in the order of 100 picoseconds) from a high value, when the device is driven into breakdown, to a low value after the avalanche shock front has travelled through the n-region. As soon as the voltage falls, the current rises as the trapped plasma is swept out of the device. Figure 10.7 shows typical voltage and current waveforms obtained when the device is connected to a suitable circuit for maintaining the oscillations (Evans, 1969). The waveforms demonstrate that at a characteristic frequency the Trapatt acts as a negative resistance: the fundamental Fourier components of the two waveforms are in antiphase. Clorfeine *et al.* (1969) have shown that the value of this negative resistance is

$$R \simeq -0.4 \frac{V_b}{I_0} ,$$

(10.9)

where V_b is the breakdown voltage and I_0 is the average current flowing through the device. It is because of this negative resistance that the Trapatt is capable of giving out microwave power.

The Trapatt is a large signal device which has to be driven into the trapped plasma mode because thermal noise fluctuations cannot be relied upon to get the oscillations started. Instead, the device is usually triggered by an Impatt oscillator.

Note that the trapatt is inherently slower than the Impatt because the process of removing the relatively slow carriers in the trapped plasma takes longer than the extraction of carriers moving at their scattering-limited velocity.

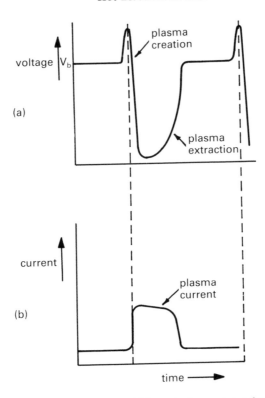

Fig. 10.7 – (a) Terminal voltage and (b) terminal current waveforms in the Trapatt. $V_b \approx 100$ volts is the breakdown voltage.

10.3.3 The Baritt

The Baritt is not unlike the Impatt and the Trapatt in that it contains two junctions, but now the arrangement is such that one of the junctions is forward-biased. Figure 10.8(a) shows the structure and biasing of a p^+–n–p^+ Baritt. The n-region is very thin, usually of the order of 10 microns. Coleman and Sze (1971) were the first to demonstrate that such a device shows a negative resistance at microwave frequencies, enabling it to sustain oscillation and give out power in the r.f. frequency range. Instead of semiconductor junctions, their device contained two metal-semiconductor (Schottky) barriers, but the principles of operation are essentially the same in the two cases.

The potential and electric field profiles in a Baritt diode under different biasing conditions are shown schematically in Figs. 10.8(b) and 10.8(c). In equilibrium, there is a zero bias across both junctions and no current flows in the device. As the applied voltage is increased, the left-hand junction becomes reverse-biased, the right-hand junction becomes slightly forward biased, and the current, which is constant throughout the device, is equal to the reverse leakage current

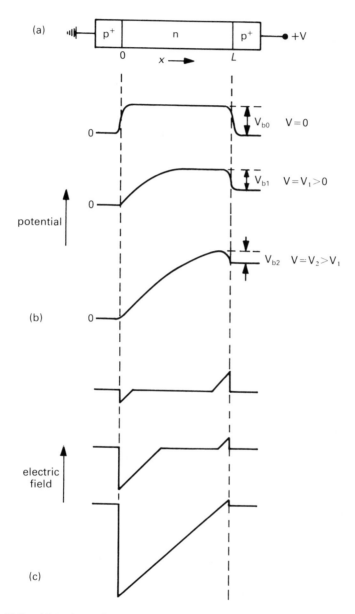

Fig. 10.8 – (a) A p^+–n–p^+ Baritt structure. (b) Three potential profiles across the diode, and (c) the corresponding electric field profiles.

of the left-hand junction (the n-region is assumed to be too wide at this stage for minority carriers injected across the forward-biased junction to be 'collected' by the reverse-biased junction). On further increasing the applied voltage, the depletion region of the reverse-biased junction extends further into the n-region, until punch-through eventually occurs.

Figure 10.9 shows a representative current-voltage characteristic for the Baritt. The current in the region A–B is the reverse leakage current. Between B and C punch-through has occurred and the current rises rapidly as the voltage is increased because the minority carriers injected across the forward-biased junction are swept across the n-region by the large electric field there. Eventually, the current becomes space-charge limited and increases much less rapidly with increasing voltage (region C–D). A fourth region occurs beyond C–D where the current again rises rapidly owing to avalanche breakdown in the vicinity of the metallurgical junction on the left, where the electric field is very large. A detailed discussion of the current mechanism in barrier injection devices has been given by Sze *et al.* (1971).

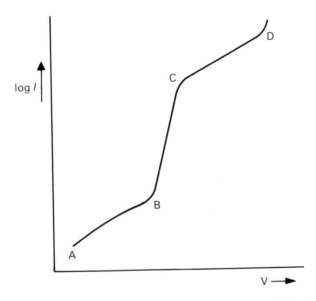

Fig. 10.9 – Sketch of the current-voltage characteristic of a Baritt diode.

The biasing of a Baritt oscillator is such that punch-through obtains, corresponding to the region B–C in Fig. 10.9, but breakdown does not occur. Assuming that the injected carriers are travelling at their scattering limited velocity v_s, the transit time across the n-region is $\tau_t = L/v_s$. Now suppose that the device is driven by a small-signal voltage source with an angular frequency $\omega_0 = 2\pi/T_0$, where the period $T_0 = 4\tau_t/3$. One cycle of this voltage waveform is shown in Fig. 10.10(a). As the voltage rises above the mean value, a 'packet' of charge is injected across the forward-biased junction into the n-region. This charge packet may be approximated as a spike or delta function which appears in the n-region at the time the voltage reaches its peak value. The charge pulse is then swept across the n-region in a time τ_t, and the current rises to a constant level above its previous value. After the transit of the charge, the current falls to its original level. The current waveform is illustrated in Fig. 10.10(b).

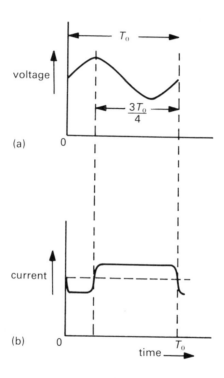

Fig. 10.10 – Schematic of (a) the voltage and (b) the current waveforms of a Baritt oscillator.

It is evident from Fig. 10.10 that at frequencies around $\omega_0/2\pi = \frac{3}{4}(v_s/L)$ (which has a value of 7.5 GHz in a device with L = 10 microns and $v_s = 10^5$ m/s) the Baritt shows a negative resistance. Coleman (1972) has derived an expression for the impedance of the device. The power output from the Baritt is on the order of 10 mW, and the Q-factor is relatively high at about 40. Under the right conditions the oscillator is self-starting owing to thermal noise fluctuations.

10.4 AVALANCHE NOISE

As we have seen, avalanche multiplication takes place in regions of high electric field, where the free charge carriers have sufficient kinetic energy to ionize the atoms of the host lattice. The ionization coefficient for electrons, α, is defined as the mean number of ionizing collisions per unit distance of drift of the electron. An analogous definition applies for the ionization coefficient of holes, β. In general, α and β take different numerical values, and both show a strong dependence on the electric field.

An ionizing impact liberates a hole and an electron, which are then themselves free to drift through the semiconductor in opposite directions under the influence of the electric field, liberating further hole–electron pairs as they go. The avalanche current is the sum of the primary current and the secondary, tertiary, ... currents produced by impact ionization.

In the general case where the ionization coefficients of holes and electrons are not equal, the algebra of the multiplication process is rather lengthy. However, the special case where $\alpha = \beta$ is considerably simpler, and is the one we adopt below in order to discuss the principles of current multiplication and avalanche noise.

Since the electric field, and hence the ionization coeffient α, may vary over the space-charge region, it is convenient to define

$$\bar{\alpha} = (1/L_a) \int_0^{L_a} \alpha(x)\, dx \tag{10.10}$$

as the average value of α over the avalanche zone of length L_a. Each mobile carrier (hole or electron) crossing the avalanche zone will liberate on average $\bar{\alpha} L_a$ pairs of holes and electrons. The total distance drifted by the constituents of such a pair is L_a, since they drift in opposite directions, and thus each pair will produce on average $\bar{\alpha} L_a$ more pairs. Therefore, if I_0 is the primary current, the current after multiplication is

$$\begin{aligned} I_M &= I_0 \left[1 + \bar{\alpha} L_a \left\{ 1 + \bar{\alpha} L_a (1 + \bar{\alpha} L_a [\ldots]) \right\} \right] \\ &= I_0 \left[1 + \bar{\alpha} L_a + (\bar{\alpha} L_a)^2 + (\bar{\alpha} L_a)^3 + \ldots \right] \\ &= I_0 M \ , \end{aligned} \tag{10.11}$$

where

$$\begin{aligned} M &= 1 + \bar{\alpha} L_a + (\bar{\alpha} L_a)^2 + (\bar{\alpha} L_a)^3 + \ldots \\ &= 1/(1 - \bar{\alpha} L_a) \end{aligned} \tag{10.12}$$

is the multiplication factor. When $\bar{\alpha} L_a$ is unity, M goes to infinity and avalanche breakdown occurs. Since α is a very rapid function of the electric field, only a very small change in applied voltage is required to drive the device into breakdown, thus accounting for the very abrupt nature of the breakdown characteristic.

The ionizing events underlying the multiplication process occur randomly, thus creating noise in the multiplied current. At low frequencies, well below the avalanche frequency, the noise shows a 'white' spectrum which varies as the third power of the multiplication factor. The key to understanding the low-frequency noise fluctuations in the avalanche current is that any hole–electron pair created in the avalanche zone will on average be multiplied by the factor M. The spectral density of the avalanche noise is determined by considering the change in the electron (or hole) current in an element of distance dx due to hole–electron pair creation by impact ionization. The change in the hole current is

$$dI_p = \alpha(x)I_M\,dx \; , \tag{10.13}$$

where I_M, given in equation (10.11), is the total current flowing through the avalanche region. I_M is independent of position. Now, the element of current in equation (10.13) shows full shot noise and hence has a spectral density $2q\alpha(x)I_M\,dx$ (excluding the dc component). As the current flowing through the device due to hole–electron pair creation in the element dx is $M\,dI_p$, the spectral density of the associated avalanche noise is

$$dS_{av}(\omega,x) = 2q\alpha(x)M^2 I_M\,dx \; . \tag{10.14}$$

Assuming that the pair-creation events throughout the device are independent, the spectral density of the noise in the total avalanche current is the integral of equation (10.14) plus the shot noise contribution associated with the primary current I_0:

$$\overline{S_a(\omega)} = 2q[M^2 I_M \int_0^{L_a} \alpha(x)\,dx + M^2 I_0]$$

$$= 2qM^3 I_0 \; , \tag{10.15}$$

where we have used the results in equations (10.11) and (10.12).

The expression in equation (10.15) for the spectral density of the avalanche noise when the ionization coefficients of the holes and electrons are equal was derived by Tager (1964). It was generalized later by McIntyre (1966) to include those cases where the hole and electron coefficients, though not equal, vary with electric field in such a way that $\beta = k\alpha$, where k is a scaling constant. McIntyre found that if the primary current consists entirely of holes, the spectral density of the noise is

$$\overline{S_a(\omega)} = 2qM^3I_0 \left[1 + \left(\frac{1-k}{k} \right) \left(\frac{M-1}{M} \right)^2 \right]$$

$$\simeq 2qM^3I_0/k \; , \tag{10.16a}$$

and when the primary current consists entirely of electrons,

$$\overline{S_a(\omega)} = 2qM^3I_0 \left[1 - (1-k) \left(\frac{M-1}{M} \right)^2 \right]$$

$$\simeq 2qM^3I_0k \; . \tag{10.16b}$$

The approximations in these expressions hold when $M \gg 1$. Notice that the expressions in equations (10.16) are symmetrical in that one transforms into the other when k is replaced by $1/k$.

Taking the spectral level in equation (10.15) as the norm, it is apparent from equations (10.16) that the noise is reduced when the primary current consists of holes and the ionization coefficient for holes is greater than that for electrons (i.e. $k > 1$), and also when the primary current consists of electrons and the ionization coefficient for electrons is greater than that for holes (i.e. $k < 1$). Experimental confirmation of the behaviour predicted by equations (10.16) has been obtained by several investigators, including Melchior and Anderson (1965), Melchior and Lynch (1966), Baertsch (1966, 1967), and Conradi (1972).

It is implicitly assumed in the derivation of equations (10.15) and (10.16) that the number of ionizing collisions per carrier transit of the avalanche region is very large. This allows the avalanche multiplication to be treated as a continuous spatial process. Van Vliet and Rucker (1978, 1979) have relaxed the condition of continuity, and have treated the multiplication process as a discrete phenomenon. The details of their theory are beyond the scope of the present discussion, but it is interesting to note that in modern avalanche devices, the number of ionizing collisions per carrier transit is often small, in the region of 2 or 3 (Lukaszek et al., 1976). Van Vliet and Ruckers's theory can handle such cases, and in the limit of a large number of ionizing collisions, their result agrees with that of McIntyre.

10.5 NOISE IN TRANSIT-TIME AMPLIFIERS

10.5.1 Impatts
Amplification at microwave frequencies can be achieved with an Impatt by

inserting the device in a resonant cavity which is coupled, by means of a transmission line and circulator, to input and output transmission lines. At the resonance frequency the diode shows a negative resistance, and hence the reflection coefficient at the cavity is greater than unity. Thus the wave emerging from the cavity has a greater amplitude than that entering it; that is, the device performs as an amplifier.

The sources of noise in such a circuit are thermal noise in the input line, thermal noise in the bulk resistance of the diode, and multiplication noise in the avalanche current. Of these three noise sources, the last is the predominant one. Hines (1966) has proposed a small-signal theory of avalanche current noise in Impatt amplifiers which is applicable at high and low frequencies. In the theory the ionization coefficients of holes and electrons are treated as equal. This constraint is relaxed by Gummel and Blue (1967), who treat the same problem as Hines but adopt a rather more general approach. As they point out, this is necessary when discussing a material such as silicon, where the hole and electron ionization coefficients differ by a factor of 10. Hines's theoretical predictions have been confirmed by Haitz and Voltmer (1966), who made noise measurements on micro-plasma-free guard-ring avalanche diodes.

Hines analysed the avalanche-noise behaviour of the Impatt amplifier on the basis of the equivalent circuit shown in Fig. 10.11. In the circuit external to the diode, L_L is a tuning inductor and R_L is the load resistance. The noise current flowing in the external circuit, $i_n(t)$, is the sum of three components. One of these is the carrier conduction current, $i_{no}(t)$, which is a fictitious noise current in the sense that it is the noise current which would flow if it were possible to keep the electric field in the avalanche region constant at the critical value necessary to maintain a steady avalanche. In practice, the charge carriers in a

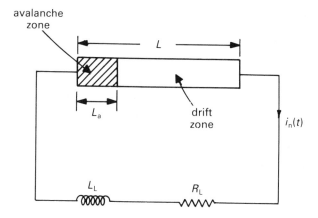

Fig. 10.11 – Circuit for analysis of avalanche current noise in the external circuit of an Impatt reflection amplifier.

high-current density avalanche will induce an ac field in the avalanche zone. A noise current ecomponent, $i_{ne}(t)$, which is correlated with $i_{n0}(t)$, is associated with this field. By linear superposition, the total avalanche conduction noise current is $i_{n0}(t) + i_{ne}(t)$. Thus, $i_{ne}(t)$ is the second of the three components contributing to the noise current in the external circuit. The third component is the displacement current in the avalanche zone.

The expression derived by Hines for the power spectral density of $i_n(t)$ is complicated, involving a number of parameters including the avalanche angular frequency ω_a, the avalanche current, I_M, the total impedance of the circuit in Fig. 10.11, and the derivative of the ionization coefficient with respect to the field. At low frequencies, below the avalanche frequency, the noise shows a 'white' spectrum which is *inversely proportional* to the dc avalanche current, I_M. Physically, this inverse dependance may be interpreted as being due to a decrease in the impedance of the avalanche as the current rises. At high frequencies, above the avalanche frequency, the power spectral density of the noise falls rapidly with increasing frequency, varying approximately as ω^{-4}. There is some structure on the curve in this region, characterized by the width of the drift region; and the power spectral density is *proportional* to I_M. At the avalanche frequency, which itself varies as the square root of I_M, the power spectral density of the noise current may show a peak if I_M is sufficiently large. Representative curves for the power spectral density of $i_n(t)$ as a function of frequency, illustrating the features described above, are shown in Fig. 10.12.

Hines has shown that the noise figure of an Impatt is given by the expression

$$F = 1 + \frac{\overline{S_{i_n}(\omega)} \cdot R_L}{Gk\theta} ,$$

(10.17)

where $\overline{S_{i_n}(\omega)}$ is the power spectral density of the noise current $i_n(t)$, G is the gain of the amplifier, and θ is the absolute temperature. With parameter values appropriate to a frequency of 10 GHz, the noise figure evaluated from equation (10.17) is 40.5 dB. Kuvas (1972) measured the noise figure of Impatts fabricated from silicon, germanium and GaAs, at 6 GHz. In silicon he found a noise figure of 34 dB, somewhat lower than the theoretical value, and the Ge and GaAs devices showed noise figures of 29 dB and 26 dB, respectively. The result for Ge is the same, within the limits of experimental error, as that measured and reported by Rulison et al. (1967) for a Ge device at microwave frequencies (8–10 GHz). At millimetre-wave frequencies (25–40 GHz), Weller et al. (1973, 1974) found that a noise figure of 26 dB could be attained with a GaAs diode used as a negative resistance element in a reflection amplifier. In the 50 GHz range of frequencies, Okamoto (1975) has found minimum noise figures of 27.5 dB and 38 dB, respectively, for GaAs and Si Impatts. In general, GaAs shows a better noise performance and a higher power capability than either Ge or Si.

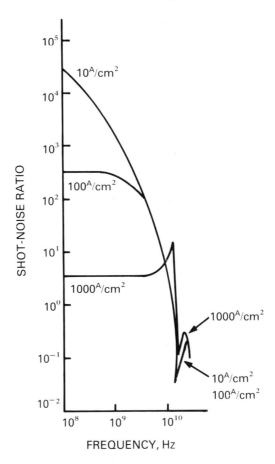

Fig. 10.12 – The calculated power spectral density of the noise current in the external circuit of an Impatt reflection amplifier, normalized to the shot noise spectrum $2qI_M$. (After Gummel and Blue (1967); © 1967 IEEE.)

10.5.2 Trapatts

Haddad *et al.* (1970) reported that little was known about the noise performance of avalanche diodes operating in the Trapatt mode of oscillation. The preliminary results that were available at that time indicate that the Trapatt amplifier is quite noisy.

It appears that since then there has been little progress in quantifying the noise in Trapatt amplifiers, presumably because these are appropriate to large-signal applications where noise is not a critical factor. It is also true that the detailed physics of Trapatt operation is very complex, requiring computer simulations (Mains *et al.*, 1980; Evans, 1970) to investigate the many sets of device waveforms and modes that are possible with different circuit/diode configurations. Until the physics of the Trapatt is more fully understood, it would seem that a detailed theory of the noise characteristics of Trapatt amplifiers cannot be given.

10.5.3 Baritts

The absence of avalanche breakdown in the space-charge region of a Baritt diode means that the device is inherently less noisy than either the Impatt or the Trapatt. Coleman and Sze (1971) reported a noise figure of 15 ± 1 dB for a metal–semiconductor–metal Baritt device, while Björkmann and Snapp (1972) measured noise figures of 10 dB and 11 dB, respectively, for p^+–n–p^+ and p^+–n–ν–p^+ structures.

There are two sources of noise in the Baritt. One is noise associated with the carriers injected across the junction into the space-charge region; this is space-charge-smoothed shot noise. The second noise source, known as diffusion noise, is due to random fluctuations in the velocity of carriers crossing the depletion region. The diffusion noise predominates over the shot noise at high current densities, when the latter component is substantially reduced through the mechanism of space-charge smoothing.

An analysis of the shot noise in a Baritt was developed by Haus et al. (1971). They derived an expression for the power spectral density of the open-circuit voltage fluctuations by assuming that full shot noise is injected across one of the junctions into the space-charge region. The injected shot noise is smoothed because the charge carriers create a correlated electric field fluctuation which modulates the bias current at the boundary. When this modulation-noise current is added to the injected noise current, the result is a reduction in the power spectral density of the noise in the external circuit. This net reduction in noise is a manifestation of the phenomenon of space-charge smoothing.

The noise figure, F, of a high gain amplifier is conveniently expressed in terms of the noise measure, M, as follows:

$$F = 1 + M . \tag{10.18}$$

The theoretical expression for the noise measure derived by Haus et al. for the space-charge smoothed shot noise, for the condition in which the transit angle, $\omega L/v_s$, is such that the maximum negative resistance is achieved, is

$$M = \frac{q J_0}{k\theta} \frac{v_s \epsilon}{\alpha^2} \{1 + (\omega \epsilon/\alpha)^2\}^{-\frac{1}{2}} , \tag{10.19}$$

where J_0 is the current density through the device, v_s is the scattering limited velocity, ϵ is the relative permittivity of the material, and α, the barrier modulation parameter, is the ratio of the current density to the electric field at the injecting boundary. α depends on the doping level. The noise measure calculated from equation (10.19) can be substantially less than unity, depending on the values of the parameters in the expression. Indeed, Haus *et al.* suggested that noise measures of 10^{-2} and lower are possible. Unfortunately, when diffusion noise is included in the analysis, these remarkably small values for the noise measure are no longer predicted by the theory.

Diffusion noise is essentially thermal noise of the hot carriers. It arises because the carriers, although nominally drifting at the scattering-limited velocity, actually show a distribution of velocities whose average is the scattering-limited value. The velocity distribution is Maxwellian-like, and appropriate to the temperature of the hot carriers. A general analytical method for handling Johnson noise of hot carriers is described in §10.7 in connection with noise in Gunn diodes.

When space-charge smoothed shot noise and diffusion noise are included in the analysis of the Baritt, Statz *et al.* (1972) found that the noise measure can be very much larger than in the absence of diffusion noise. This is illustrated in Fig. 10.13, showing the calculated noise measure as a function of current density at a frequency of 10 GHz, for the case where the transit angle is such as to maximize

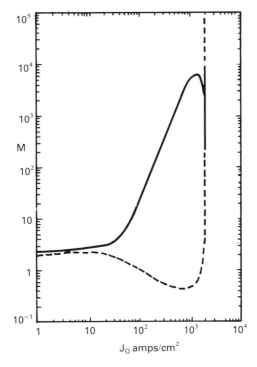

Fig. 10.13 – Calculated noise measure M as a function of the current density J_0. The broken line is for shot noise alone and the solid line is for shot noise and diffusion noise. (After Statz *et al.* (1972); © 1972 IEEE.)

the negative resistance. The two curves in the figure show the noise measure when shot noise alone is present (broken line) and when shot noise and diffusion noise are both included in the calculation (solid line). The effect of the space-charge smoothing on the shot noise is exemplified by the dip in the broken curve at the higher current densities; but this has little significance with regard to the total noise, which can be several orders of magnitude higher, and is due to the diffusion component.

Despite the relatively high contribution from diffusion noise at high current densities, the Baritt diode is still a low-noise microwave device. Under appropriate circuit conditions, a noise measure of about 10 dB at frequencies up to 20 GHz can be achieved. Several numerical models of noise in a Baritt have been developed (Sjölund, 1972; Christie and Stewart, 1975) whose computed results show reasonable agreement with experiment. One conclusion from such models is that Baritt devices with metal–semiconductor junctions show larger noise measures than those with p–n junctions. The difference depends on the current density and frequency, but is usually 3 dB or less.

10.6 PHYSICS OF THE TRANSFERRED-ELECTRON OSCILLATOR

Certain semiconductor materials, principally GaAs and InP, show a region of negative differential mobility in their velocity/field characteristic (Fig. 10.2). A specimen of such a semiconductor in the form of a short bar with ohmic contacts shows oscillations in the current when the applied voltage exceeds some critical level (Gunn, 1964). This behaviour is attributed to the transfer of electrons from a state of high mobility to one of lower mobility.

The process of electron transfer can be understood from the energy–momentum diagram for the semiconductor. Those materials capable of exhibiting Gunn oscillations possess a central conduction-band valley (whose minimum is the conduction-band edge defining the upper limit of the band gap), surrounded by symmetrically placed satellite valleys whose energy minima are higher than the minimum of the central valley. This energy-band structure is illustrated in Fig. 10.14, where just one of the satellite valleys is shown.

The current oscillations come about because the mobility of electrons in the satellite valleys is very much less than that of electrons in the central valley. Thus, at low values of the electric field, essentially all the conduction electrons are in the central valley, they drift with a velocity which is proportional to the field, and show a high mobility. As the field rises, more kinetic energy is imparted to the electrons, until, at some threshold field related to the difference in energy between the minima in the satellite valleys and the minimum in the central valley, there is a high probability of finding a substantial proportion of the electrons in the satellite valleys. These electrons are then in the low mobility state. On further increasing the field, more of the electrons make the transfer from the high to the low mobility state, which means that with the field rising,

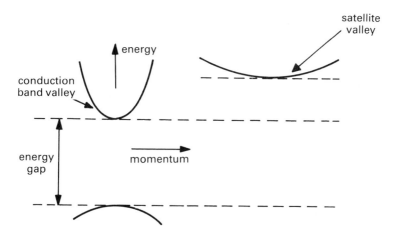

Fig. 10.14 – Sketch of the energy band structure of a semiconductor with a negative differential mobility.

the average drift velocity of the electron population is falling. This accounts for the region of negative differential mobility in Fig. 10.2. When the field is sufficiently high for essentially all the electrons to be in the satellite valleys, the velocity saturates at the scattering-limited value.

It is implicit in this description of the electron transfer process that the electric field is constant throughout the semiconductor. In fact, Gunn oscillations are associated with a field which is highly non-uniform. The reason for the non-uniform field distribution is to be found in the behaviour of an accumulation of charge when the differential mobility is negative: unlike the case of positive mobility, when dielectric relaxation occurs and the charge accumulation decays away, the charge grows rapidly. This implies that, when the differential mobility is negative, a uniform distribution of charge is very unstable. It is this instability which underlies the operation of the Gunn diode.

Consider a uniformly doped specimen of semiconductor such as GaAs, which shows a negative differential mobility when the field exceeds the critical field E_c. When the applied voltage is small, so that the field within the material is uniform and below E_c, the specimen behaves as a conventional resistance with an ohmic current-voltage characteristic. As the voltage is raised to a level where the field equals E_c, the current takes its maximum value I_c. At the same time the differential mobility is on the point of changing from a positive to a negative value. When this happens it usually occurs in a local region near the negative contact. In this local region, the electrons drift through the semiconductor more slowly than those in front and behind the region. This causes an accumulation of charge behind and a deficit of charge ahead of the region and, because of the positive feedback type of action of the negative differential mobility, these concentrations of opposite charges undergo very rapid growth to form a dipole layer. The field at the centre of this layer, or *domain,* is very high, in excess of E_c, and this is compensated by a reduction in the field below E_c in the regions outside the layer. Now, the domain drifts through the semiconductor at the scattering-limited velocity, which means that, if the domain is to retain its shape, then the electrons outside it must be drifting at the same velocity. This constrains the field outside the domain to take a value consistent with this velocity, bearing in mind that the electrons of interest here are experiencing a positive differential mobility. The current, I_d, flowing when the domain is travelling through the device is less than the maximum current I_c by a factor equal to the ratio of the scattering-limited velocity to the peak drift velocity. When the domain arrives at the anode, it collapses and a new domain is formed at the cathode. The cycle then repeats. Thus, the current alternates between two levels, I_d and I_c, with a repetition frequency approximately equal to the reciprocal of the time taken by the domain to drift through the device.

The mechanism of high-field domain formation and the subsequent transit of the domain from cathode to anode is the basis of the Gunn oscillator, which shows a negative resistance at r.f. frequencies. An alternative mode of oscillation is also possible, in which the high-field region is prevented from forming by the application of a carefully controlled r.f. voltage, which sweeps the field rapidly through the region of negative differential mobility. This prevents the accumulation of charge and maintains a more or less uniform field throughout the device. Thus, the current-voltage characteristic of this mode, which is known as the Limited Space-charge Accumulation mode of oscillation (LSA), are determined by the velocity-field characteristic, since the current is $qnv(E)A$, where n is the density of electrons, A is the area of the device and $v(E)$ is the field-dependent drift velocity. It follows that, in the regions of negative differential mobility, the diode shows a negative differential resistance at its terminals, and hence can be used for amplification. One advantage of the LSA mode is that it is potentially capable of producing a high power output at microwave frequencies. Moreover, Gunn diodes with uniform field distributions are expected to be less noisy than those with non-uniform fields, because of the higher mobility in the former.

10.7 NOISE IN GUNN OSCILLATORS

Of the two main contributors to the noise in a Gunn oscillator, one, which is fundamental, is a non-equilibrium form of Johnson noise in the microwave band, originating in the random velocity fluctuations of the hot carrier population, whereas the second is associated with $1/f$ noise in the bias voltage (or current).

The noise figures of the early GaAs devices, whose (doping $\times$ length) product was less than about 10^{12} cm^{-2} (subcritical doping), were high at around 30 dB (Thim and Barber, 1966). An improvement was found with supercritically doped

GaAs diodes, which showed noise figures in the region of 10 to 15 dB (Perlmann et al., 1970; Magarshack et al., 1974). A further reduction in the noise figure, to less than 8 dB, was achieved using InP (Baskaran and Robson, 1972). There have been predictions of even lower noise figures from computer simulations: Källbäck (1972, 1973) arrived at values around 3 dB for GaAs and InP, and Sitch and Robson (1976) calculated noise measures of 7 dB for GaAs and 4 dB for InP. In these computations, certain constraints were placed on the injection of charge at the cathode.

10.7.1 Johnson noise

When a domain travels from the cathode to the anode, there are regions in front of and behind the narrow depletion layer constituting the domain, which are essentially resistive. Since the width of the domain is small, the total length of the resistive regions is very nearly constant throughout the transit of the domain, and equal to the length of the semiconductor. The hot carriers in the undepleted regions outside the domain are the source of the Johnson noise at the terminals.

There are two mechanisms which give rise to the noise: one is the random intraband fluctuations of the carriers due to their elevated temperature, and the other is the fluctuations caused by interband transitions, that is, by carriers hopping randomly between the energy bands associated with the high and low mobility states. The noise from these mechanisms was originally treated by Shockley et al. (1966), who introduced the 'impedance field method' of calculation. As the method is important generally to the understanding of non-equilibrium Johnson noise, we now digress briefly in order to examine it in some detail, before continuing with the discussion of Johnson noise in the Gunn diode.

A relatively simple example of the impedance field method has already been introduced, in our analysis of thermal noise in a resistor in thermal equilibrium with its surroundings (§ 2.8). The treatment can be generalized to apply to non-equilibrium situations, according to the following argument. For simplicity we assume a one-dimensional geometry, but an extension to three dimensions is quite straightforward.

Consider a one-dimensional resistor of length L and cross-sectional area A, and suppose that the velocity of a particular electron after a collision is $u(t)$. The equation of motion for the particle (see Appendix 3, equation (A3.7)) is

$$\frac{m\,du(t)}{dt} = -\frac{qu(t)}{\mu} + (p_2 - p_1)\,\delta\,(t), \qquad (10.20)$$

where m is the mass of the electron, μ is the mobility, which differs from the low-field value if the carriers are hot, and p_1 and p_2 are the values of the momentum immediately before and after the collision, respectively. Thus, $(p_2 - p_1)$ is the momentum that has been imparted to the electron by the collision. On

Fourier transforming both sides of equation (10.20), the transform of $u(t)$ is found to be

$$U(j\omega) = \frac{(\mu'/q)(p_2 - p_1)}{(1 + j\omega\bar{\tau}_f)} , \qquad (10.21)$$

where $\bar{\tau}_f = \mu m/q$ is the mean free time between collisions and can be identified with the dielectric relaxation time, τ_1, in § 2.8.

If the electron is in the elemental strip between x and $x + dx$, the (transform of the) open-circuit voltage pulse at the terminals of the resistor due to the collision, can be expressed in terms of $U(j\omega)$ as follows:

$$V(j\omega, x) = qU(j\omega)H(j\omega, x), \qquad (10.22)$$

where $H(j\omega, x)$ is the linear system function relating the microscopic current at x to the open-circuit voltage at the terminals. Naturally, $H(j\omega, x)$ depends on the device under consideration and the condition in which it is being operated. The dimensions of $H(j\omega, x)$ are those of electric field divided by current. For the three-dimensional case, Shockley *et al.* designated the quantity corresponding to $H(j\omega, x)$, the 'impedance-field vector', which they represented as Z.

The power spectral density of the open-circuit noise voltage due to the velocity fluctuations between x and $x + dx$ is given by Carson's theorem (§ 2.6, equation (2.41)). From the transform of the pulse shape function given by equations (10.21) and (10.22), we have

$$d\overline{S_v(\omega, x)} = 2|H(j\omega, x)|^2 \frac{\mu^2 \overline{(p_2 - p_1)^2}}{(1 + \omega^2\bar{\tau}_f^2)} \cdot \frac{nA}{\bar{\tau}_f} dx , \qquad (10.23)$$

where the mean rate of collisions has been set equal to

$$\nu = \frac{nA}{\bar{\tau}_f} dx . \qquad (10.24)$$

Note that in general the electric field, and hence the mobility and $\bar{\tau}_f$, are functions of position. Now, p_1 and p_2 are independent, and so $\overline{(p_2 - p_1)^2} = \overline{p_2^2} + \overline{p_1^2}$, since each has a zero mean value, and therefore, on setting $\overline{p^2} = \overline{p_1^2} = \overline{p_2^2}$, equation (10.23) can be written as

$$d\overline{S_v(\omega, x)} = 4|H(j\omega, x)|^2 q^2 nA \left\{ \frac{\overline{p^2}\mu^2}{q^2\bar{\tau}_f(1 + \omega^2\bar{\tau}_f^2)} \right\} dx . \qquad (10.25)$$

On comparing the term in parenthesis with equation (A3.11) in Appendix 3 we see that it is equal to one quarter of the power spectral density of the velocity fluctuations associated with a single particle. But this defines the diffusion of u at angular frequency ω (see Appendix 3):

$$D_u(\omega) = \frac{\overline{p^2}\mu^2}{q^2\overline{\tau_f}(1 + \omega^2\overline{\tau_f^2})} \ . \tag{10.26}$$

Thus, the power spectral density of the open-circuit voltage fluctuations due to velocity fluctuations throughout the device, obtained by integrating equation (10.25), is

$$\overline{S_v(\omega)} = \int_0^L |H(j\omega,x)|^2 \, 4q^2nAD_u(\omega)\,dx \ . \tag{10.27}$$

Equation (10.27) is the one-dimensional version of the *diffusion-impedance field noise formula* originally derived by Shockley *et al.* It is important to note that in its derivation, no reference is made to equilibrium conditions. Thus, the formula is valid irrespective of whether the assembly of electrons is in thermal equilibrium with its surroundings, provided $D_u(\omega)$ is interpreted in the general sense discussed in Appendix 3. It is easy to check that the formula gives the correct result for the equilibrium case, for then $H(j\omega,x)$ equals the resistance per unit length, and, for frequencies below the reciprocal of the mean free time, $D_u(\omega)$ in equation (10.26) can be identified with the diffusion constant $D = \mu k\theta/q$ (see Appendix 3), from which it follows that

$$\overline{S_v(\omega)} = 4k\theta R \ , \tag{10.28}$$

as required.

We now return to the discussion of Johnson noise in the Gunn diode. If the impedance of the undepleted regions outside the domain is $Z = R + jX$, then the impedance-field vector is simply $H(j\omega, x) = Z/L$, where L is the total length of the device. From equation (10.27), the power spectral density of the thermal noise voltage at frequencies below the collision frequency is

$$\overline{S_v(\omega)} = \frac{4q^2nAD}{L} |Z|^2 \ , \tag{10.29}$$

where D is the diffusion constant of the hot carriers outside the domain. If we neglect the reactive part of the impedance, then equation (10.29) can be expressed in terms of the resistance $R = L/(nq|\mu|A)$, where μ is the mobility of the hot carriers:

$$\overline{S_v(\omega)} = 4k\theta_{eff}R \ , \tag{10.30}$$

where

$$\theta_{eff} = \frac{qD}{|\mu|k} \tag{10.31}$$

is the effective noise temperature. It is important to realize that θ_{eff} is no more than a convenient measure of the noise; it does not represent the temperature of the hot electron population.

Now, the noise measure of an oscillator is defined as

$$M = \theta_{\text{eff}}/\theta_0 \tag{10.32}$$

where the reference temperature θ_0 is usually taken as 290 K. Thus, from equation (10.30), the noise measure of the thermal noise in the Gunn oscillator is

$$M = qD/|\mu|k\theta_0 \ . \tag{10.33}$$

Taking as illustrative values $D = 400$ cm^2/sec and $|\mu| = 3000$ cm^2/V sec, the noise measure in equation (10.33) is

$$M = 5.33 \simeq 7 \text{ dB} \ .$$

The simple result in equation (10.33) was first derived by Thim (1971). It represents the minimum noise measure of the oscillator, and it indicates why the LSA oscillator is potentially less noisy than the Gunn oscillator with propagating domains: the modulus of the mobility throughout the uniform field device is considerably higher than the mobility in the undepleted regions either side of the domain in a Gunn diode.

The simplistic treatment of hot-electron noise given above disregards a number of factors, including noise associated with the formation and growth of domains. Robson (1974) gives a more comprehensive treatment of hot-electron noise, in a paper which he describes as 'largely tutorial in content'. In particular, he deals with the diffusion of charge in the domain, and he shows that the noise measure depends on the transit angle, θ_T, passing through a minimum when $\theta_T \simeq 2\pi$. He also shows that the noise measure varies rapidly with the nL product. Again the curve passes through a minimum, at a value of nL which increases with increasing bias level. (As an example, at a bias field of 10 kV/cm, the value of nL at the minimum in M is approximately 10^{11} cm^{-2}.) As a result of diffusion, the minima are several decibels above the corresponding minimum values calculated from equation (10.33). More recently, Constant (1976) and Rees (1977) have discussed theories of noise in transferred-electron devices at some length.

As well as AM noise, the thermal velocity fluctuations of the hot carriers also give rise to FM noise. Haus et al. (1973) calculated the magnitude of the effect, and found an r.m.s. frequency deviation of 1 to 2 Hz. Hobson (1967) has also discussed a thermal noise mechanism, in his case influencing the nucleation of domains, which produces FM noise. Hobson's mechanism is believed to represent a fundamental limit to the frequency stability of a Gunn oscillator.

10.7.2 1/f noise
The 1/f noise in Gunn devices originates in traps at the surface, within the bulk

material, and at the contact interface (Hashiguchi and Ohkoshi, 1971). It seems reasonable to assume that by careful fabrication techniques, the density of the traps can be reduced and, along with it, the $1/f$ noise. Kotani (1976) has found that thermocompression bonding produces a high level of $1/f$ noise. This bonding technique involves high temperatures and stresses, with the result that lattice imperfections are created which act as trapping centres and, in so doing, increase the $1/f$ noise. According to Kotani, the FM noise is increased significantly by the presence of these imperfections, which influence the nucleation of the domain and its transit velocity.

The dc current (or voltage) in a Gunn oscillator exhibits $1/f$ noise fluctuations. Faulkner and Meade (1968) have shown that $1/f$ fluctuations in the bias level are correlated with the FM noise, with a correlation coefficient typically falling between 0.6 and 0.8. Meade (1971) has proposed a theory to account for the observed ratio of FM noise to $1/f$ noise in the dc level, based on the triangular-domain model of Butcher, Fawcett and Hilsum (1966). The details of the argument are beyond the scope of the present discussion.

REFERENCES

R. D. Baertsch (1966), Noise and ionization rate measurements in silicon photo-diodes, *IEEE Trans. Elect. Dev. (Corresp.)*, **ED**–**13**, 987.

R. D. Baertsch (1967), Noise and multiplication measurements in InSb avalanche photodiodes, *J. Appl. Phys.*, **38**, 4267–4274.

D. J. Bartelink and D. L. Scharfetter (1969), Avalanche shock fronts in p–n junctions, *Appl. Phys. Lett.*, **14**, 320–323.

S. Baskaran and P. N. Robson (1972), Noise performance of InP reflection amplifiers in Q band, *Elect. Lett.*, **8**, 137–138.

G. Björkmann and C. P. Snapp (1972), Small-signal noise behaviour of companion p^+-n-p^+ and $p^+-n-\nu-p^+$ punchthrough microwave diodes, *Elect. Lett.*, **8**, 501–503.

P. N. Butcher, W. Fawcett and C. Hilsum (1966), A simple analysis of stable domain propagation in the Gunn effect, *Brit. J. Appl. Phys.*, **17**, 841–850.

J. E. Carroll (1974), *Physical Models for Semiconductor Devices*, Edward Arnold, Chapter 8.

J. Christie and J. A. C. Stewart (1975), Small-signal Baritt noise measures, *IEEE Trans. Elect. Dev.*, **ED**–**22**, 836–841.

A. S. Clorfeine, R. Ikola and L. S. Napoli (1969), A theory for the high-efficiency mode of oscillation in avalanche diodes, *RCA Review*, **30**, 397–421.

D. J. Coleman Jr. (1972), Transit-time oscillations in Baritt diodes, *J. Appl. Phys.*, **43**, 1812–1818.

D. Coleman and S. Sze (1971), A low noise metal-semiconductor-metal (MSM) microwave oscillator, *Bell Syst. Tech. J.*, **50**, 1695–1699.

J. Conradi (1972), The distribution of gains in uniformly multiplying avalanche photodiodes: experimental, *IEEE Trans. Elect. Dev.*, **ED**–19, 713–718.

E. Constant (1976), Noise in microwave, injection, transit time and transferred-electron devices, *Physica*, **83B**, 24–40.

W. J. Evans (1969), Circuits for high-efficiency avalanche diode oscillators, *IEEE Trans. Microwave Theory and Tech.*, **MTT**–17, 1060–1067.

W. J. Evans (1970), Computer experiments on Trapatt diodes, *IEEE Trans. Microwave Theory and Tech.*, **MTT**–18, 862–871.

E. A. Faulkner and M. L. Meade (1968), Flicker noise in Gunn diodes, *Elect. Lett.*, **4**, 226–227.

W. Fawcett, A. D. Boardman and S. Swain (1970), Monte Carlo determination of electron transport properties in gallium arsenide, *J. Phys. Chem. Solids*, **31**, 1963–1990.

M. Gilden and H. E. Hines (1966), Electronic tuning effects in the Read microwave avalanche diode, *IEEE Trans. Elect. Dev.*, **ED**–13, 169–175.

H. K. Gummel and J. L. Blue (1967), A small-signal theory of avalanche noise in Impatt diodes, *IEEE Trans. Elect. Dev.*, **ED**–14, 569–580.

J. B. Gunn (1964), Microwave oscillations of current in III-V semiconductors, *IBM J. Res. Development*, **8**, 141–159.

G. I. Haddad, P. T. Greiling and W. F. Schroeder (1970), Basic principles and properties of avalanche transit-time devices, *IEEE Trans. Microwave Theory and Tech.*, **MTT**–18, 752–772.

R. H. Haitz and F. W. Voltmer (1966), Noise studies in uniform avalanche diodes, *Appl. Phys. Lett.*, **9**, 381–383.

S. Hashiguchi and T. Ohkoshi (1971), Determination of equivalent circuit parameters describing noise from a Gunn oscillator, *IEEE Trans. Microwave Theory and Tech.*, **MTT**–19, 686–691.

H. A. Haus, H. Statz and R. A. Pucel (1971), Noise measure of metal–semiconductor–metal Schottky-barrier microwave diodes, *Elect. Lett.*, **7**, 667–668.

H. A. Haus, H. Statz and R. A. Pucel (1973), Noise in Gunn oscillators, *IEEE Trans. Elect. Dev.*, **ED**–20, 368–370.

M. E. Hines (1966), Noise theory for the Read type avalanche diode, *IEEE Trans. Elect. Dev.*, **ED**–13, 158–163.

G. S. Hobson (1967), Source of F. M. noise in cavity controlled Gunn-effect oscillators, *Elect. Lett.*, **3**, 63–64.

R. L. Johnson, B. C. De Loach and B. G. Cohen (1965), A silicon microwave diode oscillator, *Bell Syst. Tech. J.*, **44**, 369–372.

B. Källbäck (1972), Noise properties of the injection limited Gunn diode, *Elect. Lett.*, **8**, 476–477.

B. Källbäck (1973), Noise performance of GaAs and InP injection limited diodes, *Elect. Lett.*, **9**, 11–12.

C. Kittel (1970), *Introduction to Solid State Physics*, 4th edition, Wiley.

M. Kotani (1976), Design fabrication of low noise Gunn diodes with consideration of a thermocompression bonding effect, *IEEE Trans. Elect. Dev.*, **ED–23**, 567–572.

R. L. Kuvås (1972), Noise in Impatt diodes: intrinsic properties, *IEEE Trans. Elect. Dev.*, **ED–19**, 220–233.

W. A. Lukaszek, A. van der Ziel and E. R. Chenette (1976), Investigation of the transition from tunnelling to impact ionization multiplication in silicon p–n junctions, *Solid State Elect.*, **19**, 57–71.

J. Magarshack, A. Rabier and R. Spitalnik (1974), Optimum design of T. E. amplifier devices in GaAs, *IEEE Trans. Elect. Dev.*, **ED–21**, 652–654.

R. K. Mains, N. A. Masuari and G. I. Haddad (1980), Theoretical investigations of Trapatt amplifier operation, *IEEE Trans. Microwave Theory and Tech.*, **MTT–28**, 1070–1076.

R. J. McIntyre (1966), Multiplication noise in uniform avalanche diodes, *IEEE Trans. Elect. Dev..*, **ED–13**, 164–168.

M. L. Meade (1971), Relationship between F. M. noise and current noise in a cavity-controlled Gunn effect oscillator, *The Radio and Elect. Eng.*, **41**, 126–132.

H. Melchior and L. K. Anderson (1965), Noise in high speed avalanche photodiodes, presented at the 1965 Int. Electron. Devices Meeting, Washington, D. C.

H. Melchior and W. T. Lynch (1966), Signal and noise response of high speed germanium avalanche photodiodes, *IEEE Trans. Elect. Dev.*, **ED–13**, 829–838.

T. Misawa (1971), Semiconductors and semimetals, Vol. 7 (ed. R. K. Willardson and A. C. Beer), Academic Press, Chapter 7.

H. Okamoto (1975), Noise characteristics of GaAs and Si Impatt diodes for 50-GHz range operation, *IEEE Trans. Elec. Dev.*, **ED–22**, 558–565.

B. S. Perlmann, C. L. Upadhyayula and R. Marx (1970), Wide band reflection type transferred electron amplifiers, *IEEE Trans. Microwave Theory and Tech.*, **MTT–18**, 911–922.

H. J. Prager, K. K. N. Chang and S. Weisbrod (1967), High power, high efficiency silicon avalanche diodes at ultra high frequencies, *Proc. IEEE*, **55**, 586–587.

W. T. Read (1958), A proposed high frequency negative resistance diode, *Bell Syst. Tech. J.*, **37**, 401–446.

H. D. Rees (1977), Intrinsic noise of transferred-electron amplifiers, *Solid State and Elect. Dev.*, **1**, 165–179.

P. N. Robson (1974), Low-noise microwave amplification using transferred-electron and baritt devices, *The Radio and Elect. Eng.*, **44**, 553–567.

R. L. Rulison, G. Gibbons and J. G. Josenhans (1967), Improved performance of Impatt diodes fabricated from Ge, *Proc. IEEE (letters)*, **55**, 223–224.

W. Shockley, J. A. Copeland and R. P. James (1966), The impedance field method of noise calculation in active semiconductor devices, in *Quantum Theory of*

Atoms, Molecules, and the Solid State, Academic Press, New York, pp. 537–563.

J. E. Sitch and P. N. Robson (1976), Noise measure of GaAs and InP transferred electron amplifers, *IEEE Trans. Elect. Dev.,* **ED–23**, 1086–1094.

A. Sjölund (1972), Small-signal noise analysis of p^+-n-p^+ Baritt diodes, *Elect. Lett.,* **9**, 2–4.

H. Statz, R. A. Pucel and H. A. Haus (1972), Velocity fluctuation noise in metal–semiconductor–metal diodes, *Proc. IEEE (letters),* **60**, 644–645.

S. M. Sze, D. J. Coleman Jr. and A. Large (1971), Current transport in metal–semiconductor–metal (MSM) structures, *Solid State Elect.,* **14**, 1209–1218.

A. S. Tager (1964), Current fluctuations in a semiconductor under the conditions of impact ionization and avalanche breakdown, *Fiz. Tver. Tela.,* **6**, 2418–2427; translation in *Sov. Phys. – Solid State,* **6**, 1919–1925 (1965).

H. W. Thim (1971), Noise reduction in bulk negative resistance amplifiers, *Elect. Lett.,* **7**, 106–108.

H. Thim and M. Barber (1966), Microwave amplification in a GaAs bulk semiconductor, *IEEE Trans. Elect. Dev.,* **ED–13**, 110–114.

K. M. van Vliet and L. M. Rucker (1978), Theory of avalanche noise, in *Noise in Physical Systems, Proc. of the 5th International Conference on Noise, Bad Nauheim, Fed. Rep. of Germany, March 13–16, 1978* (ed. D. Wolf), pp. 333–336.

K. M. van Vliet and L. M. Rucker (1979), Theory of carrier multiplication and noise in avalanche diodes – part 1: one carrier processes, *IEEE Trans. Elect. Dev.,* **ED–26**, 746–751.

K. P. Weller (1973), A study of millimetre-wave GaAs Impatt oscillator and amplifier noise, *IEEE Trans. Elect. Dev.,* **ED–20**, 517–521.

K. P. Weller, A. B. Dreeben, H. L. Davis and W. M. Anderson (1974), Fabrication and performance of GaAs p^+-n junction and Schottky-barrier millimetre Impatt's, *IEEE Trans. Elect. Dev.,* **ED–21**, 25–31.

11

Quantum mechanics and noise

11.1 INTRODUCTION

In many applications of electronic devices the frequencies of interest fall in or below the microwave range, and the operating temperature, θ, is well above the cryogenic region. Under these conditions a quantum of energy, hf, where h is Planck's constant and f is the frequency, is very much less than the thermal energy $k\theta$ and quantum mechanical effects have a negligible influence on the noise in the system. Thus, the spectral density of the 'low frequency, high temperature' thermal noise voltage fluctuations in a resistor R is given by the 'classical' Nyquist theorem

$$\overline{S_v(\omega)} = 4k\theta R \ , \tag{11.1}$$

which is independent of Planck's constant.

At high frequencies and/or low temperatures quantum effects can no longer be ignored. In these situations the condition $hf \gtrsim k\theta$ may prevail and quantum statistics must then be invoked in order to describe the noise. The result for the spectral density of the fluctuations in a resistor is a generalized version of equation (11.1), which contains Planck's constant and which reverts to the expression given above when the condition $hf \ll k\theta$ is satisfied. The quantum mechanical form of equation (11.1) is pertinent to the analysis of thermal noise in masers and lasers, as discussed in §11.6. (It is also important in connection with noise in Josephson junction devices. In view of their unique physical characteristics, however, Josephson devices are not included in the present chapter but are treated separately in Chapter 12).

Quantum mechanical arguments lead not only to a modification of Nyquist's 'classical' theorem, but also to the profound conclusion that all linear amplifiers must show some noise. The existence of a quantum limit to amplifier noise can be understood in terms of the zero-point energy of an harmonic oscillator, since this is the basis for deriving the quantum version of equation (11.1). However, the zero-point energy term is a particular consequence of the uncertainty principle, which is a general quantum mechanical principle with no restrictions on the models or systems to which it applies. It should therefore be possible to employ the uncertainty principle to prove quite generally that there is a fundamental, quantum mechanical lower limit to the noise that a linear amplifier can exhibit. This is indeed the case. The argument is devoloped in §11.3, following immediately after the preparatory discussion of the uncertainty principle itself, which is given below.

11.2 THE UNCERTAINTY PRINCIPLE

The motion of a particle can be described classically in terms of its position coordinates and momentum, and predictions can be made about the future state of the particle based on deterministic laws of dynamics. The situation is different in quantum mechanics. In this case, it is not possible to specify simultaneously two observable quantities such as the momentum, p, and position, x, of a particle to an accuracy better than that allowed by the Heisenberg uncertainty principle. According to the principle, if Δp and Δx are the uncertainties in the measurements of p and x, then

$$\Delta p \, \Delta x \geqslant h/4\pi \ . \tag{11.2}$$

The uncertainty in a variable z, represented by the symbol Δz, may be thought of as the r.m.s. value about the mean of the variable found from a series of measurements; that is, Δz is an abbreviated notation for the average $\sqrt{\overline{\delta z^2}}$, where δz is the deviation from the mean in a single measurement. Pairs of quantities which are constrained by the uncertainty principle are known as conjugate variables. An example of another conjugate pair is the energy, E, of a system and the time, t, at which it possesses this energy. The uncertainty in these quantities is expressed as

$$\Delta E \, \Delta t \geqslant h/4\pi \ . \tag{11.3}$$

Although the uncertainty principle is a quantum mechanical phenomenon, it has a classical analogue (Gabor, 1946) which may provide an intuitive understanding of the above inequalities. Consider an harmonic signal of angular frequency ω_0, which is gated over the interval $(-T/2, T/2)$ to produce the pulse, $x(t)$, shown in Fig. 11.1(a). Suppose that this pulse has been obtained from a measurement, and that we wish to specify the time of arrival and the frequency of the harmonic signal.

If

$$x(t) = \cos \omega_0 t \quad \text{for } |t| \leqslant T/2$$
$$0 \quad \text{elsewhere} \quad , \tag{11.4}$$

then the Fourier transform of $x(t)$ is

$$X(j\omega) = \int_{-T/2}^{T/2} \cos \omega_0 t \exp -j\omega t \, dt$$

$$= \frac{\sin [(\omega - \omega_0)T/2]}{(\omega - \omega_0)} + \frac{\sin [(\omega + \omega_0)T/2]}{(\omega + \omega_0)} . \qquad (11.5)$$

(a)

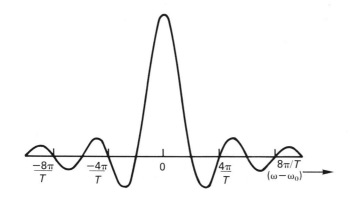

(b)

Fig. 11.1 – (a) Truncated harmonic signal and (b) its Fourier transform.

For angular frequencies around ω_0, the second term on the right is negligible and $X(j\omega)$ is closely approximated by

$$X(j\omega) = \frac{\sin [(\omega - \omega_0)T/2]}{(\omega - \omega_0)} , \qquad (11.6)$$

which is plotted in Fig. 11.1(b) as a function of $(\omega - \omega_0)$. The curve shows a principal maximum centred on $\omega = \omega_0$, indicating that the major spectral components of the signal are clustered around the frequency of the harmonic wave. But the peak has a finite width, making it impossible to determine the harmonic frequency precisely. In other words, the width of the peak constitutes an uncertainty in the measurement of the frequency.

A measure of this uncertainty is the half-width of the peak, $\Delta\omega$, between the first two zeros on either side of ω_0 :

$$\Delta\omega = 2\pi/T . \tag{11.7}$$

Since $\Delta\omega$ is inversely proportional to the pulse duration, T, the spread in frequency decreases as the observation time increases. But the length of the pulse constitutes a measure of the uncertainty in its time of arrival, which can only be reduced by sacrificing information about the frequency. Clearly, equation (11.7) is a form of uncertainty relationship, which occurs independently of any quantum mechanical considerations. However, it may be expressed in the same terms as equation (11.3) by introducing Planck's constant and formulating the uncertainties

$$\Delta E = h \, \Delta\omega/2\pi$$
$$\Delta t = T \tag{11.8}$$

where ΔE and Δt are the uncertainties in the energy and the time of arrival. The result of combining these conditions with equation (11.7) is

$$\Delta E \, \Delta t = h , \tag{11.9}$$

which clearly is consistent with the inequality in the Heisenberg uncertainty principle.

11.3 QUANTUM LIMIT TO AMPLIFIER NOISE

Electromagnetic radiation exhibits a duality in its behaviour, appearing to be either corpuscular in character, as in the photo-electric effect, or wave-like as in interference phenomena. The aspect of this 'split personality' that is observed in practice depends entirely on the experiment that is being conducted. Conceptually, it would certainly be simpler to describe all experiments involving electromagnetic radiation exclusively in terms of either a particle model or a wave model, but this is not possible and the necessity of retaining the dual representation is well recognized and almost universally accepted.

The corpuscular description of the field is due to Einstein (1905), who introduced the quantum of electromagnetic field energy, or photon. According to Einstein, the total energy E of the radiation field at frequency f is

$$E = nhf , \tag{11.10}$$

where n is the total number of photons in the field, each with an energy hf. Whilst bearing in mind equation (11.10) and the corpuscular description, suppose that the field is now regarded as a wave which may be assigned a phase

$$\phi = 2\pi ft . \tag{11.11}$$

We thus have two equations, each of which represents a different aspect of the field's character. One of these equations involves the energy E and the other the time t. As we have seen, these two quantities are a conjugate pair, satisfying the uncertaintly principle expressed in equation (11.3). Writing

$$\Delta E = hf \, \Delta n$$
$$\Delta \phi = 2\pi f \, \Delta t ,$$

(11.12)

it follows that

$$\Delta n \, \Delta \phi \geqslant 1/2 .$$

(11.13)

This result is an embodiment of the dual character of the radiation field, showing that measurements on the particulate and wave-like aspects of the field are not independent: information about one is gained at the expense of the other.

The form of the uncertainty principle in equation (11.13) can be used to establish the existence of a quantum limit to linear amplifier noise (Heffner, 1962). It is convenient to construct the argument in connection with the maser, which is an acronym employed here as a generic term standing for molecular amplification by stimulated emission of radiation and covers lasers, irasers etc. For the present purpose, it is sufficient to leave aside the microscopic details of maser action, which are dealt with in the following section, and to concentrate on the simple picture in Fig. 11.2. Here the incident radiation with spectral density S_1 passes through an absorbing material of thickness L (representing the maser) and emerges with spectral density

$$S_2 = S_1 \exp - (\alpha L) ,$$

(11.14)

where α is the coefficient of absorption at the frequency of the radiation. When the maser is functioning as an amplifier, α takes a *negative* value and the intensity

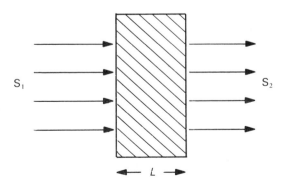

Fig. 11.2 – Electromagnetic radiation passing through an absorbing medium of thickness L.

of the output is higher than that of the input, i.e. the system shows gain. Assuming that the number of photons at the input and output within a given time are n_1 and n_2, respectively, then

$$n_2 = Gn_1 , \tag{11.15}$$

where $G > 1$ is the power gain of the system. As the amplification is a coherent process, the phases at the input and output, ϕ_1 and ϕ_2, are equal or differ by a constant phase shift ϕ_0:

$$\phi_2 = \phi_1 + \phi_0 . \tag{11.16}$$

Imagine now that the amplifier is followed by a detector, as shown in Fig. 11.3, which is 'ideal' in the sense that it is the best possible within the constraints of the uncertainty principle. Thus, the detector will provide a measure of the number of photons, n_2, and the phase, ϕ_2, of the radiation field output from the amplifier with errors satisfying the uncertainty condition

$$\Delta n_2 \, \Delta \phi_2 = 1/2 . \tag{11.17}$$

Therefore, if the amplifier were noiseless, the *measurement* errors in the number of photons and the phase at the input would satisfy the condition

$$\Delta n_1 \, \Delta \phi_1 = \frac{1}{G} \Delta n_2 \, \Delta \phi_2 = \frac{1}{2G} . \tag{11.18}$$

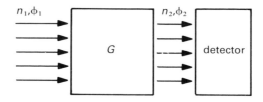

Fig. 11.3 — Amplifier with power gain G, followed by an 'ideal' detector.

But this is impossible because it contradicts the uncertainty principle, and hence the hypothesis that the amplifier is noiseless cannot be valid. The amplifier must introduce some uncertainty into the measurement, or in other words it will always show some noise.

Although this reasoning establishes the existence of a minimum level of noise in the amplifier, it does not provide an estimate of that level. This can, however, be obtained from an extension of the argument in which a matching condition between the amplifier and the detector is determined, and the level of the quantum fluctuations in the radiation field (Gabor, 1950) is employed as the minimum noise condition. The procedure is discussed in detail in Appendix 7, where the minimum power of the output noise in a frequency interval df centred on the signal frequency f_0 is shown to be

$$\Delta P_{min} = (G-1) h f_0 \, df \; . \tag{11.19}$$

On referring this to the input, by dividing by the gain G, it is clear that in the limit of high G the minimum detectable signal power is $hf_0 df$.

This is interesting, because the available noise power from a resistor in the low temperature limit (as discussed in the following section) is $\frac{1}{2}hf df$, arising from the zero point energy term. Since this is a factor of 2 below the minimum detectable level, the zero-point energy is not observable. By way of contrast, under 'classical' operating conditions the available noise power is always observable, at least in principle. However, when quantum effects are significant the amplifier itself is subject to quantum mechanical contraints which set a lower limit on the detectable power at the input.

The discussion given above is quite general in that it applies to any linear amplifier whose performance is characterized by equations (11.15) and (11.16). The maser has already been identified as such an amplifier, and indeed it could be said that it is the 'perfect' example of the type; for theoretically, the minimum detectable power at the input of the maser is exactly that implied by equation (11.19), which as we have seen represents the best performance that can be achieved. The action of the maser and the noise mechanisms in the device are discussed later in § 11.5 and § 11.6.

11.4 NYQUIST'S THEOREM AND QUANTUM MECHANICS

In his original treatment of thermal noise, Nyquist (1928) invoked an argument involving the exchange of energy between two electrical conductors connected together in equilibrium at temperature θ via a lossless transmission line. By assuming the law of equipartition, in which each degree of freedom is assigned an energy $k\theta$, he arrived at the expression in equation (11.1) for the open-circuit voltage fluctuations in a resistor R (see Appendix 2). When quantum mechanical effects are negligible, that is, when the frequency and temperature are such that a quantum of energy hf is much less than the thermal energy $k\theta$, Nyquist's theorem is valid. Otherwise, when $hf \gtrsim k\theta$, a generalized form of the theorem must be used.

Nyquist himself recognized the problem and proposed that, instead of $k\theta$ as in the equipartition law, the average energy per degree of freedom should in general be set equal to

$$\frac{hf}{[\exp\left(\dfrac{hf}{k\theta}\right) - 1]} \; ,$$

which is indistinguishable from $k\theta$ when $hf \ll k\theta$. This is the average energy of an harmonic oscillator, except that it excludes the zero-point energy term, $\frac{1}{2}hf$. When this term is included, the spectral density of the voltage fluctuations takes the general form

$$\overline{S_v(\omega)} = 4R \left\{ \tfrac{1}{2}hf + \frac{hf}{[\exp{(hf/k\theta)} - 1]} \right\}$$

$$= 2Rhf \coth{(hf/2k\theta)} \; , \tag{11.20}$$

which reduces to the familiar Nyquist theorem in equation (11.1) when quantum effects are negligible. The available noise power in a frequency interval df is easily shown from equation (11.20) to be

$$\Delta P_{\text{avail}} = \left\{ \tfrac{1}{2}hf + \frac{hf}{[\exp{(hf/k\theta)} - 1]} \right\} df \; . \tag{11.21}$$

Some doubts about the inclusion of the zero-point energy term in equations (11.20) and (11.21) have been expressed by MacDonald (1962), who points out that its presence suggests the possibility, in principle, of extracting power from the fluctuations at the absolute zero limit. This, he submits, is inadmissible.

MacDonald's argument would appear to threaten seriously the inclusion of the zero-point energy term in expressions for thermal noise. But it relies on the concept of available power being interpreted in the 'classical' sense, in which the fluctuations are assumed to be feeding into a noiseless load. This is not the case, however, when the finite value of h is significant because, then, the load, like the source, is subject to quantum mechanical constraints which ensure that it too produces noise. (The existence of a quantum mechanical lower limit to the noise in a system has been discussed for the case of a linear amplifier in § 11.3.) The effect of this in the low-temperature limit is to compensate the flow of noise power from the source resistor into the load by an equal but opposite flow from the load to the source. Thus, the zero-point energy fluctuations in the source resistor are not observable, nor is there any possibility of achieving a net flow of noise power out of the device, which could then be used to drive another system. This argument highlights a principle which is of fundamental importance in quantum mechanics: the behaviour of a physical quantity can only be specified when the means by which it is observed is taken into account.

Evidently the inclusion of the zero-point energy term in expressions like equation (11.20) describing spontaneous fluctuations does not imply physically untenable behaviour in the low-temperature limit. It is fitting to remark that this conclusion does not detract from the genuine concern surrounding the issue when it was raised by MacDonald in the early 1960s. and to point out that his paper poses a number of interesting questions concerning the application of quantum mechanical formulae to problems involving thermal noise (Brownian movement) and irreversibility.

From the discussion so far, it would seem that the zero-point energy term is included in the expressions for the noise on an ad hoc basis. The fact that it does not lead to undesirable consequences at zero temperature is not proof that it should be present. A general proof of equation (11.20) has, however, been derived by Callen and Welton (1951) from an argument involving the emission and absorption of energy quanta by electrons in the resistor. Their analysis is interesting because it does not rely on the arbitrary introduction of the zero-point energy term, and indeed no specific reference is made to the concept of a zero-point energy in their proof, although the term $\frac{1}{2}hf$ appears in their final result.

In most applications, the effect of the zero-point energy on the noise is vanishingly small. However, in the case of thermal noise in the maser it is important to take account of the zero-point energy term, at least in the theoretical analysis, because the result then has a simple physical interpretation which would not otherwise be the case (see § 11.6). With regard to *measurements* of the spectral density of thermal noise in the region where quantum modifications to Nyquist's theorem should be apparent, none are known to the author at present. This situation may be rectified soon due to an experiment being conducted at the University of Florida (van der Ziel, 1981), involving a Hanbury–Brown–Twiss type microwave circuit operating at 100 kHz. The aim is to measure thermal noise under conditions where $hf \gtrsim k\theta$ and thus to provide experimental evidence for the form of its power spectrum when quantum mechanical effects are significant.

11.5 ABSORPTION AND EMISSION OF RADIATION BY MATTER

The action of a maser involves the interaction of electromagnetic radiation with matter in the form of a population of atoms or molecules with at least two energy levels. In order to avoid circumlocution, we shall refer to the constituents of the active material simply as molecules. As the acronym suggests, *stimulated emission,* in which a downward transition between energy levels is induced by the presence of the radiation, is the mechanism responsible for the amplification process. This mechanism is essentially quantum mechanical in nature, with no clear-cut classical interpretation. On this account, maser action is fairly unusual in that it is a macroscopic manifestation of a quantum mechanical phenomenon.

There are three types of radiative transition between the energy levels of a system. The two that have yet to be introduced are *spontaneous emission* and *absorption.* (Non-radiative transitions may also occur, involving the transport of energy in the form of lattice vibrations or phonons, but these are not relevant to the present discussion). Both of these processes contribute to the noise in the maser, since both involve random, independent transitions between levels. Indeed, these are the mechanisms which alone are responsible for the inherent thermal noise fluctuations of the system. As we shall show in § 11.6.1, the expression for the spectral density of the thermal noise contains two terms, one of which can be identified with spontaneous emission and the other with absorption. Before examining the noise in more detail, it is instructive to consider briefly the basic physical mechanisms underlying the operation of the maser, following an argument due to Einstein.

11.5.1 Energy exchange in a two-level system

For simplicity, imagine that a single molecule has associated with it just two energy states, E_1 and E_2. Figure 11.4 shows the two-level system, with $E_2 > E_1$.

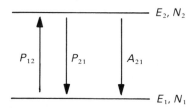

Fig. 11.4 – Two-state system of energy levels.

A downward transition will give rise to the emission of radiation at a single frequency, f_{12}, satisfying the condition

$$E_2 - E_1 = hf_{12} \ , \tag{11.22}$$

where h is Planck's constant. Now consider a large number, N, of such molecules within an enclosure, some of which will be in the lower energy state with the remainder in the higher state. If N_1 and N_2 are the numbers in the lower and upper states, respectively, then

$$N = N_1 + N_2 = \text{constant} \ , \tag{11.23}$$

and, provided the population is in thermal equilibrium with its environment at absolute temperature θ, the relative occupancy of the two states is given by the ratio

$$\frac{N_2}{N_1} = \exp - \{(E_2 - E_1)/k\theta\} \ . \tag{11.24}$$

(This result was actually derived for classical systems by Boltzmann. It is satisfactory for our present purpose but in other circumstances it may be necessary to replace it with Bose–Einstein or Fermi–Dirac statistics.)

The condition of thermal equilibrium is maintained by the continual exchange of energy with the environment. Assuming that this energy is in the form of radiation, which implies that the population of molecules is immersed in a radiation field, the exchange can only occur at the frequency f_{12} corresponding to the energy difference $(E_2 - E_1)$. It follows that equation (11.24) can be written in the alternative form

$$\frac{N_2}{N_1} = \exp - (hf_{12}/k\theta) \ . \tag{11.25}$$

In thermal equilibrium, the number of molecules in each of the two energy states remains statistically constant: on average there can be no accumulation or depletion of numbers in either level. This means that, apart from statistical fluctuations, the net rate of transitions between the two levels is zero, i.e. the mean number of upward transitions per unit time is exactly balanced by the mean number of downward transitions. Such behaviour is illustrative of a general condition known as the *principle of detailed balance* (Onsager, 1931), which requires that the average net rate of transition between any pair of energy levels in a system is identically zero. (Note that in a system with three or more levels, this condition is stronger than any imposed by thermodynamics.)

Suppose that $C_\uparrow$ is the probability per unit time that a given molecule in the lower energy state will make a transition to the higher state, and let $C_\downarrow$ be a similar probability for a downward transition. In order to satisfy the principle of detailed balance, the condition

$$C_\uparrow N_1 = C_\downarrow N_2 \tag{11.26}$$

must hold, since this represents equality between the upward and downward rates. Notice that, as $N_2 < N_1$ from equation (11.25), we must have $C_\uparrow < C_\downarrow$. This inequality is accommodated by writing (Einstein, 1917)

$$C_\uparrow = P_{12} = B_{12} \overline{S_R(\omega_{12})} df \tag{11.27a}$$

$$C_\downarrow = P_{21} + A_{21} = B_{21} \overline{S_R(\omega_{12})} df + A_{21} \ , \tag{11.27b}$$

where the A and B coefficients, or Einstein coefficients as they are often called, are constants which may depend on frequency. Their values depend on the type of molecule and the pair of levels involved in the transition process. A_{21} is the coefficient of spontaneous emission, and B_{12}, B_{21} are, respectively, the coefficients of stimulated emission and absorption. Note that the corresponding probabilities, P_{12} and P_{21}, are both proportional to the power of the radiation field, $\overline{S_R(\omega_{12})} df$, in a frequency interval df at frequency $f_{12} = \omega_{12}/2\pi$. Einstein made the important observation that, as the system is in thermal equilibrium, the radiation field is that of a black body, with energy density given by the Planck radiation law:

$$\overline{S_R(\omega_{12})} = \rho_m \frac{\hbar\omega_{12}}{\left\{\exp\left(\dfrac{\hbar\omega_{12}}{k\theta}\right) - 1\right\}}, \tag{11.28a}$$

where $\hbar = h/2\pi$, and ρ_m is the volume density of radiation modes within the enclosure per unit frequency interval:

$$\rho_m = \frac{2\,\omega_{12}^2}{\pi c^3}. \tag{11.28b}$$

In this expression c is the speed of light.

It is interesting to consider for a moment the forms postulated by Einstein for the transition probabilities in equations (11.27). The absorption term, P_{12}, is proportional to the energy density of the radiation, or equivalently to the number of photons in the field with frequency f_{12}, which is intuitively reasonable since the more photons that are present the greater the chance of an upward transition in a given time interval. But if this argument is accepted, then an emission process must exist which also depends on the intensity of the radiation, otherwise it would be impossible to maintain the condition of thermal equilibrium. This accounts for the presence of the stimulated emission term. And the spontaneous emission term is necessary to satisfy the inequality between $C_\uparrow$ and $C_\downarrow$, as shown below.

Equations (11.26) and (11.27) lead to the condition

$$B_{12} N_1 \overline{S_R(\omega_{12})}\,\mathrm{d}f = [B_{21}\,\overline{S_R(\omega_{12})}\,\mathrm{d}f + A_{21}]\,N_2. \tag{11.29}$$

In the limit of high temperature, $\overline{S_R} \to \infty$ and $N_2/N_1 \to 1$, from which it follows that

$$B_{12} = B_{21}, \tag{11.30a}$$

i.e. the probabilities of absorption and stimulated emission are the same. Thus, if $C_\downarrow$ is to be greater than $C_\uparrow$ as required, A_{21} must take a value which is greater than zero. From the above two equations, this value is

$$\begin{aligned}
A_{21} &= \overline{S_R(\omega_{12})}\left(\frac{N_1}{N_2} - 1\right) B_{21}\,\mathrm{d}f \\
&= \left(\frac{8\pi f_{12}^2}{c^3}\right) h f_{12}\, B_{12}\,\mathrm{d}f. \tag{11.30b}
\end{aligned}$$

It is apparent from equation (11.30b) that the A and B coefficients are not independent, and that their ratio varies as the cube of the frequency.

Equations (11.30) provide a statement of the relative values of the A and B coefficients but their absolute values, and hence the upward and downward transition probabilities, remain unspecified. In order to determine the constant factor necessary to rectify this situation, an experimental measurement of either the A or B coefficients must be performed. A direct measure of the spontaneous emission coefficient, for example, is obtained from the rate of decay of the number of molecules in the upper state after the system has been excited by an external source of monochromatic radiation at frequency f_{12}. The decay is exponential with a time constant equal to A_{21}^{-1}, as is easily established by constructiong the equations for the rates at which the numbers of molecules in the two levels change. Typically, the time constant is on the order of 10 ns, which in the context of a measurement presents few difficulties with currently available electronic instrumentation.

The two-level system discussed above is the simplest possible multi-level scheme. In a practical maser three or more levels are generally employed in order to facilitate population inversion, although it is possible to achieve this condition with only two levels, as was demonstrated with the original ammonia maser described by Gordon *et al.* (1954, 1955) and Shimoda *et al.* (1956).

A complication which is sometimes encountered in practice is that two or more quantum states have exactly the same energy. The system is then said to be *degenerate,* and degeneracy factors must then be introduced into the analysis in the appropriate places. This leads to quantitative changes, but the essential physics of the argument given above is retained.

11.5.2 Maser action

It is clear from equation (11.24) that in equilibrium the lower level is always more densely populated than the higher level. The obvious consequence of this is that the absorption term on the left of equation (11.29) is greater than the stimulated emission term on the right. Indeed, for levels which are separated by a few $k\theta$ or more, stimulated emission in equilibrium is negligible compared with the other two transition processes. This situation can be reversed, with dramatic effect, by a process known as *pumping.* In practice there are various means of pumping the system, the details of which we shall not dwell on here. The important point is that radical changes in the populations of the levels can be achieved, so much so that under certain circumstances the condition $N_2 > N_1$, representing *population inversion,* will be satisfied. Although the system is then no longer in equilibrium, the occupancy of the levels can be described by analogy with equation (11.24), in terms of a negative temperature $-\theta_m$:

$$\frac{N_2}{N_1} = \exp\frac{(E_2 - E_1)}{k\theta_m} \geqslant 1 \ . \tag{11.31}$$

Of course, θ_m is not the actual temperature of the system, but merely a parameter characterizing the occupancy of its levels. Note that when θ_m is infinite the system is on the threshold of population inversion with equal numbers in the two levels, and when $\theta_m = 0$ all the molecules are in the upper state and the lower state is empty.

When population inversion obtains, the energy radiated in stimulated emission exceeds that absorbed by the system. The implication of this is startlingly simple but enormously important. If a small harmonic signal in the form of a radiation field at frequency f_{12} is injected into the system, a large number of transitions due to stimulated emission will ensue, each of which contributes a photon of frequency f_{12} to the field. Thus the system acts as an amplifier. Moreover, as each downward transition adds a *fixed* amount of energy into the existing radiation field, the phase of the radiated energy must be exactly the same as that of the signal. (This argument can be readily appreciated if two sine waves with the same frequency but different phases are added together. The resultant intensity is a function of the phase difference, a conclusion which is familiar in the context of interference experiments. Therefore, in the maser, all the components of the radiation field must have the same phase if each contributes the same energy to the field.) The importance of the phase coherent output cannot be too strongly emphasized. With the advent of the maser, a phase coherent source with high intensity became available for the first time, and since then the optical maser (or laser) in particular has been employed in a diverse range of applications. An example of such an application is holography, where phase coherence is an essential requirement. The progress that has been made in this field to date would not have been possible in the absence of the laser.

11.6 NOISE IN THE MASER

Thermal noise and shot noise are the principal types of noise inherent in the maser. The former arises from random fluctuations associated with spontaneous emission and absorption, and is significant when the maser is operated close to the threshold of population inversion. Well above the threshold, the emitted radiation obeys Poisson statistics and the dominant type of noise is then shot noise in the output flux of photons.

11.6.1 Thermal noise
Below the threshold of population inversion, the active material in a maser which is immersed in a radiation field with frequency f_{12} absorbs power from the field. Since power dissipation is characteristic of a resistive material, the equivalent circuit element of a maser is a conductance, G_{mas} (or, in the dual circuit, a resistance G_{mas}^{-1}). Now, any resistance in thermal equilibrium with its environment at absolute temperature θ shows thermal noise whose power

spectral density, in general, is given by the quantum mechanical version of Nyquist's theorem. Thus, the spectral density of the thermal noise in a maser is

$$\overline{S_{mas}(\omega_{12})} = 4\,G_{mas}\left\{\frac{\hbar\omega_{12}}{2} + \frac{\hbar\omega_{12}}{\left[\exp\left(\dfrac{\hbar\omega_{12}}{k\theta}\right) - 1\right]}\right\}, \quad (11.32)$$

where $\omega_{12} = 2\pi f_{12}$ is the angular frequency corresponding to the energy difference between the two active levels in the system. In order to progress further, it is necessary to express G_{mas} in terms of the parameters of the maser.

Imagine that the active material of the maser, consisting of molecules with two energy states, is connected to a signal source with internal conductance G_s. The connection is via a (lossless) transmission line which is terminated by the maser, whose input impedance is matched to the source. Figure 11.5 shows the equivalent circuit of the complete system. (In reality, other elements such as a

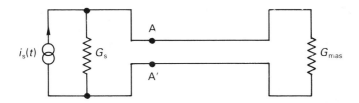

Fig. 11.5 — Equivalent circuit of the maser. The source is represented by the current generator $i_s(t)$ with internal conductance G_s, and A, A' are the input terminals of a transmission line connecting the source to the active medium.

circulator may also be required, but these are not relevant to the present discussion.) The terminals A, A' in the figure represent the input of the transmission line, where the admittance is purely real and equal to the source conductance G_s. From simple circuit considerations it is readily shown that the power absorbed by the maser, that is in the conductance G_{mas}, is

$$P_{abs} = P_{av}\frac{G_{mas}}{G_s}, \quad (11.33a)$$

where P_{av} is the available power from the source. But, according to equations (11.27), the power absorbed by the maser in the presence of an incident radiation field with power P_{av} is

$$P_{abs} = P_{av}\,B_{12}\,\rho_m\,\hbar\omega_{12}\,(N_1 - N_2), \quad (11.33b)$$

where the absence of the spontaneous emission term implies that the induced emission and absorption terms are dominant. This is so because the power in the incident radiation is assumed to be considerably greater than that of the black body radiation field which alone is present in equilibrium. On comparing equations (11.33), the conductance of the maser is found to be

$$G_{mas} = B_{12}\,\rho_m\hbar\omega_{12}(N_1 - N_2)G_s \ . \tag{11.34}$$

In the absence of pumping, N_1 and N_2 take their equilibrium values and G_{mas} is positive. However, when a pump signal is applied to achieve population inversion, equation (11.34) shows that the conductance of the maser goes negative, indicating that in this condition the active medium of the maser emits more power than it absorbs.

In order to determine the thermal noise in the maser, we need only consider the equilibrium case, in which there is no pump signal. From equations (11.32) and (11.34) the spectral density of the thermal noise is

$$\overline{S_{mas}(\omega_{12})} = 4B_{12}\,\rho_m\hbar\omega_{12}\,G_s \left\{ \frac{\hbar\omega_{12}}{2} + \frac{\hbar\omega_{12}}{\left[\exp\left(\dfrac{\hbar\omega_{12}}{k\theta}\right) - 1\right]} \right\} . \tag{11.35}$$

Now, the relative occupancy of the levels in equilibrium is given by the Boltzmann law

$$\frac{N_2}{N_1} = \exp -\frac{\hbar\omega_{12}}{k\theta} \ , \tag{11.36}$$

which when substituted into equation (11.35) yields

$$\overline{S_{mas}(\omega_{12})} = 2B_{12}\rho_m\,G_s(\hbar\omega_{12})^2\,(N_1 - N_2) + 4B_{12}\rho_m(\hbar\omega_{12})^2\,G_s N_2 . \tag{11.37a}$$

Note that the first term on the right of this expression arises from the zero-point energy, and would be absent if this were omitted from the quantum mechanical version of Nyquist's theorem. However, since it is included here, a simpler form for the spectral density of the noise is

$$\overline{S_{mas}(\omega_{12})} = 2B_{12}\,\rho_m G_s\,(\hbar\omega_{12})^2\,(N_1 + N_2)$$

$$= 2B_{12}\,\rho_m G_s(\hbar\omega_{12})^2\,N \ , \tag{11.37b}$$

where N is the total number of active molecules in the system.

As N is constant, irrespective of whether the system is in equilibrium, equation (11.37b) should be valid even when a pump signal is present. Moreover, as van der Ziel (1981) has pointed out, the terms containing N_1 and N_2 in the final expression for the output noise can be identified with the true absorption and true spontaneous emission noise, respectively. This physically reasonable interpretation would not be possible if the zero-point energy term were omitted from equation (11.32), suggesting strongly that its presence is correct.

11.6.2 The noise figure

When the maser acts as an amplifier, it is no longer in thermal equilibrium, although as we have seen equations (11.37) may still be used to describe the thermal noise at the output of the system. If this noise generator is referred to the input, by dividing by the power gain G of the amplifier, the noise figure can be immediately expressed as

$$F = 1 + \frac{\overline{S_{\text{mas}}(\omega_{12})}}{G \, \overline{S_s(\omega_{12})}} \, , \qquad (11.38)$$

where

$$\overline{S_s(\omega_{12})} = 4 \, G_s \left\{ \frac{\hbar\omega_{12}}{2} + \frac{\hbar\omega_{12}}{\left[\exp\left(\dfrac{\hbar\omega_{12}}{k\theta} \right) - 1 \right]} \right\} \qquad (11.39)$$

is the spectral density of the noise from the source at temperature θ. (It is implicit in equation (11.38) that only thermal noise is present in the output of the maser. In practice, other types of noise, including shot noise, may also be present, but since thermal noise sets the ultimate limit to performance we examine it exclusively here.) In order to evaluate the noise figure, the gain of the amplifier must now be specified.

When the maser is pumped beyond threshold, so that population inversion obtains, the relative occupancy of the two active levels can be expressed in terms of a negative temperature $-\theta_m$, as defined in equation (11.31). In this condition the output power from the maser is

$$P_{\text{out}} = P_{\text{av}} + \hbar\omega_{12} \, B_{12} \, \rho_m \, (N_2 - N_1) P_{\text{av}} \, , \qquad (11.40)$$

where P_{av} is still the available power from the source. On dividing by P_{av}, the gain is found to be

$$G = 1 + \hbar\omega_{12} \, B_{12} \, \rho_m \, (N_2 - N_1)$$

$$= 1 + \hbar\omega_{12} \, B_{12} \, \rho_m \, N \frac{\left[\exp\left(\dfrac{\hbar\omega_{12}}{k\theta_m} \right) - 1 \right]}{\left[\exp\left(\dfrac{\hbar\omega_{12}}{k\theta_m} \right) + 1 \right]} \, . \qquad (11.41)$$

Equation (11.41) may be re-arranged to obtain the alternative form

$$\frac{\hbar\omega_{12} B_{12} \, \rho_m N}{G} = \left(1 - \frac{1}{G} \right) \frac{\left[\exp\left(\dfrac{\hbar\omega_{12}}{k\theta_m} \right) + 1 \right]}{\left[\exp\left(\dfrac{\hbar\omega_{12}}{k\theta_m} \right) - 1 \right]} \, , \qquad (11.42)$$

which is convenient for evaluating the noise figure since the expression on the left appears as a factor in the second term on the right of equation (11.38).

With the aid of equations (11.37b), (11.39) and (11.42) the noise figure may be expressed in the general form

$$F = 1 + \left(1 - \frac{1}{G}\right) \frac{\left\{\exp\left(\frac{\hbar\omega_{12}}{k\theta}\right) - 1\right\} \left\{\exp\left(\frac{\hbar\omega_{12}}{k\theta_m}\right) + 1\right\}}{\left\{\exp\left(\frac{\hbar\omega_{12}}{k\theta}\right) + 1\right\} \left\{\exp\left(\frac{\hbar\omega_{12}}{k\theta_m}\right) - 1\right\}} \qquad (11.43)$$

For low frequencies, that is when $\hbar\omega_{12} \ll k\theta$ and $\hbar\omega_{12} \ll k\theta_m$, this reduces to

$$F \simeq 1 + \left(1 - \frac{1}{G}\right) \frac{\theta_m}{\theta} , \qquad (11.44)$$

in which case the noise temperature is

$$\theta_n = \left(1 - \frac{1}{G}\right) \theta_m . \qquad (11.45)$$

Clearly, the noise figure in general depends on the degree of population inversion. In the low frequency range it behaves in a particularly simple way, expressed in equation (11.44), which shows that F reduces as the levels become more inverted.

The approximate formulation of the noise figure in equation (11.44) should be representative of the actual noise figure of the maser near the threshold of population inversion, since thermal noise is likely to be the predominant type of noise in this region. At higher levels of inversion, other sources of noise, such as shot noise, may be dominant, in which case the above expressions for the noise figure cannot be expected to hold. Nevertheless, since thermal noise sets a lower limit to the level of noise in the maser, it is interesting to examine the limiting forms of the general expression in equation (11.43).

We may safely assume that the gain of the system is considerably greater than unity (otherwise further amplification would be required which would introduce a significant amount of noise), in which case the general expression reduces to

$$F = 1 + \frac{\left\{\exp\left(\frac{\hbar\omega_{12}}{k\theta}\right) - 1\right\} \left\{\exp\left(\frac{\hbar\omega_{12}}{k\theta_m}\right) + 1\right\}}{\left\{\exp\left(\frac{\hbar\omega_{12}}{k\theta}\right) + 1\right\} \left\{\exp\left(\frac{\hbar\omega_{12}}{k\theta_m}\right) - 1\right\}} . \qquad (11.46)$$

This function takes its smallest value when $\theta_m = 0$, corresponding to full inversion of the levels. Thus, the minimum noise figure is

$$
F_{\min} = 1 + \frac{\left\{ \exp\left(\dfrac{\hbar\omega_{12}}{k\theta}\right) - 1 \right\}}{\left\{ \exp\left(\dfrac{\hbar\omega_{12}}{k\theta}\right) + 1 \right\}}
$$

$$
\simeq 1 + \frac{\hbar\omega_{12}}{2k\theta} \ , \tag{11.47}
$$

where the approximation holds when the low frequency condition $\hbar\omega_{12} \ll k\theta$ is satisfied. Another interesting case is when $\theta = \theta_m = 0$, describing a condition in which the levels are fully inverted and the source is at zero temperature. Clearly, the exponential functions in equation (11.46) are now all very large compared with unity and the noise figure equals 2. At first sight this appears to be a strange result, but not too much importance should be attached to it: it is merely a statement of the fact that the noise power referred to the input of the maser consists of two equal zero-point contributions, one from the source and one from the maser itself. As the zero-point energy is not observable (we have seen that the minimum detectable energy is $\hbar\omega_{12}$, not $\hbar\omega_{12}/2$), the noise figure in this case is of doubtful value.

11.6.3 Shot noise

Van der Ziel (1970) has suggested that a maser with significant power gain shows shot noise in the output, in addition to the thermal noise discussed above. The reason is that the pump signal necessary to achieve population inversion consists of a succession of independent random events, and hence itself shows shot noise. As the rate of emission of photons equals the pump rate, the output flux of photons also shows shot noise. Obviously, in order to increase the output power of a maser, the level of population inversion must be increased, which can only be achieved with a higher pump rate. It follows that the shot noise component in the output increases as the energy states become more inverted, which accounts for the fact that shot noise may be the dominant source of noise well above threshold.

11.7 CONCLUDING REMARKS

The maser is an extremely quiet amplifier, the limit to performance being set by thermal noise inherent in the active part of the device. In practice, the overall level of noise in the amplifier may be considerably higher than the limiting level, owing to noise in external parts of the system such as the circulator and various electronic circuits. In order to exploit the low-noise potential of the maser it is essential to keep these extraneous sources of noise as low as possible.

REFERENCES

H. B. Callen and T. A. Welton (1951), Irreversibility and generalized noise, *Phys. Rev.*, **83**, 34–40.

A. Einstein (1905), Über eine die erzeugung und verwandlung des lichtes betreffenden heuristischen gesichtspunkt (Heuristic viewpoint concerning the generation and transformation of light), *Ann. d. Phys.*, **17**, 132–148.

A. Einstein (1917), On the quantum theory of radiation, *Phys.Z.*, **18**, 121–128.

D. Gabor (1946), Theory of communication, *J. of the IEE*, **93**, 429–457.

D. Gabor (1950), Communication theory and physics, *Phil. Mag.*, **41**, 1161–1187.

J. P. Gordon, H. J. Zeiger and C. H. Townes (1954), Molecular microwave oscillator and new hyperfine structure in the microwave spectrum of NH_3, *Phys. Rev.*, **95**, 282–284; (1955), The maser – new type of microwave amplifier frequency standard and spectrometer, *Phys. Rev.*, **99**, 1264–1274.

H. Heffner (1962), The fundamental noise limit of linear amplifiers, *Proc. IRE*, **50**, 1604–1608.

D. K. C. MacDonald (1962), On Brownian movement and irreversibility, *Physica*, **28**, 409–416.

H. Nyquist (1928), Thermal agitation of electric charge in conductors, *Phys. Rev.*, **32**, 110–113.

L. Onsager (1931), Reciprocal relations in irreversible processes, *Phys. Rev.*, **37**, 405–425; ibid., **38**, 2265–2279.

K. Shimoda, T. C. Wang and C. H. Townes (1956), Further aspects of the theory of the maser, *Phys. Rev.*, **102**, 1308–1321.

A. van der Ziel (1970), Noise in solid-state devices and lasers, *Proc. IEEE*, **58**, 1178–1206.

A. van der Ziel (1981), Quantum noise effects at high frequencies and low temperatures, *Proc. 6th International Conference on Noise in Physical Systems held at the National Bureau of Standards, MD, USA, 6–10 April 1981.*

12

Josephson junction devices

12.1 INTRODUCTION

It is well known that certain metals and metallic compounds exhibit the phenomenon of superconductivity: below a transition temperature, θ_c, which varies from one superconductor to another but is always in the cryogenic region, the electrical resistance vanishes completely. The effect was discovered in 1911 by Heike Kamerlingh Onnes, who found that the resistance of mercury is zero at temperatures below 4.2 K. The absence of resistance means that a 'supercurrent' can be induced in a superconducting ring below the transition temperature, and such a current may be expected to flow at a uniform level for an indefinitely long time. Indeed, supercurrents which persist for extremely long periods (years) with no perceptible decay have been observed.

Supercurrents are a quintessentially quantum mechanical phenomenon, involving weakly bound pairs of electrons which move through the material without experiencing any collisions with the host lattice. The pair mechanism was originally proposed by Cooper (1956), whose theory was extended later by Bardeen, Cooper and Schrieffer (1957) to take into account the multiple interactions of all the electrons in the superconductor. The important conclusion from the BCS theory, as it is often called, is that only the Cooper-pair interactions are significant, the effect of all the others being merely to limit the number of states into which the Cooper pairs can be scattered.

The force binding the two electrons in a Cooper pair is extremely small, which is why superconductivity is observed only at low temperatures where the thermal energy of the electrons is less than the energy gap (i.e. the energy needed to separate the pair). The bond can also be separated by a sufficiently strong applied magnetic field, or if the current is raised too high. In either case, superconductivity is destroyed and the material then behaves as a normal conductor with the current carried by single electrons.

When two normal electrical conductors are separated by a thin insulating layer, a current can flow by the process of tunnelling. According to classical physics this is impossible, but quantum mechanical arguments show that, even if an electron has insufficient energy to surmount the barrier, there is still a non-zero probability that it will make the crossing and appear on the other side. Similarly, when two superconductors are separated by a thin insulating layer, a normal tunnel current of single electrons can flow across the junction. But in addition it is also possible for a supercurrent of Cooper pairs to tunnel through the barrier, which means that the insulator material between the two supei-conductors is itself acting as a superconductor. This remarkable behaviour, along with several other important effects, was predicted from theoretical considerations by B. D. Josephson (1962). He referred to the junction as a 'weak' superconductor because the supercurrent flow is more easily extinguished by external influences, such as an applied magnetic field, than would be the case in the absence of the insulating layer.

The effect of a magnetic field on the supercurrent through a Josephson junction (assuming that the field strength is below the level where the supercurrent is destroyed) is to introduce a spatial variation in the magnitude and direction of the current across the junction. This in turn means that the net current through the junction depends on the applied magnetic field, the functional form of the dependence being the same as that representing the Fraunhofer diffraction pattern obtained when light is passed through a single slit. Similarly, when two Josephson junctions are operated in parallel in a superconducting ring, the functional form of the current dependence on the magnetic flux is the same as that for the interference pattern obtained in a double-slit experiment. Indeed, the mechanism underlying the current variation in a multi-junction device is quantum interference, and such a device is known as a Josephson interferometer or a superconducting quantum interference device (SQUID).

The SQUID is extremely sensitive to tiny changes in the magnetic field strength, and can therefore be exploited as a high resolution magnetometer. It was conceived in the mid-1960s by Jaklevic et al. (1964). The ultimate resolution of the SQUID is set by inherent thermal noise in the Josephson junctions, and multi-junction devices have been constructed whose performance approaches this limit. So low is the noise in the SQUID that it has been employed as the sensing element in experiments designed to detect gravitational radiation (see Chapter 13). This is perhaps the most exacting problem in signal detection that has ever been undertaken, demanding meticulous care in order to achieve an exquisite sensitivity. The SQUID is a natural choice for such an application.

The noise mechanisms encountered in superconductors, Josephson junctions and SQUIDs are introduced and discussed in the latter part of this chapter. First, however, the essential physics of Josephson junction devices is outlined in order to provide a framework for supporting the subsequent sections on noise and fluctuations.

12.2 SUPERCONDUCTIVITY

A number of metals, including niobium, lead and tin, are known to be superconductors, and all have transition temperatures below 10 K. Some metallic compounds are also superconductors, notably Nb_3Sn whose transition temperature is relatively high at 18 K; and some alloys exhibit superconductivity, including Pb–Bi, which is a particularly interesting example because the element bismuth is not itself a superconductor. Superconductors are not found among the monovalent metals or the ferromagnetic and antiferromagnetic metals.

An important characteristic property of a superconductor is the complete absence of resistance at temperatures below the transition temperature θ_c. This suggests that a superconductor is no different from a 'perfect' conductor, that is a hypothetical metal with zero resistance, and indeed for many years after the discovery of superconductivity this was assumed to be the case. But a *super-conductor* below θ_c is not merely a *perfect* conductor: it is also perfectly diamagnetic, which is to say that, even in the presence of an applied magnetic field, the magnetic flux density **B** within its interior is always zero. This is known as the Meissner effect, after the experiments of Meissner and Ochsenfeld (1933), and it implies that, if a superconductor immersed in a magnetic field is cooled, the lines of induction are ejected from the material as the transition temperature is passed.

The Meissner effect is not a consequence of the zero resistance of the superconductor but is an independent phenomenon which is characteristic of superconductivity. This may be understood by considering Maxwell's equation relating the magnetic induction **B** and the field **E**:

$$\frac{\partial \mathbf{B}}{\partial t} = -\operatorname{curl} \mathbf{E} = 0 , \tag{12.1}$$

where the zero on the right appears because the absence of resistance means that the electric field is zero. According to equation (12.1), the rate of change of **B** within a perfect conductor is zero. But if this were the case in a superconductor, any magnetic flux within the material would remain there on cooling to temperatures below θ_c. As we have indicated above, this is not what is observed in practice, forcing the conclusion that zero resistance and perfect diamagnetism are two independent properties of the superconducting state. In brief, a perfect conductor may or may not be penetrated by lines of magnetic induction, whereas a superconductor never contains any magnetic flux within its interior.

An interesting corollary to this is that any current flowing along a superconductor cannot pass through the interior of the specimen but must flow on the surface. In order to see why this should be so, consider Maxwell's equation connecting the magnetic flux density **B** and the current density **J**:

$$\operatorname{curl} \mathbf{B} = \mu_0 \mathbf{J} , \tag{12.2}$$

where μ_0 is the permeability of free space, and it has been assumed that the relative permeability of the superconductor is unity. Now, as the superconductor exhibits perfect diamagnetism, the magnetic flux density within its interior is zero, hence curl **B** is also zero, and therefore from Maxwell's equation there can be no current flowing within the material. Thus, any current flow along the superconductor must be on the surface, which is permitted since the flux density outside the specimen is not constrained to be zero.

In reality, the current flow cannot be strictly confined to a sheet of zero thickness on the surface of the superconductor, for this would imply an infinite current density, which is physically unrealistic. There is in fact a finite penetration depth which depends on the superconductor but is typically on the order of 500 Å. It is the depth to which lines of magnetic flux density appear to penetrate the material. Thus, although a superconductor is said to be 'perfectly' diamagnetic, there is in fact a superficial penetration of magnetic induction. There is an analogy here between the penetration depth and the skin depth of a normal conductor.

Any satisfactory theory of superconductivity must obviously go beyond the simple argument stemming from equation (12.2) if it is to account for features such as the penetration depth. The London theory, as it is called, developed by F. London and H. London (1935) addresses the electrodynamics of superconductivity, and provides an elegant mathematical description of the superconducting state. The essential conclusions from the theory are embodied in two equations, known as the London equations, which are in close agreement with the observed electromagnetic properties of superconductors.

The London theory is not a microscopic theory of superconductivity, and is not concerned with the fundamental mechanism underlying supercurrent flow. Rather, it takes the familiar equations of electromagnetic theory and introduces a constraint which leads to a description of a physical phenomenon which happens to correspond closely to what is observed in superconductors. It was not until many years later that a transport mechanism for superconductivity was proposed.

The breakthrough in understanding superconductivity was made by Bardeen, Cooper and Schrieffer (1957), whose theory is based on the earlier ideas of Leon N. Cooper (1956). The BCS theory involves a mechanism for supercurrent flow which is quite different from that of ordinary current flow in a normal metal or even the hypothetical perfect conductor with zero resistance. Whereas the normal conduction process involves single electrons whose repeated collisions with the lattice are responsible for the resistance of the metal, the constituents of a supercurrent are pairs of weakly bound electrons which do not collide with the atoms of the lattice. The absence of collisions between the Cooper pairs and the host lattice accounts for the zero resistance of the superconductor.

The BCS theory involves a detailed quantum mechanical analysis which is beyond the scope of this discussion. However, the qualitative features of pair conduction are comparatively easy to describe.

The binding force between the electrons in a Cooper pair is due to an interaction with the atomic lattice. The negative charge on each of the electrons attracts the local, positively charged metal ions and the lattice undergoes a slight distortion, thereby creating a region of enhanced positive charge. This acts to attract the other electron, albeit very weakly, the binding energy or energy gap being approximately $4k\theta_c$. This energy is so low that separation is easily caused by the thermal fluctuations of the electrons, which is why superconductivity is observed only at cryogenic temperatures. For the same reason it can be destroyed by an excessive magnetic field or by too large a current.

According to the BCS theory, a Cooper pair can be treated as though it were a single particle located at the centre of mass of the two constituent electrons. In the absence of a current, the net momentum of each pair is zero, but when a current is flowing all the pairs show exactly the same momentum, in the direction of the current. A high degree of coherence exists between the movement of the pairs, which is the reason for the absence of collisions with the lattice, and hence accounts for the zero resistance of a superconductor.

Thus, the BCS theory provides a mechanism which explains one important aspect of superconductivity, namely the absence of resistance. A more detailed examination of the theory would reveal that it also accounts for the Meissner effect. In view of the success of the pair-conduction mechanism in explaining superconductivity, there remains little doubt that normal conduction, even in a perfect conductor, and superconduction are physically distinct, involving fundamentally different transport processes and displaying different electrodynamic behaviour.

12.3 THE JOSEPHSON JUNCTION

Electron tunnelling through an insulating barrier between two normal conductors is a well known quantum mechanical phenomenon. Even though an electron may not have sufficient energy to surmount the barrier, there is a finite probability that it will appear on the far side. In essence, the wave function of the electron penetrates the barrier, suffering a certain amount of attenuation in so doing which depends on the height and width of the potential hill separating the two conductors. An appreciable current will flow only if the barrier is very thin, a typical thickness being a few tens of angstroms.

When two superconductors are separated by a thin insulating layer, it is also possible for a tunnelling current to flow. This current consists of a normal flow of single electrons and also, under favourable conditions, a flow of Cooper pairs. The tunnelling supercurrent of paired electrons was predicted by Brian D. Josephson (1962) from a quantum mechanical analysis of the situation. His theoretical arguments have long since received experimental confirmation, and indeed Josephson junctions have been employed as the principal elements in a number of practical electronic devices. A good example of such a device is the SQUID, to be discussed later.

The flow of current through the junction is characterized by two distinct types of behaviour, known as the d.c. and a.c. Josephson effects. In the d.c. Josephson effect, a direct supercurrent can flow up to some critical level, I_0, and no voltage is developed across the junction. Thus, the insulator separating the two superconductors can itself act as a superconductor. However, the critical current I_0 through the junction is much less than the maximum current that could flow if the junction were not present, and the junction supercurrent can be extinguished with a smaller magnetic field than would be needed with a homogeneous superconductor. On this account, the junction was designated by Josephson a 'weak' superconductor.

The a.c. Josephson effect is observed when a d.c. bias voltage, V_0, is applied across the junction: the supercurrent of Cooper pairs then alternates with a frequency

$$f_0 = 2qV_0/h \, , \tag{12.3}$$

where q is the magnitude of the electronic charge and h is Planck's constant. f_0 is known as the Josephson frequency or sometimes as the Josephson plasma frequency. It may be inferred from equation (12.3) that the Josephson junction can be employed as a single-frequency oscillator or alternatively as a very precise voltage standard. Note that even for a small applied voltage, say 10 μv, the frequency is very high at approximately 50 GHz.

Equation (12.3) follows directly from Josephson's equations for the junction current. If Φ is the phase change in the wave function across the insulating layer, then the equations are

$$I = I_0 \sin \Phi \tag{12.4a}$$

and

$$\frac{d\Phi}{dt} = \frac{2q\,V_0}{\hbar} \, , \tag{12.4b}$$

where I is the current through the junction and $\hbar = h/2\pi$. On integrating the second of these equations with respect to time, the phase difference is found to be

$$\Phi = \frac{2q\,V_0}{\hbar} t + \text{const.} \, , \tag{12.5}$$

which, when substituted into equation (12.4a) shows that the frequency is as given in equation (12.3).

A Josephson junction may be fabricated in a number of ways. The structure used in the original observation of Josephson tunnelling (Anderson and Rowell, 1963) consisted of a strip of superconductor evaporated onto a substrate, with an oxide layer on the order of 10 Å thick separating it from a similar strip evaporated perpendicular to the first. More recently, other types of tunnelling junction have been constructed, including bridges and point contact junctions. Several of these devices and their relative merits have been discussed by Clarke (1973).

Generally, the tunnelling element in a practical Josephson junction is shunted by a capacitance which gives rise to hysteresis in the current-voltage characteristic. This is most undesirable in applications such as the SQUID, and is usually eliminated by shunting the junction with an appropriate resistance R. The device can then be characterized by the equivalent circuit shown in Fig. 12.1(a). Naturally, the current-voltage characteristic of the resistively shunted junction (RSJ) is important in various applications, but its derivation is not immediately obvious. On this account it is discussed below.

With the aid of equations (12.4) the (time-varying) voltage V_0 across the RSJ can be expressed as

$$V_0 = \frac{\hbar}{2q} \frac{d\Phi}{dt} = R(I - I_0 \sin \Phi) , \qquad (12.6)$$

where $I > I_0$ is the total (time-independent) current through the circuit. This equation may be solved for Φ by integrating as follows:

$$\int \frac{d\Phi}{[\alpha - \sin \Phi]} = \frac{2qRI_0}{\hbar}(t + t_0) , \qquad (12.7)$$

where t_0 is a constant of integration and $\alpha = I/I_0$. The integral on the left of

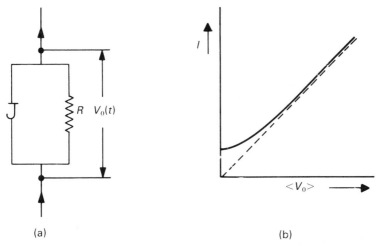

(a)   (b)

Fig. 12.1 – (a) Resistively shunted junction and (b) its current-voltage characteristic (solid line) which approaches the value RI (broken line) asymptotically at high currents.

this expression is evaluated by the well-known method involving the substitution $u = \tan(\Phi/2)$. After some straightforward algebraic manipulation this leads to the result

$$\Phi = 2 \tan^{-1} \left[\alpha^{-1} \left\{ (\alpha^2 - 1)^{\frac{1}{2}} \tan \left[(\alpha^2 - 1)^{\frac{1}{2}} \eta (t + t_0) \right] + 1 \right\} \right] , \tag{12.8}$$

where $\eta = \dfrac{qRI_0}{\hbar}$. Now, on inspecting this expression it can be seen that, although the inner tangent goes to infinity when its argument is equal to an odd multiple of $\pi/2$, the differential $d\Phi/dt$ (which is of interest here since it gives V_0) is continuous in the variable t. In fact, $d\Phi/dt$ is periodic with period

$$T = \frac{\pi}{\eta(\alpha^2 - 1)^{\frac{1}{2}}} . \tag{12.9}$$

However, the oscillation of V_0 over this interval is *not symmetrical,* and the mean voltage over the cycle, defined as

$$\langle V_0 \rangle = \frac{\hbar}{2q} \left\langle \frac{d\Phi}{dt} \right\rangle = \frac{\hbar}{2qT} \int_0^T \frac{d\Phi}{dt} \, dt = \frac{\hbar[\Phi(T) - \Phi(0)]}{2qT} \tag{12.10}$$

is non-zero. Defining the limits of a cycle as the two successive times at which the inner tangent function in equation (12.8) goes to infinity, we have $[\Phi(T) - \Phi(0)] = 2\pi$, and hence

$$\langle V_0 \rangle = \frac{\eta\hbar}{q} (\alpha^2 - 1)^{\frac{1}{2}} = R(I^2 - I_0^2)^{\frac{1}{2}} . \tag{12.11}$$

This equation is the current-voltage relation for the RSJ. Note that for currents through the RSJ which are large compared with the critical current, the mean voltage is asymptotic to RI.

The current-voltage characteristic in equation (12.11) is plotted in Fig. 12.1(b). It may be inferred from the curve that the a.c. supercurrent makes an appreciable contribution to the time-averaged current through the RSJ, even when the voltage is varying with time, at current levels several times the critical current. It follows that a change in the critical current can affect the slope of the characteristic over a considerable region of the curve. Measurements of the current-voltage characteristics of resistively shunted junctions show good agreement with equation (12.11) (Hansma et al., 1971).

12.4 JOSEPHSON INTERFEROMETERS

When a current flows in a superconducting ring a quantum mechanical phenomenon known as flux quantization occurs: the flux in the ring takes only discrete values which are the multiples of the flux quantum $\Phi_0 = h/2q \simeq 2 \times 10^{-15}$ Weber. This is exploited to achieve extremely fine resolution in two types of magnetic field sensor, both of which are superconducting quantum interference devices (SQUID's). The first type of SQUID to be developed was the d.c. SQUID, consisting of a superconducting ring containing two Josephson junctions (Fig. 12.2(a), and this was followed later by the RF SQUID in which a single junction mounted in a superconducting ring is coupled inductively to an external tank circuit (Fig. 12.2(b)).

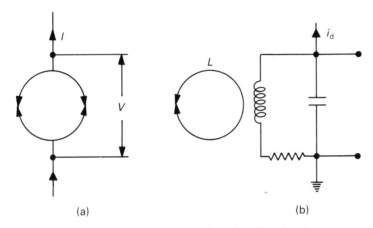

(a)   (b)

Fig. 12.2 – (a) d.c. SQUID, containing two Josephson junctions in a superconducting ring. (b) An RF SQUID, with one Josephson junction in a superconducting ring weakly coupled to a resonant tank circuit.

The behaviour of both types of SQUID can be investigated on the basis of simple theoretical models, provided it is assumed that a superconducting ring containing one or more Josephson junctions still exhibits flux quantization. It is implicit in this assumption that the inductance of the junction or junctions is negligible compared with the geometrical inductance of the ring itself. When this is not the case the problem is somewhat more complicated, as discussed by Ouboter and de Waele (1970) and Webb (1972).

12.4.1 The d.c. SQUID

The current flowing through a single Josephson junction displays an oscillatory dependence on the magnetic flux applied to the junction (Rowell, 1963). This is illustrated in Fig. 12.3 for the case of a Pb-I-Pb junction. Minima at 6.5, 13.5 and 19.5 gauss are clearly evident, separated by well-defined maxima. The functional form of the curve in the figure is $\sin x/x$, where x is proportional to the magnetic flux, and the period of the oscillation is the flux quantum Φ_0. This is analogous to the form of the Fraunhofer diffraction pattern obtained when light is passed through a single slit.

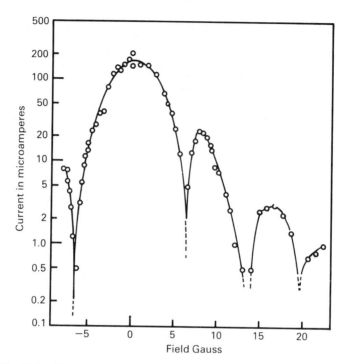

Fig. 12.3 – Experimental measurement of the field dependence of the current in a Josephson junction (from Rowell, 1963, by kind permission of the American Physical Society).

When two Josephson junctions are present in a superconducting ring, as shown in Fig. 12.2(a), the current as a function of the magnetic flux applied perpendicular to the plane of the ring is also oscillatory in nature, in this case being analogous to the interference pattern obtained when light is passed through a double slit. There are now two periodicities in the current as a function of the flux, one associated with the flux at a single junction and the other arising from the flux in the region between the junctions. This situation can be described in terms of a critical current I_0' for the double junction which shows an oscillatory dependence on the applied magnetic flux with a period equal to Φ_0 (Jaklevic *et al.*, 1964). The theory of the SQUID has been described by Clarke (1973), a summary of which is given below.

Under the usual conditions of operation the SQUID is biased with a constant current I which is greater than twice the critical current I_0 of the junctions (both junctions are assumed here to have the same critical current, but this is not essential to the argument). A voltage $V > 0$ is then developed across the ring, which is assumed to have an inductance L. When an external magnetic flux $\Phi_e < \Phi_0/2$ is applied, a screening current $I_s = \Phi_e/L$ must flow in order to maintain the flux in the material of ring at zero. If Φ_e is increased to

$\Phi_e = \Phi_0/2$ the screening current increases to $I_s = \Phi_0/2L$. As the external flux is further increased, it becomes energetically favourable for the flux through the ring to make a transition between quantum states (bearing in mind flux quantization) in such a way that the screening current reverses direction. Thus, when $\Phi_e = \Phi_0/2+$ the screening current is $I_s = -\Phi_0/2L$, and over the range $\Phi_0/2 < \Phi_e \leqslant 3\Phi_0/2$ I_s increases until at the upper end of the range I_s again reverses direction and the whole process keeps on repeating. The screening current as a function of the external flux takes the form of a saw-tooth curve, as shown in Fig. 12.4(a). Note that the transitions occur when $\Phi_e = (n + \frac{1}{2})\Phi_0$, where $n = 0, 1, 2, \ldots$.

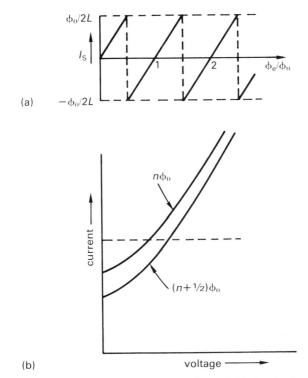

(a)

(b)

Fig. 12.4 – (a) Screening current versus ϕ_e/ϕ_0 and (b) sketch of current-voltage characteristics of the d.c. SQUID.

On referring to Fig. 12.2(a) it is apparent that when a screening current I_s is flowing in the ring, the net currents flowing through the two junctions differ by $2|I_s|$. Therefore, if the current through one junction is $I_0 - 2|I_s|$, then the current through the other is I_0, which is the critical current of the junction. It follows that the critical current of the d.c. SQUID, I_0', is less than $2I_0$ by $2|I_s|$:

$$I_0' = 2I_0 - 2|I_s| \ . \tag{12.12}$$

This result indicates that the critical current of the d.c. SQUID oscillates with the applied flux, as stated above. The maximum reduction in the critical current is equal to Φ_0/L and occurs when $\Phi_e = (n + \frac{1}{2})\,\Phi_0$, whereas the minimum reduction is zero, occurring when $\Phi_e = n\,\Phi_0$.

The current-voltage characteristics of the d.c. SQUID at these extrema are sketched in Fig. 12.4(b). When the bias current I is greater than $2I_0$, as shown in the figure, the voltage across the device also oscillates in a triangular fashion as Φ_e is increased. If R is the resistance of each of the junctions, the voltage modulation depth is simply

$$\Delta V = 2\,|I_s|_{max}\,\frac{R}{2} = \frac{\phi_0 R}{2L}\,, \tag{12.13a}$$

where $2\,|I_s|_{max} = \Phi_0/L$ is the maximum reduction in the critical current of the SQUID. For $R = 5\Omega$ and $L = 10^{-9}$ H, V is approximately equal to $5\mu v$.

If the variation of voltage with external flux is approximated as a triangular wave, then the output voltage produced by a small change in applied flux, $\delta\Phi_e$, is

$$\delta V \simeq \frac{R}{L}\delta\phi_e\,. \tag{12.13b}$$

For the values used above, this gives $\delta V/\delta\Phi_e \simeq 5 \times 10^9$ V/Wb.

In most applications the d.c. SQUID is used as a null detector in a negative feedback loop: a small change in the applied flux produces a change in the voltage which is detected and used to generate a nulling flux that is fed back to the system. For a SQUID with an inductance $L = 10^{-9}$ H it is possible to achieve a resolution better than $10^{-4}\,\Phi_0/\sqrt{Hz}$, corresponding to a voltage change across the junctions of 1 nV/$\sqrt{Hz}$.

12.4.2 The a.c. SQUID

The RF SQUID, illustrated in Fig. 12.2(b), consists of a single Josephson junction in a superconducting ring which is coupled inductively to a resonant tank circuit. The resonance frequency of the tank circuit is in the region of 30 MHz. An RF current at the resonance frequency drives the tank circuit at such a level that the peak current induced in the SQUID is just above the critical current of the junction. The output is the voltage developed across the tank circuit, which is an oscillatory function of the magnetic flux applied perpendicular to the plane of the superconducting ring, with a period equal to ϕ_0.

A detailed discussion of the RF SQUID is given by Clarke (1973), who includes in his review a discussion of practical designs and their performance as well as a number of references on the operation of the device.

12.5 NOISE IN JOSEPHSON JUNCTION DEVICES

In many applications of Josephson junction devices, the limit to performance is set by noise in the system. Thus, in the SQUID for example it is important to eliminate all extraneous noise, such as may be generated by electrical machinery and fluctuations in the earth's magnetic field. This can usually be achieved by careful screening of the cryostat containing the SQUID, and ensuring that the device itself is mounted as rigidly as possible since even the minutest movement in a small magnetic field may generate relatively large signals. Electrical interference on the leads entering the cryostat due to radio transmissions can also be a problem and must be removed by suitable filtering.

Once all the externally generated noise has been removed or suppressed, one is left with the inherent noise in the Josephson device itself and also the noise generated in the associated electronic circuitry. Various types of inherent noise are discussed below.

12.5.1 Shot noise in a Josephson junction

Noise in the a.c. Josephson effect, that is when the current flowing through the junction is greater than the critical current, was investigated theoretically by Scalapino (1967) and Stephen (1969). A Langevin treatment of the noise was given by Stephen, who found that the system is unstable with respect to phase fluctuations. This instability has the effect of broadening the bandwidth of the radiation emitted from the junction so that, instead of appearing at a single frequency, it is spread into a Lorentzian line shape.

The Langevin treatment also predicts a fluctuation in the current flowing through the junction. It is not, however, necessary to solve the differential equations relating the variables in the problem in order to derive the expression for the power spectral density of the noise. Instead, a simple shot-noise argument can be employed, which leads to the correct result (van der Ziel and Chenette, 1978).

The current flowing across the junction consists of normal electrons, the magnitude of the charge on each being q, and Cooper pairs with charge of magnitude $2q$. If i is the total current through the junction, then

$$i = i_n + i_p = i_{n1} - i_{n2} + i_{p1} - i_{p2} \ , \qquad (12.14)$$

where $i_n = i_{n1} - i_{n2}$ is the net current of normal electrons and $i_p = i_{p1} - i_{p2}$ is the net current of paired electrons across the junction. (It is implicit that there is a current flow of normal and paired electrons in both directions across the junction.) Now, each of the four current components on the right of equation (12.14) shows full shot noise, since they are all independent, and hence the power spectral densities of i_n and i_p, respectively, are

$$\overline{S_{i_n}(\omega)} = 2q(I_{n1} + I_{n2}) \tag{12.15a}$$

and

$$\overline{S_{i_p}(\omega)} = 4q(I_{p1} + I_{p2}) , \tag{12.15b}$$

where the upper case I is the mean value of the corresponding lower case i. Note that in equation (12.15b) the double electronic charge on the Cooper pairs has been included, giving rise to a prefactor which is twice that in the previous expression for normal electrons.

By considering the probabilities of a charge carrier moving across the barrier in either direction the following results can be proved:

$$I_{n1} + I_{n2} = I_n \coth(qV/2k\theta) \tag{12.16a}$$

and

$$I_{p1} + I_{p2} = I_p \coth(2qV/2k\theta) , \tag{12.16b}$$

where V is the height of the barrier, which in the present case is the voltage across the junction, and θ is the absolute temperature. Again, note the extra factor of 2 appearing in equation (12.16b), associated with the double charge on the Cooper pairs. It follows from equations (12.15) and (12.16) that the power spectral density of the total fluctuation in the current flowing across the junction is

$$\overline{S_i(\omega)} = \overline{S_{i_n}(\omega)} + \overline{S_{i_p}(\omega)}$$

$$= 2qI_n \coth(qV/2k\theta) + 4qI_p \coth(2qV/2k\theta) . \tag{12.17}$$

When $|qV| \gg k\theta$ the coth functions in this expression are unity, giving

$$\overline{S_i(\omega)} = 2qI_n + 4qI_p . \tag{12.18}$$

Thus, in this limit, the expressions for the noise in the currents of normal electrons and Copper pairs appear in the familiar shot-noise form. At the other extreme, when $|qV| \ll k\theta$, the coth functions can be approximated by the reciprocal of their arguments, which gives

$$\overline{S_i(\omega)} = 4k\theta I/V , \tag{12.19}$$

where $I = I_n + I_p$. This result will be recognized as the expression for thermal noise in a conductance I/V. It is apparent that as V approaches zero this conductance, and hence also the noise, can become very large (Kanter and Vernon, 1970).

12.5.2 Thermal noise in the d.c. Josephson effect

When the temperature of a Josephson junction is close to the transition temperature it is possible for thermal fluctuations — arising from equilibrium energy
exchange with the environment — to affect the phase relationship between the
two sides of the junction. This results in a fluctuating voltage appearing across
the junction with a non-zero mean value. The phenomenon has been investigated
theoretically by Ivanchenko and Zil'berman (1968) on the basis of a kinetic
theory argument, and also by Ambegaokar and Halperin (1969) who solved the
differential equations for the phase change across the junction. In the latter
treatment, an expression for the mean voltage is obtained which in general must
be integrated numerically in order to arrive at a solution. However, the authors
give several analytical forms for the solution which are valid under different
limiting conditions.

12.5.3 Noise in the RSJ

Noise in the resistively shunted Josephson junction is important in connection
with determining the ultimate sensitivity of SQUIDs and other Josephson
junction devices. Provided the tunnelling current is small compared with the
current through the shunt resistor, shot noise in the current crossing the
junction is negligible and the only significant source of noise is thermal fluctuations in the shunt. These fluctuations produce a noise voltage across the RSJ
when the device is driven from a constant current source.

The spectral density of the voltage fluctuations across a current-biased RSJ
has been treated theoretically by Likharev and Semenov (1972) and Vystavkin
et al. (1974) on the assumptions that the junction has zero capacitance and that
the thermal noise in the shunt resistor is in the classical limit, i.e. the Josephson
frequency satisfies the condition $hf_0 \ll k\theta$, where θ is the absolute temperature,
h is Planck's constant and k is Boltzmann's constant. These assumptions have
been relaxed by Koch *et al.* (1980), who show that the limiting noise level as
the absolute temperature approaches zero is set by the zero point fluctuations
in the shunt resistor.

The approach of Koch *et al.* is to set up the inhomogeneous, non-linear
Langevin equation for the phase, in which the thermal fluctuations are represented
as part of the source term. In the quantum limit (i.e. $\theta = 0$) at a frequency $\omega/2\pi$
which is much less than the Josephson frequency, they find that the spectral
density of the voltage noise across the RSJ is

$$\overline{S_v(\omega)} = 2qI_0^2R^3/V \ , \qquad (12.20)$$

where R is the shunt resistance, I_0 is the critical current of the junction, and V is
the mean voltage across the device. Expressions for the spectral density of the
voltage fluctuations under various other limiting conditions can also be obtained
from the theory.

In addition to producing a voltage noise across the device, the thermal fluctuations have another effect: they give rise to a phenomenon known as 'noise rounding' in the current-voltage characteristic of the RSJ. In the absence of noise, the characteristic is as given in equation (12.11). This may be expressed in terms of the normalized voltage $\widetilde{V} = V/I_0R$ and the normalized current $\widetilde{I} = I/I_0$ as follows:

$$\widetilde{I} = (\widetilde{V}^2 + 1)^{\frac{1}{2}} , \tag{12.21}$$

a sketch of which is shown in Fig. 12.5 (solid line). When noise is present the current-voltage characteristic is modified, taking a form which is illustrated qualitatively by the broken line in the figure. Koch *et al.* have investigated quantitatively the noise rounding of the current-voltage characteristic in various circumstances, by employing numerical techniques for solving the non-linear differential equations involved in the problem. Experimental confirmation of their noise theory, including the noise rounding phenomenon, has been obtained recently using PbIn-Ox-Pb junctions with CuAl shunt resistances of the order of 0.1 Ω (Koch *et al.*, 1981a).

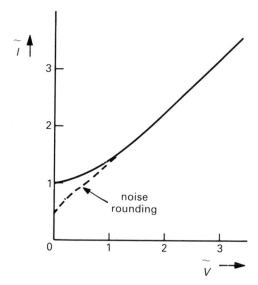

Fig. 12.5 – Current-voltage characteristic of the RSJ in the absence of noise (solid line) and when noise is present (broken line).

12.5.4 Noise in the d.c. SQUID

The theoretical treatment of thermal noise in the RSJ, outlined above, is relevant to the case of the d.c. SQUID, which in essence consists of two RSJs in a super-conducting ring. On the basis of the theory, Koch *et al.* (1981b) have found that the limit to sensitivity is set by the zero point energy fluctuations in the resistors shunting the junctions, and that under optimum conditions these fluctuations give rise to an equivalent noise energy per unit frequency equal to $\hbar$. Assuming optimum coupling to the input, this corresponds to a noise temperature for the amplifier approximately equal to $hf/k \ln 2$, where f is the frequency of the signal. As this noise temperature is equal to the limiting value set by the uncertainty principle, the theoretical sensitivity of the SQUID in the low temperature limit is equal to that of an ideal linear amplifier.

Much effort has been devoted recently to fabricating a d.c. SQUID with a sensitivity approaching the theoretical limit. Ketchen and Voss (1979) have achieved a device operating at 4.2 K with an energy sensitivity per hertz equal to $5h$, which is an improvement of an order of magnitude or more over earlier devices (Clarke et al., 1975; Hollenhorst and Gifford, 1979).

12.5.5 Noise in the RF SQUID

The absolute limit to the sensitivity of an a.c. SQUID at zero temperature is set by the uncertainty principle. The absolute limiting noise flux as well as various other factors limiting the performance of practical devices are discussed by Kurkijärvi (1981) and Leggett (1981), both of whom give a number of recent, relevant references.

REFERENCES

V. Ambegaokar and B. I. Halperin (1969), Voltage due to thermal noise in the d.c. Josephson effect, *Phys. Rev. Lett.*, **22**, 1364–1366.

P. W. Anderson and J. M. Rowell (1963), Probable observation of the Josephson superconducting tunnelling effect, *Phys. Rev. Lett.*, **10**, 230–232.

J. Bardeen, L. N. Cooper and J. R. Schrieffer (1957), Theory of superconductivity, *Phys. Rev.*, **108**, 1175–1204.

J. Clarke (1973), Low-frequency applications of superconducting quantum interference devices, *Proc. IEEE*, **61**, 8–19.

J. Clarke, W. M. Goubau and M. B. Ketchen (1975), Thin film d.c. SQUID with low noise and drift, *Appl. Phys. Lett.*, **27**, 155–156.

L. N. Cooper (1956), Bound electron pairs in a degenerate Fermi gas, *Phys. Rev.*, **104**, 1189–1190.

P. K. Hansma, G. I. Rochlin and J. N. Sweet (1971), Externally shunted Josephson junctions: generalized weak links, *Phys. Rev.*, **43**, 3003–3014.

J. H. Hollenhorst and R. P. Gifford (1979), High sensitivity microwave SQUID, *IEEE Trans. Mag.*, **MAG-15**, 474–477.

Yu. M. Ivanchenko and L. A. Zil'berman (1968), Destruction of Josephson current by fluctuations, *JETP Lett.*, **8**, 113–115. (Originally published in *ZhETF Pis. Red.*, **8**, 189–192, 1968.)

R. C. Jaklevic, J. Lambe, A. H. Silver and J. E. Mercereau (1964), Quantum interference effects in Josephson tunnelling, *Phys. Rev. Lett.*, **12**, 159–160.

B. D. Josephson (1962), Possible new effects in superconductive tunnelling, *Phys. Lett.*, **7**, 251–253.

H. Kanter and F. L. Vernon Jr. (1970), Current noise in Josephson point contacts, *Phys. Rev. Lett.*, **25**, 588–590.

M. B. Ketchen and R. F. Voss (1979), An ultra-low-noise tunnel junction d.c. SQUID, *Appl. Phys. Lett.*, **35**, 812–815.

R. H. Koch, D. J. Van Harlingen and J. Clarke (1980), Quantum-noise theory for the resistively shunted Josephson junction, *Phys. Rev. Lett.*, **45**, 2132–2135.

R. H. Koch, D. J. Van Harlingen and J. Clarke (1981a), Quantum noise in Josephson junctions and SQUIDs, *Proc. of the Sixth Int. Conf. on Noise in Physical Systems, held at the National Bureau of Standards, Gaithersburg, MD, USA, April 6–10 1981*, pp. 359–363.

R. H. Koch, D. J. Van Harlingen and J. Clarke (1981b), Quantum noise theory for the d.c. SQUID, *Appl. Phys. Lett.*, **38**, 380–382.

J. Kurkijärvi (1981), Ultimate sensitivity of an a.c. SQUID, *Proc. of the Sixth Int. Conf. on Noise in Physical Systems, held at the National Bureau of Standards, Gaithersburg, MD, USA, April 6–10 1981*, pp. 373–375.

A. J. Leggett (1981), Quantum tunnelling and noise in SQUIDs, *Proc. of the Sixth Int. Conf. on Noise in Physical Systems, held at the National Bureau of Standards, Gaithersburg, MD, USA, April 6–10 1981*, pp. 355–358.

K. K. Likharev and V. K. Semenov (1972), Fluctuation spectrum in superconducting point junctions, *JETP Lett.*, **15**, 442–445. (Originally published in *ZhETF Pis. Red.*, **15**, 625–629, 1972.)

F. London and H. London (1935), The electromagnetic equations of the supraconductor, *Proc. Roy. Soc. Lond.*, **A149**, 71–88.

W. Meissner and R. Ochsenfeld (1933), Eine neuer effeckt eintritt der supraleitfahigkeit, *Naturwiss.*, **21**, 787–788.

R. de Bruyn Ouboter and A. Th. A. M. de Waele (1970), Superconducting point contacts weakly connecting two superconductors, in *Progress in Low Temperature Physics*, vol. 6 (editor C. J. Gorter), Amsterdam, London, North-Holland, pp. 243–290.

J. M. Rowell (1963), Magnetic field dependence of the Josephson tunnel current, *Phys. Rev. Lett.*, **11**, 200–202.

D. J. Scalapino (1967), Proc. of the Symposium on the Physics of Superconducting Devices, University of Virginia, USA (unpublished).

M. J. Stephen (1969), Noise in the a.c. Josephson effect, *Phys. Rev.*, **182**, 531–538.

A. van der Ziel and E. R. Chenette (1978), Noise in solid state devices, in *Advances in Electronics and Electron Physics*, **46**, 313–383.

A. N. Vystavkin, V. N. Gubankov, L. S. Kuzmin, K. K. Likharev, V. V. Migulin and V. K. Semenov (1974), S-c-S junctions as non-linear elements of microwave receiving devices, *Phys. Rev. Appl.*, **9**, 79; see also (1975), 'Non-Josephson' radiation from the cavity containing a superconducting point contact junction, *IEEE Trans. Mag.*, **MAG-11**, 834–837.

W. W. Webb (1972), Superconducting quantum magnetometers, *IEEE Trans. Mag.*, **MAG-8**, 51–60.

13

Gravitational radiation detectors

13.1 INTRODUCTION

Einstein's (1916) general theory of relativity leads to the prediction that accelerating masses radiate gravitational waves which propagate at the speed of light. According to the theory, the gravitational field at any point is a tensor quantity, and the potential is therefore specified by ten numbers. (There are alternatives to the tensor theory, notably the Brans–Dicke (1961) theory which is a modification of Einstein's general relativity and which predicts a mixture of tensor and scalar radiation. In connection with the present discussion on detection, however, the distinction between the different types of gravitational radiation is immaterial.) Unlike electromagnetic waves, which are generated by dipole sources, the lowest mass multipole that can generate gravitational radiation is a quadrupole. The fluctuating quantity constituting the radiation is the curvature of space-time. One may imagine, like waves on the surface of the sea, gravitational waves propagating through the universe, generated by highly energetic celestial sources. These may include stellar collapse, supernovae, pulsars, collisions between black holes, and rapidly rotating binary stars.

The gravitation field equations, whose oscillatory solution constitutes the theoretical evidence for the existence of gravitational radiation, were known to Einstein in 1915. At that time there was no question of attempting to detect gravity waves, and so provide a check on the theory, because no means existed for conducting a sophisticated experiment with sufficient sensitivity to respond to the extremely low energy fluxes that are predicted theoretically. It was not until 1958 that the first efforts were made to build a gravity wave detector. The pioneer in the field was Joseph Weber, who designed an antenna for receiving gravitational waves which is familiarly known as the 'Weber bar'. Weber took great pains to isolate his detectors from interference due to spurious acoustic, electromagnetic and seismic effects, and in addition employed a coincidence counting technique involving two detectors – one at the University of Maryland and the other at the Argonne National Laboratory near Chicago, which is a baseline of 1000 km — intended to minimize the effect of local sources of noise at either of the sites. Some ten years after he began his search, Weber published a series of papers in which he reported observing simultaneous pulse signals at his detectors (Weber, 1969, 1970), which he attributed to gravitational radiation generated by an extra-terrestrial source. He also observed that the rate of arrival of the pulses varied with the time of day, and on examining the frequency of occurrence over a period of several months as a function of sidereal time (which is time measured relative to the background stars, and differs from solar time by about 4 minutes a day) he found peaks at 12 hourly intervals. This can be explained in terms of the directional response of the detectors, which shows a maximum when the signal is incident broadside-on and a minimum in the axial direction. As the detectors (whose axes in Weber's experiment were in an east—west alignment) rotate with the Earth, which is essentially transparent to gravitational radiation, the 12 hourly peaks in the response would occur if the radiation were incident from a particular direction. According to Weber, the signals he observed were coming from the centre of the galaxy, or possibly from a direction 180° away from this, which, as it happens, contains the Crab nebula.

The energy in the signals reported by Weber was immensely higher than the theoretical levels that had been predicted. Indeed, many physicists expected that he would observe nothing at all. Assuming the source were at the centre of the galaxy, the signals from his detectors correspond to gravitational energy fluxes which could only be generated by enormously energetic events in which mass is converted into gravitational radiation at an embarrassingly high rate. It has been estimated (Hawking, 1972) that each of Weber's pulses would require the conversion of about thirty solar masses into radiation. This could be accommodated by the collision of two black holes, but, even though this may be a possible mechanism for a single pulse, the rate of conversion of mass into radiation needed to explain the frequency of Weber's pulses is so high that the entire mass of the galactic centre would be radiated away in a time which is about 1/1000 the age of the universe. There are obvious difficulties with this deduction. They could perhaps be circumvented by assuming either that we are living in a privileged age when the galactic centre is unusually active in producing gravitational radiation or that the Earth is in a privileged position close to a local source of gravitational radiation. There is little convincing evidence to support either hypothesis.

The implications of Weber's claims aroused considerable interest amongst theoreticians and experimentalists alike, and stimulated a dozen or so groups around the world to embark on the quest for gravitational radiation, with the aim of confirming his observations. The independent confirmation never materialized and even now, more than a decade later, Weber's signals remain unexplained. But wherever they originated, it is generally accepted that the pulses were not due to gravitational radiation. However, it is important to recognize that Weber's initiative opened up a new field of astronomy and as a result gravitational-wave detectors are being constructed today which are far more sensitive than the original Weber bar. The sensitivity of these new detectors is approaching that needed to detect signals at the theoretically estimated levels, and thus there exists the possibility that in the near future gravitational radiation will be detected unequivocably, thereby confirming Einstein's original prediction.

13.2 GRAVITATIONAL-WAVE DETECTORS

Gravitational and electromagnetic radiation have certain features in common, but whereas EM waves interact only with electronic charge or current, gravitational waves interact with all forms of energy and matter. Thus, the sensitivity of a gravity wave detector scales with both its mass and size, suggesting that the largest available solid body should be employed as an antenna in order to maximize the chance of detection. The most massive object available to mankind is the Earth itself, and this has been seriously considered as a possible gravitational-wave detector. A gravitational wave passing through the Earth would excite the quadrupole modes of vibration, with eigenfrequencies in the region of cycles per hour. From seismometer records of the Earth's activity, Weber (1967) has established an upper limit for the energy flux of gravitational radiation incident at the Earth of 3×10^4 $Wm^{-2}Hz^{-1}$ at a frequency of 0.3 mHz. Of course, as a detector of gravitational radiation the Earth is very noisy, suffering from fairly intensive seismic and meteorological disturbances. The Moon is a much quieter body and on this account has been considered as an alternative to the Earth as a very-low-frequency resonant detector. Some interest has also been shown in laser ranging between the Earth and the Moon or artificial satellites, designed to detect small changes in the distance between the bodies due to a fluctuating gravitational field. A serious problem with the ranging technique is low sensitivity.

Theoretical arguments suggest that the gravitational radiation from the final stage of stellar collapse, terminating in the formation of a black hole, or from the collision of two black holes, takes the form of a pulse of approximately 1 millisecond duration. The frequency components of such a pulse are largely contained in the kilohertz region. In order to detect such pulses, Weber (1960, 1966) constructed an antenna consisting of an aluminium alloy cylinder approximately 1.5 m long and 0.6 m in diameter, weighing about 1½ tons, and with a longitudinal resonance frequency of 1661 Hz. Piezoelectric ceramic transducers were cemented around the periphery of the bar at the middle which responded to the longitudinal mode of oscillation. The bar was suspended by a single loop of wire, as shown in Fig. 13.1(a), and was housed in an evacuated chamber in order to eliminate acoustic interference. The whole assembly was supported on anti-vibration mountings consisting of alternating layers of rubber and steel. This type of detector shows an extremely sharp resonance with a Q-factor of the order of 10^5 or higher. Mass-resonators following Weber's basic design, but with

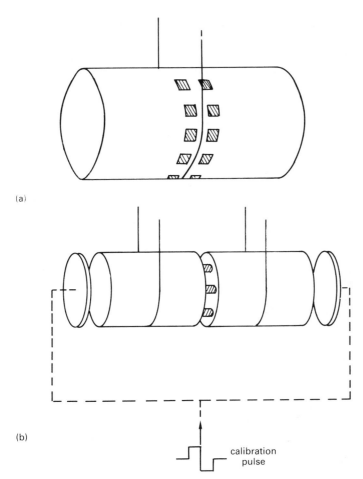

(a)

(b)

calibration
pulse

Fig. 13.1 − (a) The Weber bar with the transducers (hatched) around the periphery of the aluminium alloy cylinder. (b) The split bar with transducers sandwiched between two aluminium alloy cylinders. The capacitive plates shown at either end of the split bar are used for applying a calibration pulse to the system.

improved sensitivity obtained by cooling to liquid helium temperatures, are still being used today (see, for example, Gifford *et al.*, 1981). An alternative arrangement also exists, suggested by Peter Aplin at the University of Bristol and discussed by Gibbons and Hawking (1971), in which the bar is split into two identical cylinders with the transducers sandwiched between them (Fig. 13.1(b)). Drever (1971) was one of the first to operate such a detector. The split bar shows better electromechanical coupling than the Weber bar but a considerably lower Q-factor, of about 1000. The analysis described below can be used to predict and compare the performance of the two systems.

When a gravitational wave is incident broadside-on to either a Weber bar or a split bar, the fundamental longitudinal mode of oscillation is excited, provided the signal has frequency components at the resonance frequency of the detector. This response may be explained in simplistic terms if the gravity wave is regarded as a fluctuation in the Newtonian gravitational constant. (The pattern of forces in a tensor wave and a scalar wave are quite different, but this is not important in the present qualitative argument.) As the wave passes through the bar, the centre of mass remains essentially at rest (which is equivalent to saying that the bar is in free fall and that the gravitational attraction towards the centre of the Earth is insignificant) but other points in the bar experience tensile and compressive forces, with the result that the bar is set ringing. The expected amplitude of the oscillation is very small, as exemplified by Weber's experiment in which he was searching for displacements of 10^{-16} m in the end faces of the bar. This is an order of magnitude less than the radius of an electron! The reason for the small response is twofold: the incoming signals are themselves low-level (for example, the Earth orbiting the sun radiates about 1 kW of gravitational power); and the coupling between the radiation and the antenna is *extremely* weak.

When dealing with such small signals, it is inevitable that noise in the detection system will set the ultimate limit of sensitivity. In resonant-mass detectors such as the Weber bar and the split bar there are three main sources of noise, namely Brownian motion of the bar, thermal noise in the transducers, and noise in the first stage of amplification. As observed above, the trend nowadays is to reduce the noise dramatically, and hence improve sensitivity, by operating the system at liquid helium temperatures. It is also important to employ optimum filters for maximizing the output signal/noise ratio. The filtering problem associated with pulse-signal recovery from a sharply resonant, inherently noisy detector constitutes the main topic of discussion in this chapter. As a prelude, however, intended to introduce and exemplify the concepts involved, the relatively simple case of a non-resonant reactive transducer is considered with the aim of establishing the minimum pulse energy that such a sensor can detect.

13.3 MINIMUM DETECTABLE ENERGY

A schematic illustration of a non-resonant reactive transducer is shown in Fig. 13.2(a). Note that the circuit is a low-pass filter with time constant $T_1 = R_1 C_1$, corresponding to a cut-off frequency equal to $(2\pi T_1)^{-1}$. A voltage appears at the output of the transducer consisting of a signal component, $v_s(t)$, due to the input pulse, and a thermal noise component, $v_n(t)$, associated with the loss resistance of the transducer. In the absence of an input, the circuit is assumed to be in thermal equilibrium with its surroundings at temperature θ. There are two approaches — with distinctly different conclusions — to the question of the

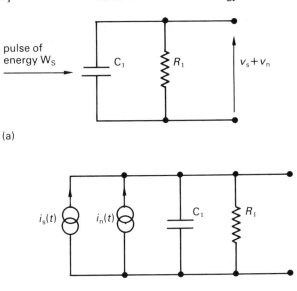

(a)

(b)

Fig. 13.2 — (a) Non-resonant reactive transducer with pulse input and (b) an equivalent circuit showing parallel signal and noise current generators.

minimum energy that can be detected with the circuit. The first argument goes as follows.

The signal/noise ratio at the output can be expressed in terms of the energy stored in the capacitor due to the input pulse. As the energy in the incident pulse is absorbed by the circuit, the voltage across the capacitor builds up to a peak value, v_p, at which point the stored energy due to the signal is

$$W_s = C_1 v_p^2 / 2 \ . \tag{13.1}$$

The mean energy stored in the capacitor due to the thermal noise in the resistor is

$$W_n = C_1 \overline{v_n^2} / 2 = k\theta / 2 \ , \tag{13.2}$$

where $\overline{v_n^2}$ is the mean-square value of the thermal noise voltage and the second equality follows from the equipartition requirement. On defining the signal/noise ratio as $v_p^2 / \overline{v_n^2}$, it follows immediately that

$$\text{signal/noise} = \frac{v_p^2}{\overline{v_n^2}} = \frac{2 W_s}{k\theta} \ , \tag{13.3}$$

from which we deduce that the minimum detectable energy in the pulse is $k\theta / 2$.

The above argument takes no account of the spectral shapes of the signal and the noise, and one is tempted to think that by judiciously enhancing signal components and rejecting noise by appropriate filtering at the output, an improvement in the minimum detectable energy could be achieved. However, if a *passive* filter at the same temperature as the transducer is employed for the purpose, no improvement can be realized because the filter is subject to the same thermodynamic laws as the transducer itself. Thus, the noise inherent in the filter will always ensure that the minimum detectable energy is no better than $k\theta/2$.

But if this thermodynamic constraint were removed, and thermal noise in the output circuit were eliminated or at least substantially reduced, the minimum detectable energy could be made less than $k\theta/2$. There are several ways of achieving this condition, one being to cool the output circuit below the ambient temperature, and another, involving a delay line, has been suggested by Maeder (1971). A third, and perhaps more appropriate, approach is to use an *active* filter (Faulkner and Buckingham, 1972), which is not constrained by the second law of thermodynamics. The active filter approach is appealing in its simplicity, combined with the fact that it allows almost any realizable rational system function to be implemented without introducing a significant amount of noise.

Assuming that an appropriate active filter is applied to the output terminals of the transducer, the argument for the minimum detectable energy takes a different form from that for the equilibrium case given above. Now it is necessary to take into account the spectral content of the signal and the noise and, as will become apparent, it is especially important to recognize the effect that the filtering action of the transducer has on the higher frequency components in the pulse.

An equivalent circuit for the transducer is shown in Fig. 13.2(b), in which the current generators $i_s(t)$ and $i_n(t)$ represent the pulse signal and thermal noise in R_1, respectively. The signal, which we take as a single, square pulse of duration $T_s \ll T_1$ and height i_m, is sketched in Fig. 13.3(a) and its power spectrum is shown in Fig. 13.3(b). Note that most of the energy in the pulse is contained in the (angular) frequency range below π/T_s. The total noise power within this bandwidth is

$$\overline{i_n^2} = \frac{2k\theta}{R_1 T_s} , \tag{13.4}$$

which is simply the Nyquist expression for the power spectral density of the thermal noise in R_1 multiplied by the (signal) bandwidth $(2T_s)^{-1}$. Now, when the short pulse condition $T_s \ll T_1$ is satisfied, the peak signal voltage across the capacitor is

$$v_p = i_m R_1 T_s/T_1 , \tag{13.5}$$

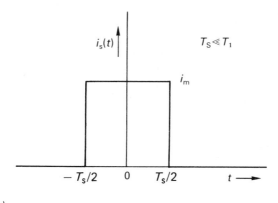

(a)

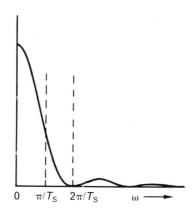

(b)

Fig. 13.3 – (a) The single pulse input and (b) its power spectrum (the units on the ordinate are arbitrary).

which from equation (13.1) gives a stored energy

$$W_s = i_m^2 R_1 T_s^2 / 2T_1 \ . \tag{13.6}$$

Thus, on defining the signal/noise ratio as $i_m^2 / \overline{i_n^2}$, we have

$$\text{signal/noise} = \frac{i_m^2}{\overline{i_n^2}} = \frac{W_s}{k\theta} \frac{T_1}{T_s} \ . \tag{13.7}$$

According to this result, the minimum detectable energy is $(T_s/T_1)k\theta$, which is smaller than the value of $k\theta/2$ predicted from equation (13.3) by a factor of $2T_s/T_1$.

The reason for the improvement is as follows. The bandwidth of the signal is very much broader than the bandwidth of the transducer, which acts as a low-pass filter as may be seen from the equivalent circuits in Fig. 13.2. Therefore the high frequency components of the signal are attenuated by the transducer, which has the effect of reducing the peak voltage across the capacitor. As may be seen from equation (13.5), the peak-square voltage falls as the *square* of the bandwidth. The noise is also reduced by the low-pass filtering, but the noise power scales in *proportion* to the bandwidth. There is thus a net degradation in the signal/noise ratio when the high frequency components of the signal and noise are removed. This corresponds to the situation in the first argument given above, leading to equation (13.3). When the high frequency components of the signal and noise are retained, a substantial increase in signal/noise ratio can be achieved, as indicated by the second analysis, culminating in equation (13.7).

Clearly, in order to realize the improvement in signal/noise ratio predicted in equation (13.7), it is necessary to broaden the bandwidth of the output circuit without introducing a significant amount of noise. Assuming that this is accomplished with an active filter, as suggested above, the system may be represented as shown in Fig. 13.4. The system function $H_T(s)$, where s is the complex angular frequency, is the current-to-voltage transfer function for the RC combination representing the transducer,

$$H_T(s) = \frac{R_1}{(1 + sT_1)} \, , \tag{13.8}$$

and $H_A(s)$ is the system function of the active filter connected to the output of the transducer. For the moment we shall assume that the active filter is noiseless,

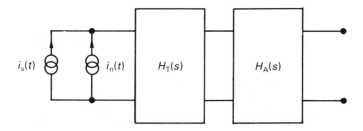

Fig. 13.4 – System diagram of the reactive transducer and active filter, with system functions $H_T(s)$ and $H_A(s)$, respectively.

but later, in dealing with gravitational antennas, noise from the output electronics will be included in the analysis. A suitable form for $H_A(s)$ is

$$H_A(s) = \frac{(1 + sT_1)}{(1 + sT_2)} \, , \tag{13.9}$$

where $T_1 \gg T_2 \simeq T_s$. (Strictly, the function in equation (13.9) cannot be realized since it has infinite bandwidth. However, this difficulty is alleviated by a minor modification to the form of $H_A(s)$ given here, which has little effect on the present argument). The overall system function is the product of equations (13.8) and (13.9):

$$H(s) = H_T(s) H_A(s)$$

$$= \frac{R_1}{(1 + sT_2)} \quad , \tag{13.10}$$

and thus the complete system acts as a low-pass filter, but with a cut-off frequency much higher than that of the transducer alone. This increase in bandwidth is exactly the condition that had to be achieved.

Evidently, an active filter used in conjunction with the non-resonant transducer can produce a substantial enhancement in the signal/noise ratio. The technique also has application in connection with sharply resonant gravitational-wave detectors, where similar signal/noise gains can be made.

13.4 SIGNAL RECOVERY FROM A GRAVITATIONAL ANTENNA

Both the split bar and the Weber bar are highly resonant mechanical systems which are coupled to piezoelectric ceramic transducers whose electrical characteristics may be represented by a parallel RC combination. In order to discuss the response of the antenna to a gravitational wave, an equivalent circuit must be developed for the complete electro-mechanical system.

13.4.1 Equivalent circuit of a resonant-mass antenna

An idealized representation of a Weber bar or split bar consists of a pair of masses connected by a spring, as illustrated in Fig. 13.5. This system shows a single resonance frequency which may be identified with the longitudinal mode of oscillation of an actual antenna. The equivalent circuit of the idealized detector which is appropriate to the present discussion is a series LCR combination. When the transducers are coupled into this circuit, the complete equivalent circuit of the gravitational-wave antenna is as shown in Fig. 13.6. This circuit was originally published by Weber (1969) and has since been discussed in a general account of gravitational-wave detection by Fellgett and Sciama (1971). A detailed analysis of the circuit, incorporating a discussion of active filters to enhance the signal/noise ratio, has been given by Buckingham and Faulkner (1972), whose treatment forms the basis of the argument outlined below.

In the equivalent circuit of Fig. 13.6, the circuit elements with subscript B represent the bar, $v_s(t)$ represents the signal coupled into the bar due to the incident pulse of gravitational energy, and C_T, G_T are the *actual* capacitance and

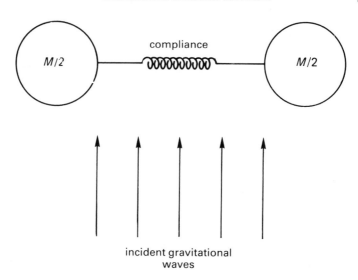

Fig. 13.5 – Idealized resonant-mass detector.

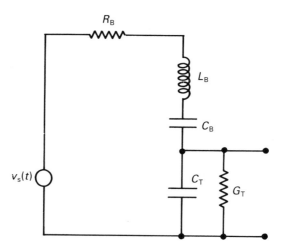

Fig. 13.6 – Equivalent circuit of a resonant-mass gravitational-wave antenna.

conductance of the piezoelectric transducer. The electromechanical coupling between the bar and the transducer is represented by the parameter β, defined as

$$\beta = \frac{\text{electrical energy stored in } C_T \text{ (open circuit)}}{\text{elastic energy stored in the mechanical system}}$$

at any instant. Over the frequency range of interest, the loss angle of a typical ceramic (lead zirconate titanate or barium titanate) transducer is very small, and thus the coupling constant is given by the ratio

$$\beta = C_B/C_T < 1 . \qquad (13.11)$$

The actual value of β depends on the properties of the piezoelectric transducer and the degree of electromechanical coupling between it and the bar, but even under optimum conditions it is not possible to achieve values of β greater than about 0.3. In the split bar, the electromechanical coupling is so tight that β takes a value close to the theoretical maximum, in the region of 0.1, but in the Weber bar, where the transducers are only loosely coupled to the mechanical oscillations, β is several orders of magnitude less than this.

In order to complete the equivalent circuit in Fig. 13.6, the circuit elements representing the bar must be expressed in terms of the parameters of the bar/transducer system, and the voltage signal generator must be calibrated in terms of an input force, $f_s(t)$, applied externally to the bar. If ω_0 is the angular resonance frequency of the bar and Q_B is the mechanical Q-factor with the transducer terminals short-circuited, then

$$C_B = \beta C_T$$
$$L_B = (\omega_0^2 \beta C_T)^{-1} \qquad (13.12)$$
$$R_B = (\omega_0 \beta Q_B C_T)^{-1} .$$

It remains to find the scaling factor between the input force, $f_s(t)$, and the equivalent voltage generator $v_s(t)$. This is achieved by equating the energies imparted to the respective systems in the two cases. Thus, an impulse of momentum $p = \int f_s(t)dt$ imparts to the masses in Fig. 13.5 an energy $W_M = p^2/2M$; and similarly, an impulse of flux $\phi = \int v_s(t)dt$ imparts to the inductor in the equivalent circuit an energy $W_L = \phi^2/2L_B$. On equating W_L to W_M the scaling factor between voltage and force is found to be

$$v_s(t) = f_s(t) \sqrt{\frac{L_B}{M}} = \frac{f_s(t)}{\omega_0 \sqrt{\beta C_T M}} . \qquad (13.13)$$

Assuming that β, Q_B and ω_0 are known from simple measurements performed on the bar, equations (13.12) and (13.13) are sufficient to completely specify the equivalent circuit of Fig. 13.6, and through them an absolute theoretical limit can be set on the sensitivity of the antenna.

13.4.2 The gravitational pulse

At the time when Weber was beginning his search the characteristics of gravitational signals were unknown, and this still holds true today. Until such time as gravitational radiation is detected unequivocably, the detailed structure of the waveform will remain obscure. However, theoretical arguments suggest that extra-terrestrial sources, such as a stellar collapse, produce double pulses of gravitational radiation about 1 millisecond long, with zero total area. An example of such a wave is illustrated in Fig. 13.7(a), which shows a square pulse like that in Fig. 13.3(a) followed by a similar pulse of opposite polarity. This waveform provides an adequate basis for discussion, since the detailed shape of the pulse is not a critical factor in the argument.

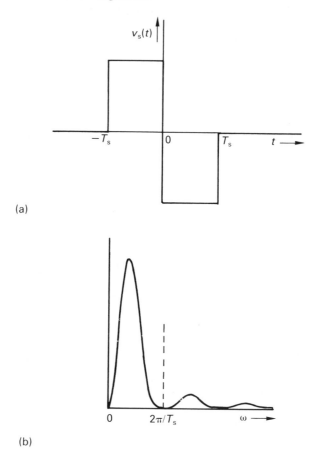

(a)

(b)

Fig. 13.7 – (a) A square double pulse with zero total area and (b) its power spectrum (the units on the ordinate are arbitrary).

The power spectrum of the double pulse is shown in Fig. 13.7(b). Note that the spectrum shows a peak which is displaced from the origin, falling at a frequency approximately equal to $1/2T_s$. It is clear from this spectral shape that by broadening the bandwidth of the output circuit, as described in § 13.3, much the same improvement in signal/noise ratio can be achieved in this case as in the previous case of the non-resonant transducer and the single pulse. But this is not an optimum procedure: because the spectrum of the double pulse peaks at a non-zero frequency, it is possible to achieve an even better signal/noise ratio by connecting the output of the transducer to an active filter whose transfer function is such that the complete detection system (bar included) shows a bandpass response. This argument is developed below.

13.4.3 Noise generators in the equivalent circuit

Figure 13.8(a) shows the equivalent circuit of the resonant-mass antenna, including the thermal noise generators associated with Brownian motion in the bar and losses in the transducer. By the use of Thévenin's theorem, the noise current generator, $i_{nT}(t)$, can be referred to the input, as shown in Fig. 13.8(b). The spectral densities of the two input noise voltage generators are

$$\overline{S_{v_{nB}}(\omega)} = 4k\theta R_B \tag{13.14}$$

and

$$\overline{S_{v_{nT}}(\omega)} = 4k\theta G_T |Z_B|^2 , \tag{13.15}$$

where Z_B is the impedance of the $L_B C_B R_B$ series combination:

$$Z_B = R_B + j \left(\omega L_B - \frac{1}{\omega C_B}\right)$$

$$\simeq R_B + 2j L_B (\omega - \omega_0) \tag{13.16}$$

The approximation here is valid in the region around resonance.

The need for an active filter giving an overall bandpass response can be appreciated when the approximate expression in equation (13.16) is substituted into equation (13.15), to give

$$\overline{S_{v_{nT}}(\omega)} = 4k\theta G_T \{R_B^2 + 4L_B^2 (\omega - \omega_0)^2\} . \tag{13.17}$$

When this is integrated over frequency, it is clear that the inductive term gives a contribution to the noise which increases as the *cube* of the bandwidth. Since the peak-square value of the signal varies as the *square* of the bandwidth (as in the previous case of a single pulse and a non-resonant transducer), there will come a point, on increasing the width of the passband, beyond which the signal/noise ratio begins to fall. There is therefore an optimum bandwidth for the overall system which maximizes the signal/noise ratio.

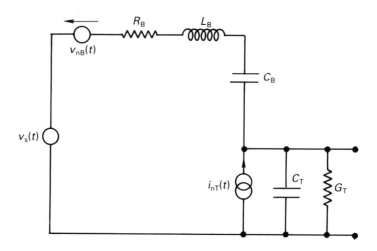

(a)

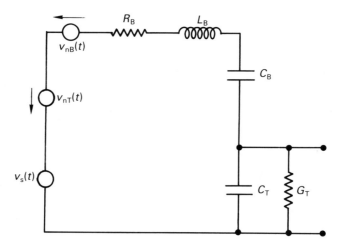

(b)

Fig. 13.8 – (a) Equivalent circuit of the antenna showing noise generators $v_{nB}(t)$ and $i_{nT}(t)$, representing Brownian motion in the bar and noise in the transducer, respectively, and (b) its Thévenin equivalent circuit with both noise generators referred to the input.

13.4.4 The optimum filter

The input impedance of the filter connected to the output of the transducer can of course take any value, but there is an advantage to be had in selecting a charge amplifier with zero input impedance. This would be a natural choice anyway for a piezoelectric sensor, and in the present case, as may be seen from the circuits in Fig. 13.8, it short-circuits the transducer and thus prevents the noise peaking at a different frequency from the signal. The noise in the amplifier can be represented in the usual way by a series voltage generator and a parallel current generator at the input, whose spectral densities can be expressed in terms of equivalent noise conductances which are in general functions of frequency. However, in a practical charge amplifier over the frequency range of interest, the series voltage generator is negligible and the equivalent noise conductance, G_{nA}, of the parallel noise current generator is independent of frequency. The effect of this noise current generator on the overall signal/noise ratio is included in the analysis by referring it, using Thévenin's theorem, to the input side of the circuit as a series voltage generator, $v_{nA}(t)$, whose power spectral density is

$$\overline{S_{v_{nA}}(\omega)} = 4k\theta\, G_{nA}\,|Z_B|^2 \;. \tag{13.18}$$

This argument is analogous to the one that was applied above to the noise current generator $i_{nT}(t)$ representing noise in the transducer.

The complete equivalent circuit of the antenna, including all the noise generators and the charge amplifier, is shown in Fig. 13.9 as a systems diagram. The overall system function is a dimensionless voltage-to-voltage ratio, given by the product

$$H(s) = H_B(s)\,H_A(s) \;, \tag{13.19}$$

where $H_B(s) = 1/Z_B(s)$ is the voltage-to-current transfer function of the bar and $H_A(s)$ is the current-to-voltage transfer function of the amplifier. We are now at a stage where the overall system function, $H(s)$, must be specified. It has already

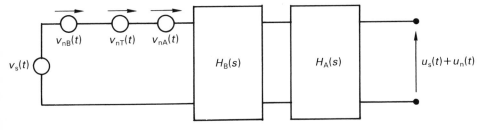

Fig. 13.9 — Systems diagram of the antenna showing the three noise generators associated with the bar, the transducer and the amplifier; and the signal and noise output voltages $u_s(t)$ and $u_n(t)$.

been argued that for optimum performance $H(s)$ should be a bandpass function with a bandwidth sufficiently broad to include all the significant frequency components of the signal. The bandwidth of the (short-pulse) signal is very broad compared with the bandwidth of the bar, and so one function of the active filter must be to broaden the bandwidth of the overall system. There is an analogy here with the non-resonant transducer, where the active filter extended the pass band in order to avoid rejection of the high frequency components in the signal.

A suitable form for $H(s)$ is the fourth-order bandpass function

$$H(s) = \frac{A \omega_0^2 s^2}{Q_m^2 \left(s^2 + \frac{s\omega_0}{Q_m} + \omega_0^2\right)^2} \, , \tag{13.20}$$

where A is a constant and $Q_m < Q_B$ is a measure of the overall bandwidth, whose optimum value is to be determined. According to equation (13.19), the system function, $H_A(s)$, of the active filter which gives this overall response is the ratio $H(s)/H_B(s) = H(s)Z_B(s)$, and hence from equations (13.16) and (13.20) it follows that

$$H_A(s) = \frac{A R_B Q_B \left(s^2 + \frac{s\omega_0}{Q_B} + \omega_0^2\right) s\omega_0}{Q_m^2 \left(s^2 + \frac{s\omega_0}{Q_m} + \omega_0^2\right)^2} \, . \tag{13.21}$$

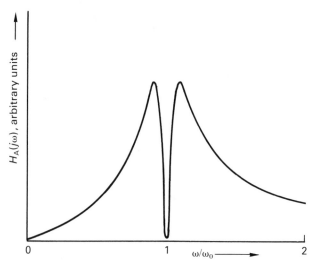

Fig. 13.10 – Frequency dependence of the function $|H_A(j\omega)|$ derived from equation (13.21), with $Q_m = 5.5$ and $Q_B = 1000$. These Q-factors are appropriate to the split bar.

A possible scheme for implementing a filter with this system function has been proposed by Buckingham and Faulkner (1972). The filter shows a deep, narrow notch in its frequency response at the resonance frequency of the bar, as sketched in Fig. 13.10. Thus, the noise from the antenna, which is concentrated in a narrow band around the frequency of the mechanical resonance, is strongly attenuated by the filter, but the spectral components of the relatively broadband signal fall mainly within the passbands on either side of the notch.

The purpose of the filter is to enhance the output signal/noise ratio, the optimum case being the one in which the signal/noise ratio is a maximum. It remains for us now to determine the value of Q_m which gives this condition.

13.4.5 The signal/noise ratio

The systems diagram in Fig. 13.9 can be represented in the alternative form shown in Fig. 13.11, where $H(s)$ is the transfer function given in equation (13.20). The mean-square value of the noise output from $H(s)$ is given by the integral

$$\overline{u_n^2} = \frac{1}{2\pi} \int_0^\infty [\overline{S_{v_{nB}}(\omega)} + \overline{S_{v_{nT}}(\omega)} + \overline{S_{v_{nA}}(\omega)}] \, |H(j\omega)|^2 \, d\omega \ ,$$

(13.22)

where we have set $s = j\omega$, and the power spectral densities are given by equations (13.14), (13.15) and (13.18). After some fairly straightforward manipulation, this evaluates to

$$\overline{u_n^2} = \frac{A^2 k\theta}{2\beta C_T Q_m} \left[\frac{1}{Q_B} + \frac{1}{\beta Q_B^2 Q_T} + \frac{3}{\beta Q_T Q_m^2} \right] \ ,$$

(13.23)

where Q_T is defined by the relation

$$Q_T = \frac{\omega_0 \, C_T}{(G_T + G_{nA})} \ .$$

(13.24)

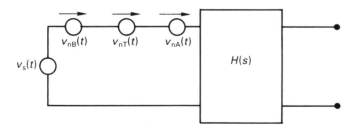

Fig. 13.11 – Equivalent circuit of the complete detection system.

Note that, apart from ω_0, the parameters in this expression relate exclusively to the transducer and the amplifier, not to the bar.

Although the double pulse with its spectral peak shifted from the origin provided the basis for deducing that $H(s)$ should be a bandpass function, we may calculate the peak value of the signal at the output, u_{sp}, by considering a single step function input of magnitude v_m. Now, the impulse response of the system is the inverse transform of the function $H(s)$ in equation (13.20), and therefore the step response is the inverse transform of the function $(1/s)H(s)$. This is an oscillatory signal, with frequency equal to the resonance frequency of the bar, whose envelope peaks at a time $\Delta t = 2Q_m/\omega_0$ after the step is applied to the input. The peak value is

$$u_{sp} \simeq A v_m/Q_m e \ , \tag{13.25}$$

corresponding to a peak energy stored in the bar equal to

$$W_s = \beta C_T v_m^2/2 \ . \tag{13.26}$$

The signal/noise ratio at the output is the peak-square signal divided by the mean-square noise, which from equations (13.23), (13.25) and (13.26) can be expressed as

$$\text{signal/noise} = \frac{u_{sp}^2}{\overline{u_n^2}} = \frac{4\,W_s}{k\theta\,e^2} \left\{ \frac{Q_m}{Q_B} + \frac{3}{\beta Q_m Q_T} \right\}^{-1} , \tag{13.27}$$

where the condition $\beta Q_T Q_B \gg 1$ – which applies to both the Weber bar and the split bar – has been employed in order to simplify the result. The optimum filter is determined by maximizing this expression for the signal/noise ratio with respect to Q_m. The optimum value of Q_m is easily found to be

$$Q_{m\,\text{opt}} = \sqrt{3 Q_B/\beta Q_T} \ , \tag{13.28a}$$

giving

$$(\text{signal/noise})_{\max} = \frac{2\,W_s}{k\theta\,e^2} \sqrt{\frac{\beta Q_T Q_B}{3}} \tag{13.28b}$$

corresponding to a

$$\text{minimum detectable energy} = \left(\frac{k\theta}{2} \right) \frac{e^2 \sqrt{3}}{\sqrt{\beta Q_T Q_B}} \ . \tag{13.28c}$$

Taken in conjunction with the expression for the time resolution of the optimized system, namely

$$\Delta t_{opt} = \frac{2 Q_{m_{opt}}}{\omega_0} , \tag{13.29}$$

equations (13.28) completely specify the signal-detection properties of the optimized antenna.

13.5 COMPARISON OF THE WEBER BAR WITH THE SPLIT BAR

In order to evaluate the expressions in equations (13.28), a numerical value must be assigned to Q_T. For the purpose of the comparison, we take a value of $Q_T = 10^3$, which should be achievable with ceramic transducers and a well-designed amplifier. Approximate values for β and Q_B are shown in Table 13.1, which also includes the optimal value of Q_m, the minimum detectable energy and the time resolution of the Weber bar and the split bar when each is working in an optimized condition.

It is apparent from the table that when each detector is optimized, the minimum detectable energy with the split bar is a factor of 10 better than that with the Weber bar. This is because the tighter electromechanical coupling in the split bar more than compensates for the degradation in performance due to the higher losses in this system. The theoretical performance of the split bar predicted from equations (13.28) has been confirmed experimentally by Drever et al. (1973).

Table 13.1 shows another aspect of performance where the Weber bar and the split bar differ significantly. This is in the time resolution of the two systems. With the optimum value of Q_m shown in the table, the time of arrival of a pulse at the Weber bar cannot be specified to an accuracy of better than 1 second, which compares unfavourably with the value of 1 millisecond for the optimized split bar. In coincident counting experiments involving pairs of detectors, as

Table 13.1

Comparative performance of the optimized detection systems.

	$\omega_0/2\pi$, Hz	Q_B	β	Q_T	$Q_{m\,opt}$	min. detect. energy	Δt_{opt}, sec
Weber bar	1660	10^5	10^{-5}	10^3	5480	$0.2k\theta$	1
Split bar	1660	10^3	10^{-1}	10^3	5.48	$0.02k\theta$	10^{-3}

conducted by Weber, a time resolution of the order of milliseconds is highly desirable since this is the timescale of a pulse transit over the 1000 km baseline between the detectors. In order to achieve this with the Weber bar, the value of Q_m must be reduced from the optimum value of 5500 by a factor of 1000. Under this conditon, however, which is far removed from the optimum, the signal/noise ratio is reduced by a factor of 500 which, in terms of the minimum detectable energy, means that the Weber bar shows a penalty of 5000 compared with the split bar.

13.6 THE CURRENT STATUS OF GRAVITATIONAL-WAVE DETECTION

Soon after Weber's original claims were reported, a number of investigators constructed gravitational-wave antennas which were similar to but not identical with the Weber bar (e.g. Braginskii et al. 1972; Tyson, 1973; Garwin and Levine, 1973) or were of the split bar type (Aplin, 1972; Drever et al., 1973, Allen and Christodoulides, 1975). Several of the experiments had a higher sensitivity than Weber's bars, and some involved a single detector rather than two or more detectors working in coincidence. (A single detector is sufficient to prove the *absence* of gravitational pulses above some specified threshold level. In order to prove the *presence* of such pulses, two detectors and a coincident counting technique are necessary to eliminate false alarms due to spurious local interference.) Negative or at best marginal results were obtained. A review of these experimental investigations has been given by Drever (1977).

By 1975 the intense effort to detect gravitational pulses with the first generation detectors (i.e. room temperature, aluminium bars) was essentially complete. It became clear that if gravity waves were to be detected a second generation of detectors, with higher sensitivities, would have to be developed. In view of the analysis given above, demonstrating the advantages of the split bar over the Weber bar, one might have supposed that the split bar would have been the basis for further development, but in fact this was not the case. The main reason for this is that the configuration of the transducers in the split bar, combined with its intrinsic mechanical complexity, pose problems in connection with cryogenic operation which are almost insuperable.

The second-generation detectors span a wide range of frequency and fall conveniently into three categories: firstly, there are cryogenic variants of the Weber bar made from niobium and single-crystal sapphire as well as aluminium, cooled with liquid helium to temperatures between approximately 1 K and 4 K, and suitable for frequencies around 1 kHz and above; the second approach is appropriate to low frequencies, between 10^{-4} Hz and 10^{-2} Hz, and involves Doppler tracking of interplanetary spacecraft; and for intermediate frequencies, between the millihertz and kilohertz bands, room temperature laser interferometry is being exploited. Recent progress in each of these fields is outlined below.

13.6.1 Cryogenic resonant-mass detectors

The obvious advantage to be gained from low-temperature operation of a resonant-mass antenna is that thermal noise in the bar is dramatically reduced. In such a system, the factor which limits performance may well be noise in the transducers or the output electronics. Piezoelectric ceramic transducers of the sort used in the Weber bar and the split bar are unsuitable for a cryogenic system, a fact which has led to the development of very novel, quiet sensors specifically for this application. These new transducers are incompatible with the split bar, which is one reason why the existing low-temperature gravitational antennae are all basically of the Weber type. There is, nevertheless, a wide variation in their detailed design.

At Stanford a massive aluminium cylinder weighing 4800 kg is reported as being in almost continuous use at a temperature of 4.2 K (Michelson and Tabor, 1981). The signals are recovered from the antenna via an inductively coupled, superconducting transducer mounted on an end-face of the bar, whose output is fed into a SQUID (superconducting quantum interference device) amplifier. Thus, in this case, the low electrical and mechanical losses of superconductors are exploited in an effort to minimize noise in the sensors. The mechanical Q-factor of the unloaded bar is $Q_B = 2 \times 10^6$, and the minimum detectable energy of the system is reported to be 3×10^{-25} J. This is a factor of approximately 300 better than the performance of the split bar given in Table 13.1, an improvement which arises primarily from the reduction of noise due to the cooling. (The electro-mechanical coupling with the superconducting transducer is not significantly better than that in the original Weber bar).

Detectors similar to the one at Stanford are also being developed by other groups around the world, notably at the Louisiana State University and at Frascati near Rome. The sensor system used by the Louisiana group differs from that of the Stanford group in that it employs the acceleration of the end faces of the bar to modulate two resonant radio frequency cavities which form part of a superconducting parametric amplifier.

A group at Perth is constructing an antenna of niobium which will be supported superconductively. The mass of the bar is a few hundred kilograms, with a mechanical Q-factor of approximately 10^8 at the operating temperature of 4 K. Another exotic idea, which has been proposed by V. B. Braginskii from Moscow (for references see the review by Drever, 1977), is to use a relatively small bar, with a mass in the region of 10 to 100 kg, made from a single crystal of sapphire. The losses in certain single crystal materials are extremely low, and in sapphire a Q-factor of approximately 10^{13} is predicted theoretically. As we have seen in § 13.4.5, the minimum detectable energy of a resonant-mass antenna improves as the mechanical Q-factor of the bar increases. Spurred on by this knowledge, the Moscow group has already produced sapphire crystals with Q values on the order of 10^9. One of the main difficulties they face with such a high-Q system is extracting the signals without significantly damping the bar.

They propose to use a capacitive sensing system, one component of which will be a superconducting coating over the end faces of the crystal.

13.6.2 Spacecraft tracking

Doppler tracking of interplanetary spacecraft is a possible means of detecting low-frequency gravitational waves, between 10^{-4} Hz and 10^{-2} Hz. It is based on the idea that a gravity wave produces a small relative motion between the earth and a distant space vehicle, which in turn produces a fluctuation in the Doppler shift of tracking signals. The first serious attempts to exploit this effect are due to begin in May 1983, in the form of experiments on board two spacecraft, one American and one European, both of which are destined to pass over the Sun's pole. Not unexpectedly, a serious difficulty that besets the scheme is noise in the Doppler tracking system. This arises from a number of sources, including random variations of the refractive index of the solar wind along the path of the tracking signal, and similar variations in the Earth's troposphere. It may be possible to minimize the effects of these noise sources if they can be monitored sufficiently accurately.

13.6.3 Room-temperature laser interferometry

For the intermediate range of frequencies, room-temperature laser interferometry appears to show some promise. Forward (1978) has already built a gravitational-wave detector employing the technique, which is sensitive to frequencies in the region of 100 Hz. The basis of his experiment was a modified Michelson interferometer, with two optical paths at right angles to each other and three masses, one at the intersection of the optical paths and the other two at the ends of the paths. The separation of the masses was about 3 m. Forward and his group were able to detect relative movements of the masses of approximately 10^{-15} m, which is a factor of 10 worse than the Weber bar, but considering the relative simplicity of this prototype system the results are felt to be encouraging.

Like other gravitational antennae, noise is a fundamental limiting factor in the optical system. In this case, fluctuations in the number of photons emitted during the period of a gravitational pulse sets an absolute limit on the signal/noise ratio that can be expected. A discussion of this and other aspects of the laser system has been given by Drever (1977) and Thorne (1980).

13.7 CONCLUDING REMARKS

Gravitational-wave detection is now a well-established branch of astronomy, and one which is developing rapidly. In a sense this is ironic, since an unequivocal observation of a gravity wave has yet to be made. However, this has not impeded the production of a second generation of gravitational-wave antennae, with sensitivities – in the case of the cryogenic bars – two to three orders of magnitude higher than that of the original Weber bar. What is more, ideas for a third generation of detectors with sensitivities higher by about the same factor again have already been formulated by a number of investigators in the field. One of the proposals in this connection is to reduce the operating temperature of resonant-mass detectors yet further, to the region of 50 millikelvins.

This continuing activity suggests that in the not too distant future many exciting discoveries may be made which could add immensely to our understanding of the universe. If, for example, an optical and a gravitational signal were to be identified with a common source, it would be possible to measure directly the speed of propagation of gravitational radiation, which in turn would provide a test for the various theories of gravitation. Other tests would be facilitated if the polarization of gravitational waves were determined. Of course, in order to reach the stage where these possibilities could be realized, a tremendous investment in technological ingenuity is required. And perhaps it is as well to temper any undue optimism with the reminder that success in the search for gravity waves is by no means assured, despite the fact that the potential rewards are high.

REFERENCES

W. D. Allen and C. Christodoulides (1975), Gravitational radiation experiments at the University of Reading and the Rutherford Laboratory, *J. Phys. A.*, **8**, 1726–1733.

P. S. Aplin (1972), Gravitational radiation experiments, *Contemp. Phys.*, **13**, 283–293.

P. Bonifazi, F. Bordoni, G. V. Pallottino and G. Pizzella (1981), Measurements of the Brownian noise of a harmonic oscillator with mass $M = 389$ kg, *Proc. 6th Int. Conf. on Noise in Physical Systems held at the National Bureau of Standards, Gaithersburg, MD, USA, 6–10 April 1981*, pp. 298–301.

V. B. Braginskii, A. B. Manukin, E. I. Popov, V. N. Rudenko and A. A. Khorev (1972), Search for gravitational radiation of extraterrestrial origin, *Sov. Phys. – JETP Lett.*, **16**, 108–112.

C. Brans and R. H. Dicke (1961), Mach's principle and a relativistic theory of gravitation, *Phys. Rev.*, **124**, 925–935.

M. J. Buckingham and E. A. Faulkner (1972), The principles of pulse signal recovery from gravitational antennas, *Radio and Elect. Eng.*, **42**, 163–171.

R. W. P. Drever (1971), Observations on pulse response of a wide-band gravitational wave detector, presented at the 6th Int. Conf. on Gravitation and Relativity, Copenhagen, 5–10 July, 1971.

R. W. P. Drever (1977), Gravitational wave astronomy, *Quarterly J. Roy. Astron. Soc.*, **18**, 9–27.

R. W. P. Drever, J. Hough, R. Bland and G. W. Lessnoff (1973), Search for short bursts of gravitational radiation, *Nature*, **246**, 340–344.

A. Einstein (1916), Grundlage der allgemeinen relativitätstheorie (Foundation of the theory of general relativity), *Ann. d. Phys.*, **49**, 769–822.

E. A. Faulkner and M. J. Buckingham (1972), Comment on 'Can a pulse excitation smaller than kT be detected?', *Elect. Lett.*, **8**, 152–153.

P. B. Fellgett and D. W. Sciama (1971), Gravitational wave astronomy: an interim survey, *Radio and Elect. Eng.*, **42**, 391–397.

R. L. Forward (1978), Wideband laser-interferometer gravitational-radiation experiment, *Phys. Rev.*, **D 17**, 379–390.

R. L. Garwin and J. L. Levine (1973), Absence of gravity-wave signals in a bar at 1695 Hz, *Phys. Rev. Lett.*, **31**, 173–176.

G. W. Gibbons and S. W. Hawking (1971), Theory of the detection of short bursts of gravitational radiation, *Phys. Rev.*, **4**, 2191–2197.

R. P. Gifford, P. F. Michelson and R. C. Tabor (1981), Noise in resonant gravitational wave detectors, *Proc. 6th Int. Conf. on Noise in Physical Systems held at the National Bureau of Standards, Gaithersburg, MD, USA, 6–10 April 1981*, pp. 292–297.

S. W. Hawking (1972), Gravitational radiation: the theoretical aspect, *Contemp. Phys.*, **13**, 273–282.

D. G. Maeder (1971), Can a pulse excitation smaller than kT be detected?, *Elect. Lett.*, **7**, 767–769.

P. F. Michelson and R. C. Tabor (1981), Sensitivity analysis of a resonant-mass gravitational wave antenna with resonant transducer, *J. Appl. Phys.*, **52**, 4313–4319.

K. S. Thorne (1980), Gravitational-wave research: current status and future prospects, *Rev. Mod. Phys.*, **52**, 285–297.

J. A. Tyson (1973), Null search for bursts of gravitational radiation, *Phys. Rev. Lett.*, **31**, 326–329.

J. A. Tyson and R. P. Gifford (1978), Gravitational wave astronomy, *Ann. Rev. Astron. Astrophys.*, **16**, 521–554.

J. Weber (1960), Detection and generation of gravitational waves, *Phys. Rev.*, **117**, 306–313.

J. Weber (1966), Observation of the thermal fluctuations of a gravitational-wave detector, *Phys. Rev. Lett.*, **17**, 1228–1230.

J. Weber (1967), Gravitational radiation, *Phys. Rev. Lett.*, **18**, 498–501.

J. Weber (1969), Evidence for the discovery of gravitational radiation, *Phys. Rev. Lett.*, **22**, 1320–1324.

J. Weber (1970), Gravitational radiation experiments, *Phys. Rev. Lett.*, **24**, 276–279; Anisotropy and polarization in the gravitational-radiation experiments, *Phys. Rev. Lett.*, **25**, 180–184.

Appendices 1 – 7

APPENDIX 1 – THERMAL NOISE AND THE POISSON DISTRIBUTION

The noise processes encountered in solid state devices and elsewhere can often be represented as random pulse trains. This is true of thermal noise and shot noise, for example, as discussed in Chapter 2. In order to investigate the statistical properties of such processes, we shall take for definiteness the model of thermal noise described in § 2.8.

Thermal noise in a resistive material originates in the thermally induced random motion of the electrons within the device. The noise at the terminals may be represented as a random pulse train in which each pulse is associated with an 'event', constituted by an initial action, due to the transit of an electron between successive collisions with atoms of the host lattice, and a subsequent relaxation which restores the system to equilibrium. From this description, it is apparent that each electron in the assembly produces a random pulse train at the terminals as it moves through the material; and the observed noise waveform is the sum of the pulse trains associated with the total population of electrons in the device.

Let us imagine that we can follow the progress of a particular electron, say the ith, in its meanderings through the lattice. Each collision it experiences marks the onset of an 'event', and the collisions occur at random. In this context, 'random' means that the probability of finding the onset of an event occurring between t and $t + \delta t$ is very small and independent of t; and 'very small' means that the probability of finding two events within δt is negligible. If the probability of finding the onset of an event in the interval δt is $\nu_i \delta t$, where ν_i is independent of t, then the required conditions for randomness are satisfied by allowing δt to become infinitesimal. The constant ν_i may then be regarded as a probability density function; and as we shall now show, it features prominently in the Poisson distribution.

The Poisson distribution is the statistical law giving the probability, $p_i(m, t)$, of *exactly* m events (i.e., collisions in the case of our electron) occurring in an interval t, when the events themselves are randomly distributed in the sense described above. The subscript i is used here as a reminder that we are discussing a single carrier. The following simple argument leads to a differential equation from which a general expression for $p_i(m, t)$ can be deduced.

There are only two possible ways in which exactly m events can occur in the time interval $(t + \delta t)$: either m events occur in time t and none in δt or $(m - 1)$ occur in t and one occurs in δt. Therefore the probability of m events in time $(t + \delta t)$ is the sum

$$p_i(m, t + \delta t) = p_i(m, t)(1 - \nu_i \delta t) + p_i(m - 1, t)\nu_i \, \delta t \; ; \tag{A1.1}$$

and of course $\dot{p}_i(m, t)$ is zero for $m < 0$. Now, since the variable t is continuous, it is permissible to form the derivative of $p_i(m, t)$ with respect to time as follows:

$$\frac{dp_i(m, t)}{dt} = \lim_{\delta t \to 0} \frac{p_i(m, t + \delta t) - p_i(m, t)}{\delta t} , \tag{A1.2}$$

which, on being combined with the previous expression, leads to the differential equation

$$\frac{dp_i(m, t)}{dt} = -\nu_i \, p_i(m, t) + \nu_i \, p_i(m - 1, t) . \tag{A1.3}$$

When $m = 0$, this reduces to the form

$$\frac{dp_i(0, t)}{dt} = -\nu_i \, p_i(0, t) \tag{A1.4}$$

which has the solution

$$p_i(0, t) = \exp -\nu_i t , \tag{A1.5}$$

where the multiplicative constant of integration is unity, consistent with the fact that in zero time the probability of no events is unity. When $m > 0$ we look for solutions of the form

$$p_i(m, t) = f_i(m, t) \exp -\nu_i t . \tag{A1.6}$$

On substituting this expression into equation (A1.3), the function $f_i(m, t)$ is found to satisfy the differential equation

$$\frac{df_i(m, t)}{dt} = \nu_i f_i(m - 1, t) , \tag{A1.7}$$

whose solutions are of the form

$$f_i(m, t) = \alpha_m \, t^m . \tag{A1.8}$$

Bearing in mind that $f_i(0,t) = 1$, the constant of integration here is easily shown to be $\alpha_m = v_i^m/m!$, and hence it follows that the Poisson distribution is

$$p_i(m, t) = \frac{(v_i t)^m}{m!} \exp - v_i t \ . \tag{A1.9}$$

Note that when $p_i(m, t)$ is summed over all m, the result is independent of $v_i t$ and equal to unity.

The significance of the constant v_i becomes apparent when the first moment about the origin (i.e. the mean) of the distribution is formed. Denoting this by $\overline{m}$, we have

$$\overline{m} = \sum_{m=0}^{\infty} m\, p_i(m, t) = v_i t \ . \tag{A1.10}$$

Thus v_i is equal to the mean rate at which the events occur.

The variance of the distribution is the second moment about the mean:

$$\sigma^2 = \overline{(m - \overline{m})^2} = \overline{m^2} - \overline{m}^2 = \sum_{m=0}^{\infty} m^2\, p_i(m, t) - \overline{m}^2 \ . \tag{A1.11}$$

After some straightforward algebraic manipulation, the summation here can be expressed in terms of the exponential function $e^{-v_i t}$, and the variance is found to be

$$\sigma^2 = \overline{m} \ . \tag{A1.12}$$

Evidently the r.m.s. fluctuation about the mean equals the square root of the mean, a result which characterizes the Poisson distribution and is sometimes known as the 'law of large numbers'.

A parameter of interest in connection with the motion of an electron through a resistive material is the mean-free time, $\overline{\tau}_{fi}$. When $t = \overline{\tau}_{fi}$ we have by definition that $\overline{m} = 1$, and hence from equation (A1.10)

$$\overline{\tau}_{fi} = 1/v_i \ , \tag{A1.13}$$

which is consistent with our earlier interpretation of v_i as the mean rate at which collisions occur.

An alternative derivation of the mean-free time is as follows. The probability, $p_i(\tau_{fi} + \delta\tau_{fi})$, of finding a given free time lying between τ_{fi} and $(\tau_{fi} + \delta\tau_{fi})$ is equal to the joint probability of no collision in time τ_{fi} and one collision in the (small) interval $\delta\tau_{fi}$. Now the probability of a collision in $\delta\tau_{fi}$ is $v_i\delta\tau_{fi}$, and hence the joint probability is

$$p_i(\tau_{fi} + \delta\tau_{fi}) = v_i\, p_i(0, \tau_{fi})\, \delta\tau_{fi} \ , \tag{A1.14}$$

where $p_i(0, \tau_{fi})$ is the probability, given in equation (A1.5), of no event in time τ_{fi}. The function $\nu_i p_i(0, \tau_{fi}) = \nu_i \exp -(\nu_i \tau_{fi})$ on the right of equation (A1.14) is the probability density function of the free times, and therefore the mean-free time is

$$\bar{\tau}_{fi} = \nu_i \int_0^\infty \tau_{fi} \exp -(\nu_i \tau_{fi}) \, d\tau_{fi}$$

$$= 1/\nu_i \, , \tag{A1.15}$$

in agreement with equation (A1.13). The probability density function of the free times may also be employed to give the mean-square free time:

$$\overline{\tau_{fi}^2} = \nu_i \int_0^\infty \tau_{fi}^2 \exp -(\nu_i \tau_{fi}) \, d\tau_{fi}$$

$$= 2 \, \bar{\tau}_{fi}^2 \, . \tag{A1.16}$$

Thus the mean-square free time of an electron is twice the square of the mean-free time, a result which is employed in the discussion of mobility in Appendix 3. The integrals in equations (A1.15) and (A1.16) are evaluated by integrating by parts.

A thermal noise waveform observed at the terminals of a device is generated, not by the motion of just one electron, but by the individual motions of the assembly of electrons within the device. Now we have seen above that the noise waveform associated with a single electron is a pulse train in which the pulses are Poisson distributed. The noise due to all the electrons in the population is also a pulse train, being the superposition of all the waveforms generated by the individual electrons. We now examine the statistical distribution of the pulses in the total noise fluctuation at the terminals.

Consider just two electrons in the resistive material, say the ith and the jth, having probabilities $\nu_i \delta t$ and $\nu_j \delta t$, respectively, of experiencing a collision in the time interval δt. (Later on we shall be less general but more realistic by setting $\nu_i = \nu_j$ for all i and j.) The probability of the ith electron experiencing s collisions and the jth electron $(m - s)$ collisions in time t is

$$p_i(s, t) p_j(m - s, t) = \frac{\nu_i^s \nu_j^{m-s} t^m}{s! \, (m-s)!} \exp -(\nu_i + \nu_j) t \, , \tag{A1.17}$$

and hence the probability of *exactly* m collisions in time t is the sum of equation (A1.17) taken over all s from 0 to m:

$$p_{ij}(m, t) = t^m \exp -(\nu_i + \nu_j) t \sum_{s=0}^m \frac{\nu_i^s \nu_j^{m-s}}{s!(m-s)!} \, . \tag{A1.18}$$

From the binomial theorem, the summation here is equal to $(v_i + v_j)^m/m!$, and therefore

$$p_{ij}(m, t) = \frac{\{(v_i + v_j)t\}^m}{m!} \exp - (v_i + v_j)t \quad , \tag{A1.19}$$

which we recognize from equation (A1.9) as the Poisson distribution. We conclude that the sum of two Poisson processes is itself Poisson in character, with a mean rate of events which is equal to the sum of the mean rates of the two constituent processes. It follows that if N Poisson processes are superposed, the probability of finding *exactly* m events in a time t is given by the Poisson distribution

$$p(m, t) = \frac{(vt)^m}{m!} \exp - vt \quad , \tag{A1.20}$$

where

$$v = \sum_{i=1}^{N} v_i \quad . \tag{A1.21}$$

Equations (A1.20) and (A1.21) constitute the addition theorem for Poisson processes.

Statistically, the electrons in a resistor behave identically. Thus the mean rate of collisions, v_i, for an individual carrier is independent of which carrier is under examination. It follows from equation (A1.21), remembering that v_i equals the reciprocal of the mean-free time, that the mean rate of events is simply

$$v = nV/\bar{\tau}_f \tag{A1.22}$$

where n is the carrier concentration, V is the volume of the device, and $\bar{\tau}_f$ is the mean-free time of the electrons. The result in equation (A1.22) is used in the discussion of thermal noise in Chapter 2, § 2.8.

APPENDIX 2 – NYQUIST'S TREATMENT OF THERMAL NOISE IN A RESISTANCE

The power spectral density of the voltage fluctuations across a resistance in thermal equilibrium with its environment was originally derived by Nyquist (1928), who employed an argument involving thermodynamic and statistical mechanical considerations. His treatment of the problem appeared very soon after Johnson had observed the thermal fluctuations in a resistance, and the result he obtained is often referred to as Nyquist's law. Nyquist's approach to the question of thermal noise is described below.

He initially considered two electrical conductors, R_1 and R_2, each with resistance R and both at the same uniform temperature θ. With the two conductors connected in parallel as shown in Fig. A2.1, it is evident that the electromotive force due to the thermal motion of the charge carriers in R_1 will produce

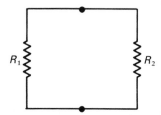

Fig. A2.1 – Equal resistances connected in parallel.

a current in the circuit, leading to an absorption of power by R_2. A similar flow of energy exists from R_2 to R_1. The power from R_1 which is absorbed by R_2 is $V_1^2/4R$ and in the reverse direction from R_2 into R_1 it is $V_2^2/4R$, where V_1^2 and V_2^2 are the mean-square values of the open-circuit e.m.f.s generated by R_1 and R_2, respectively. As the two conductors are at the same temperature the power flowing in each direction must be exactly the same, otherwise the second law of thermodynamics would be violated, and hence the mean-square voltages V_1^2 and V_2^2 are equal. This conclusion is valid irrespective of the nature of the two conductors; as Nyquist pointed out, one may be made of lead and the other of silver, or one may be metallic and the other electrolytic and the result still holds. Moreover, it holds not only for the total power exchanged between the conductors, but also for the power exchanged in any frequency band. For if this were not so it would be possible to insert a solely reactive filter (i.e. one containing only capacitors and inductors) between the conductors, whose bandpass spanned the frequency range where the imbalance in the power flow in the two directions existed. But, as R_1 and R_2 are at the same temperature, this would again lead to a violation of the second law of thermodynamics, and hence the power flows in the two directions in any frequency band must be the same. In other words, the power spectrum of the voltage fluctuations across a resistance is a universal function of R, θ and frequency f.

In order to derive this function Nyquist considered a long, lossless transmission line, terminated at either end by the conductors R_1 and R_2 (Fig.
A2.2). The characteristic impedance of the line is R, its length is l and the wave
velocity along the line is c. When the system is in thermal equilibrium there is
a flow of energy along the line from R_1 to R_2 and another flow in the reverse
direction from R_2 to R_1. These flows originate in the thermal fluctuations of the
charge carriers in R_1 and R_2, respectively, and in both cases the receiving conductor absorbs the energy that is incident upon it.

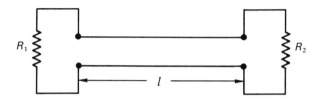

Fig. A2.2 – Lossless transmission line of length l terminated by equal resistances
R_1 and R_2.

The power delivered to the line from one of the conductors in a frequency
interval $d\omega/2\pi$ is

$$dP = \frac{1}{4R} \overline{S_V(\omega)} \frac{d\omega}{2\pi} , \qquad (A2.1)$$

where $\overline{S_V(\omega)}$ is the power spectral density of the voltage fluctuations across the
conductor in question. The time of transit along the line is l/c, and thus the total
energy delivered to the line from both conductors in a time interval of this
duration is

$$dE = \frac{l}{2Rc} \overline{S_V(\omega)} \frac{d\omega}{2\pi} . \qquad (A2.2)$$

Nyquist continued the argument by assuming that the transmission line is
now short-circuited at both ends, so that the energy on it is trapped as standing
waves. The frequencies of these waves correspond to the natural frequencies of
the line. Thus, the lowest frequency of vibration, corresponding to a voltage
wave with a node at either end of the line and no nodes between the ends, is
$c/2l$. The frequency of the next mode of vibration, in which there are nodes at
both ends and one in the middle of the line, is $2c/2l$; and in general the natural
frequencies of the line are given by $nc/2l$, where n is an integer. In a frequency
interval $d\omega/2\pi$, the number of modes of vibration is therefore $(2l/c)\,d\omega/2\pi$,
which will be a very large number compared to unity if l is allowed to become
indefinitely large. Assuming this to be the case, and identifying each mode as a
degree of freedom of the system, it is permissible to invoke the equipartition law

Appendix 2

to determine the total energy in the line. Provided quantum mechanical considerations are unimportant, the law of equipartition of energy states that on average the energy associated with each degree of freedom is equal to $k\theta$, where k is Boltzmann's constant and θ is the absolute temperature of the system. Thus, the energy in the line in the frequency interval $d\omega/2\pi$ is

$$dE = \frac{2l}{c} k\theta \frac{d\omega}{2\pi} , \qquad (A2.3)$$

and on comparing this with the previous expression for dE the power spectral density of the open-circuit voltage fluctuations is found to be

$$\overline{S_V(\omega)} = 4k\theta R , \qquad (A2.4)$$

which is Nyquist's result. It may be generalized to the case of a complex impedance by an extension of the above argument involving simple circuit considerations. Exactly the same form is then found for the spectral density of the voltage fluctuations, with R in this case representing the real part of the impedance. Thus, when the impedance is purely reactive, having no resistive component, there is no noise. This is to be expected since in the absence of resistance there can be no relaxation from a disturbed condition back towards the state of thermal equilibrium. The connections between resistance, relaxation and noise are explored in Chapter 2, where the Langevin approach to thermal noise is discussed.

At frequencies and temperatures where quantum mechanical effects are significant, that is when the quantum of energy hf is not negligibly small compared with $k\theta$, the equipartition law as stated above no longer holds. Nyquist briefly addressed this problem in his original treatment of thermal noise, and suggested that instead of $k\theta$ each degree of freedom should on average be assigned the energy

$$\frac{hf}{(e^{hf/k\theta} - 1)} , \qquad (A2.5)$$

which reduces to $k\theta$ in the 'classical' limit when $hf \ll k\theta$. The symbols h and f appearing above represent Planck's constant and frequency, respectively. The expression in equation (A2.5) gives the energy of an harmonic oscillator, except that it excludes the zero-point energy term. These days most authors agree that the zero-point term should be included in the description of the noise, as discussed more fully in Chapter 11.

REFERENCE

H. Nyquist (1928), Thermal agitation of electric charge in conductors, *Phys. Rev.*, **32**, 110–113.

APPENDIX 3 – MOBILITY, BROWNIAN VELOCITY AND DIFFUSION

The mobility of electrons (or holes) describes the macroscopic flow of the carriers in an electric field. It can be expressed in terms of certain parameters – such as the mean-free time between collisions with the lattice – which characterize the microscopic properties of the assembly of carriers. The argument is based on the classical particle model of the electron, and for simplicity a one-dimensional model of a resistor is assumed.

Suppose that an electric field, E, is applied across the resistor. Let the field be small, so that it does not significantly change the equilibrium velocity distribution of the electrons. An average drift velocity equal to μE will be imparted to the electron population by the field, where μ is the mobility.

In order to relate μ to the parameters for the microscopic motion of the electrons, consider a collision experienced by a particular particle. After the collision, and until the next one, it undergoes an acceleration $|a| = qE/m$ in the direction of the field, where m is the mass of the particle. Therefore, if the free time is τ_f, the displacement due to the field is $|a|\tau_f^2/2$, and hence after K collisions the total displacement of the particle due to the field is

$$\frac{qEK}{2m}\overline{\tau_f^2} \ ,$$

where $\overline{\tau_f^2}$ is the mean-square free time. The mean drift velocity of the particle over the K collisions is the total displacement divided by the total time:

$$v = \frac{(qEK/2m)\overline{\tau_f^2}}{K\overline{\tau_f}} \ , \tag{A3.1}$$

where $\overline{\tau_f}$ is the mean-free time. Since $v = \mu E$, it follows that the mobility is

$$\mu = q\overline{\tau_f^2}/(2m\overline{\tau_f}) \ . \tag{A3.2}$$

Now the free times are Poisson distributed, and $\overline{\tau_f^2} = 2\overline{\tau_f}^2$, as discussed in Appendix 1. Thus, the mobility is

$$\mu = q\overline{\tau_f}/m \ . \tag{A3.3}$$

If v_θ is the thermal velocity of the particle between collisions, then the free path length, l_f, is equal to $v_\theta\tau_f$. Given that the thermal velocity and the free time are uncorrelated, it follows that

$$\overline{l_f^2} = \overline{v_\theta^2}\ \overline{\tau_f^2} \ , \tag{A3.4}$$

where $\overline{l_f^2}$ is the mean-square free path and $\overline{v_\theta^2}$ is the mean-square thermal velocity.

From the law of equipartition of energy, we have for the one-dimensional resistor in thermal equilibriun: with its surroundings

$$m\overline{v_0^2}/2 = k\theta/2 \ ,$$
(A3.5)

where k is Boltzmann's constant and θ is the absolute temperature. On combining equations (A3.2), (A3.4) and (A3.5), the mobility can be expressed as

$$\mu = q\overline{l_f^2}/(2k\theta\overline{\tau_f}) \ .$$
(A3.6)

This formulation of μ is used for deriving the thermal noise in a resistor in Chapter 2, § 2.8.

Shockley (1963) gives an extensive discussion of mobilities, mean-free times and mean-free paths, which includes an analysis based on the particle model described above.

The velocity fluctuations of the assembly of electrons are responsible for the thermal noise in a resistor. Although the current and voltage thermal fluctuations at the terminals of the device are discussed in Chapter 2, § 2.8, it is nevertheless rewarding to examine specifically the velocity fluctuations themselves.

Consider an electron which undergoes a collision with the lattice. Its momenta before and after the event are p_1 and p_2, respectively. Thus, the collision imparts a momentum $(p_2 - p_1)$ to the particle in an infinitesimal time. This corresponds to a force $(p_2 - p_1)\delta(t)$. The equation of motion is therefore

$$m\frac{du(t)}{dt} = -q\frac{u(t)}{\mu} + (p_2 - p_1)\delta(t) \ ,$$
(A3.7)

where $u(t)$ is the velocity of the particle. Equation (A3.7) is essentially Langevin's equation for the velocity of a Brownian particle.

On Fourier transforming both sides of equation (A3.7) and rearranging terms, the transform of $u(t)$ is found to be

$$U\ (j\omega) = \frac{\mu(p_2 - p_1)/q}{(1 + j\omega\overline{\tau_f})} \ ,$$
(A3.8)

where $\overline{\tau_f} = m\mu/q$ is the mean-free time given by equation (A3.3). From Carson's theorem, the power spectral density of the velocity fluctuations is

$$S_u(\omega) = 2\nu\overline{(p_2 - p_1)^2} \, |F(j\omega)|^2 \ ,$$
(A3.9)

where $\nu = 1/\overline{\tau_f}$ is the mean rate of collisions for a single particle and

$$F(j\omega) = \frac{\mu/q}{(1 + j\omega\overline{\tau_f})}$$
(A3.10)

is the transform of the pulse shape function associated with a single collision. Now, p_1 and p_2 are independent, and hence

$$\overline{S_u(\omega)} = \frac{4\,\overline{p^2}\,\mu^2/(q^2\overline{\tau_f})}{(1 + \omega^2\overline{\tau_f^2})} \ , \tag{A3.11}$$

where we have set $\overline{p^2} = \overline{p_1^2} = \overline{p_2^2}$. For an equilibrium population we have from the law of equipartition, for the one-dimensional case,

$$\overline{p^2}/2m = k\theta/2 \ , \tag{A3.12}$$

and therefore the power spectral density of the thermal velocity fluctuations is

$$\overline{S_u(\omega)} = \frac{4(\mu k\theta/q)}{(1 + \omega^2\overline{\tau_f^2})} \ , \tag{A3.13}$$

where we have employed the result for the mobility in equation (A3.3)†. It follows from the Wiener–Khintchine theorem that the corresponding auto-correlation function is

$$\overline{\phi_u(\tau)} = \overline{u^2} \exp{-(|\tau|/\overline{\tau_f})} \ , \tag{A3.14}$$

where

$$\overline{u^2} = \overline{\phi_u(0)} = (k\theta/q)\,(\mu/\overline{\tau_f}) \tag{A3.15}$$

is the mean-square value of $u(t)$.

The thermal velocity fluctuations which are characterized by equations (A3.13) and (A3.14) are responsible for the *diffusion* of the particles through the resistive material. Shockley *et al.* (1966) related diffusion to velocity in an analysis that has general application to non-equilibrium as well as equilibrium ensembles. They defined the *diffusion of u at (angular) frequency* ω as

$$D_u(\omega) = \int_0^\infty \overline{\phi_u(\tau)} \cos \omega\tau \, d\tau \ , \tag{A3.16}$$

where $\overline{\phi_u(\tau)}$ is the autocorrelation function of the velocity fluctuations, but is no longer exclusively representative of the equilibrium condition, unlike equation (A3.14): $\overline{\phi_u(\tau)}$ can now be the autocorrelation function of a non-equilibrium

---

† The velocity fluctuation $u(t)$ produces a short-circuit current fluctuation $i(t) = qu(t)/L$, where L is the length of the resistor. It follows that the power spectral density of the current fluctuation is the right-hand side of equation (A3.13) multiplied by $(q/L)^2$ times the total number of carriers. This gives Nyquist's expression, $4k\theta/R$, divided by the denominator of equation (A3.13), in accord with the derivation of $\overline{S_i(\omega)}$ in § 2.8.

ensemble. On comparing the definition in equation (A3.16) with the Wiener–Khintchine inversion integral over τ, it is apparent that $D_u(\omega)$ is equivalent to the power spectral density of the velocity fluctuations divided by four. Therefore, from equation (A3.13), we have for the particular case of thermal equilibrium that

$$D_{ueq}(0) = \mu k \theta / q \ , \tag{A3.17}$$

which is Einstein's relationship when $D_{ueq}(0)$ is identified as the more familiar diffusion constant, D. It follows from equations (A3.15) and (A3.17) that, for one-dimensional motion, $D = \overline{u^2}\,\overline{\tau_f}$.

The general formulation for $D_u(\omega)$ in equation (A3.16) is important in the analysis of noise in hot electron devices, where the charge carriers are not in thermal equilibrium with the lattice.

REFERENCES

W. Shockley (1963), Electrons and holes in semiconductors, Van Nostrand, New York, Chapters 8 and 11.

W. Shockley, J. A. Copeland, and R. P. James (1966), The impedance field method of noise calculation in active semiconductor devices, in *Quantum Theory of Atoms, Molecules and the Solid State* (Ed. P. O. Lowdin), Academic Press, New York, pp. 537–563.

Appendix 4

APPENDIX 4 – NOISE DUE TO CARRIERS CROSSING THE DEPLETION LAYER

A carrier which crosses the depletion layer of a forward-biased p–n junction causes a change in the minority carrier concentration, p_0, at the plane $x = 0$ (see Fig. 4.1). This departure from the steady state population results in two relaxation current flows, one back across the depletion layer, causing a change in I_R, and one through the N-region, causing a change in I_D. These relaxation flows are the means by which the steady state, minority carrier concentration is restored.

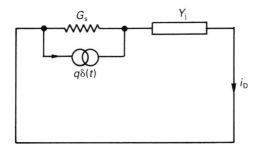

Fig. A4.1 – a.c. equivalent circuit representing the crossing of the depletion layer by an individual carrier.

The initial action of crossing the depletion layer can be represented by a current generator $q\delta(t)$, connected as shown in the a.c. equivalent circuit in Fig. A4.1. The conductance in this figure, $G_F = qI_F/k\theta$, is the reciprocal of the resistance in equation (4.6), which, as we have seen in Chapter 4, is a measure of the departure from the condition of constant quasi-Fermi levels separated by the applied voltage V throughout the depletion layer. The significance of G_F in the present context is that it represents the relaxation mechanism by which equilibrium is restored after a carrier crosses the depletion layer. Y_j in Fig. A4.1 is the junction admittance.

The current in the external circuit due to the initial action, consists of the initial action itself, $q\delta(t)$, followed by a relaxation current, $qg(t)$. Thus the total flow is

$$i_D(t) = q\{\delta(t) - g(t)\} , \tag{A4.1a}$$

or in transform terminology,

$$i_D(j\omega) = q \int_0^\infty \{\delta(t) - g(t)\} \exp -j\omega t \, dt . \tag{A4.1b}$$

The problem now is to determine $i_D(j\omega)$.

This may be achieved by applying a Thevenin transformation to the current generator and parallel conductance, G_F, in Fig. A4.1, to obtain a voltage generator whose transform is q/G_F, in series with G_F. It then follows immediately that

$$i_D(j\omega) = \frac{q\,Y_j}{G_F + Y_j} = q\left\{1 - \frac{G_F}{G_F + Y_j}\right\} , \qquad (A4.2)$$

and hence from Carson's theorem the power spectrum of the noise current in the external circuit is

$$\overline{S_D(\omega)} = \qquad\qquad\qquad |^2 , \qquad (A4.3)$$

where the mean nu_   been taken as $(I_F + I_R)/q \simeq 2I_F/q$, and

$$Y_T = \frac{}{G_F + Y_j} \qquad\qquad (A4.4)$$

is the total admittance of the circuit in Fig. A4.1. The power spectrum in equation (A4.3) of the current fluctuations due to carriers crossing the depletion layer has a simple interpretation: it is equivalent to thermal noise in the effective series resistance $1/G_F$. Now, over the frequency range of practical interest, G_F is so large that it can normally be safely ignored in junction calculations, and noise calculations are no exception to this general rule: since $G_F \gg Y_j$, the total admittance $Y_T \simeq Y_j$, and the contribution to the total noise from carriers crossing the depletion layer can be seen from equation (A4.3) to be negligible.

This conclusion is at variance with that from the corpuscular theory of van der Ziel and Becking (1958). The reason for the disparity may be understood from the following interpretation of the corpuscular theory.

If the d.c. component of $i_D(t)$ is ignored, then from equation (A4.1b),

$$\int_0^\infty g(t)\mathrm{d}t = 1 , \qquad (A4.5)$$

and hence the transform of $i_D(t)$ may be written in the form

$$i_D(j\omega) = q \int_0^\infty (1 - \exp -j\omega\tau)g(\tau)\,\mathrm{d}\tau . \qquad (A4.6)$$

This integral forms the basis of van der Ziel and Becking's analysis. They contend that it represents two delta functions separated by a time τ, corresponding to the forward and return journeys of an individual carrier across the depletion layer, and that the function $g(\tau)$, which they leave unspecified, can be interpreted as a distribution function for the time intervals τ.

With this interpretation of $g(\tau)$, they derive the spectral density of i_D by forming the spectral density of all those double-pulse events with separation τ, and integrating it over all τ, taking into account the statistical weighting function $g(\tau)$. This procedure leads to the result

$$\overline{S_D(\omega)} = 2qI_F \int_0^\infty g(\tau)|1 - \exp -j\omega\tau|^2 \, d\tau = 4qI_F \int_0^\infty g(\tau)(1 - \cos \omega\tau) \, d\tau$$

$$= 4qI_F \, \mathrm{Re} \left\{ \frac{Y_j}{G_F + Y_j} \right\} = 4k\theta \, \mathrm{Re} \, Y_T \, , \qquad (A4.7)$$

where equations (A4.1b) and (A4.2) have been employed to eliminate the integral, and the mean rate of the double pulses has been taken as I_F/q. The final result in equation (A4.7) is the one obtained by van der Ziel and Becking in their analysis. (They then set $\mathrm{Re} \, Y_T = G_j - G_0$, whereas from our definition of the total admittance we have $\mathrm{Re} \, Y_T \simeq G_j$. The difference between these two expressions is not very important in connection with the present argument).

The results in equations (A4.3) and (A4.7) differ by a substantial amount. From a mathematical point of view the reason for the difference is that the model of junction action proposed by van der Ziel and Becking, giving rise to equation (A4.7), involves the function

$$\int g(\tau)|1 - \exp -j\omega\tau|^2 \, d\tau$$

in the expression for the spectral density. In contradistinction, the expression in equation (A4.3) derives from the function

$$|1 - \int g(\tau) \exp -j\omega\tau \, d\tau|^2 \, .$$

The first thing to note about the final result in equation (A4.7) is that, according to our definition of Y_T, it is the same expression as that for the total noise associated with the real component of the junction admittance, and as such is not negligible. But does it represent the noise due to carriers crossing the depletion layer? On examining the implications of the corpuscular model, we are forced to conclude that it does not.

For the reason why, consider the essential feature of the model, namely a crossing of the depletion layer by an individual carrier, followed after a time τ by a return crossing of the same carrier. Now, by comparing equations (A4.1a) and (A4.2), the Fourier transform of $g(\tau)$, the function alleged to govern the distribution of the time intervals τ, is seen to be

$$\int_0^\infty g(\tau) \exp -j\omega\tau \, d\tau = \frac{G_F}{G_F + Y_j} . \tag{A4.8}$$

The important point about this equation is that the right-hand side is constant and essentially equal to unity up to frequencies well beyond the operating range of the junction. This is because, over this frequency range, $G_F \gg Y_j$. It follows that $g(\tau)$ is a function which is sensibly non-zero *only* for values of τ which are very much less than the usual time constants associated with a p–n junction, including the recombination time τ_R. This in turn means that the time interval τ between the forward and return journeys of a carrier crossing the depletion layer is very much less than the recombination time, τ_R. The difficulty with the corpuscular theory can now be appreciated; for, according to the model, if a carrier that has just crossed the depletion layer is to recombine in the bulk region, it must do so within a time τ of the crossing, otherwise it will return to the region from whence it came. But since $\tau \ll \tau_R$, recombination almost certainly will not occur, and thus the carrier, like all the other carriers that cross the depletion layer, will make the return journey back to the region of origin. This, however, is not in accord with our understanding of the current-flow mechanism in a p–n junction. On the basis of this contradiction, we are led to conclude that the corpuscular model of van der Ziel and Becking is not physically tenable; and thus equation (A4.7) does not represent the noise due to carriers crossing the depletion layer.

A similar argument applies to the treatment of transistor noise given by van der Ziel and Becking.

APPENDIX 5 – FORMAL SOLUTION FOR THE OUTPUT NOISE OF AN OSCILLATOR

The equation to be solved for the amplitude and phase fluctuations, $a(t)$ and $\psi(t)$, at the output of a van der Pol oscillator is equation (8.34) which, when differentiated with respect to time, becomes

$$C\frac{d^2 v_n}{dt^2} - (G_L - G_0)\frac{d(v_n + v_q)}{dt} + \frac{v_n}{L} = \frac{di_n}{dt} , \qquad \text{(A5.1)}$$

where

$$v_n \equiv v_n(t) = v_0[a(t)\cos\omega_0 t + \psi(t)\sin\omega_0 t] , \qquad \text{(A5.2)}$$

$$v_q \equiv v_q(t) = v_0[a(t)\cos\omega_0 t - \psi(t)\sin\omega_0 t] \qquad \text{(A5.3)}$$

and the remaining parameters are defined in Chapter 8. After performing the necessary differentiations of v_n and v_q and making the substitutions, equation (A5.1) can be written in the form

$$g_1(t)\sin\omega_0 t + g_2(t)\cos\omega_0 t = \frac{di_n}{dt} , \qquad \text{(A5.4)}$$

where $g_1(t)$ and $g_2(t)$ are very slowly varying functions compared with the free oscillation and are given by the expressions

$$g_1(t) = v_0 \left[C\left\{ \frac{d^2\psi}{dt^2} - 2\omega_0\frac{da}{dt} \right\} + 2\omega_0(G_L - G_0)a \right] \qquad \text{(A5.5)}$$

and

$$g_2(t) = v_0 \left[C\left\{ \frac{d^2 a}{dt^2} + 2\omega_0\frac{da}{dt} \right\} - 2(G_L - G_0)\frac{da}{dt} \right] . \qquad \text{(A5.6)}$$

The problem now is to solve equations (A5.4) to (A5.6) for $a(t)$ and $\psi(t)$, assuming that $i_n \equiv i_n(t)$ is a white noise source.

The technique employed is to multiply equation (A5.4) by $\sin\omega_0 t$ and $\cos\omega_0 t$ in turn, and then integrate over one cycle of the free oscillation. Since $g_1(t)$ and $g_2(t)$ are essentially unchanged over such a short time scale they may be treated as constants and taken outside the integral sign (this is sometimes known as the method of stationary phase). The orthogonality of the sine and cosine functions over one cycle then leads to the elimination of one or other of the terms on the left of equation (A5.4). Thus, two differential equations are constructed, which may be written in the form

$$\frac{d^2\psi}{dt^2} - 2\omega_0\frac{da}{dt} - \frac{2\omega_0^2 a}{Q_0} = \frac{\omega_0 n_1(t)}{\pi v_0 C} \qquad \text{(A5.7)}$$

and

$$\frac{d^2a}{dt^2} + 2\omega_0 \frac{d\psi}{dt} + \frac{2\omega_0}{Q_0}\frac{da}{dt} = \frac{\omega_0 n_2(t)}{\pi v_0 C} , \tag{A5.8}$$

where $Q_0 = \omega_0 C / |G_L - G_0|$ is the external Q-factor of the oscillator. The time dependent functions on the right of these expressions represent the source term, and are given by the integrals

$$n_1(t) = \int_{t-T_0/2}^{t+T_0/2} \frac{di_n}{dt'} \sin \omega_0 t' \, dt' \tag{A5.9}$$

and

$$n_2(t) = \int_{t-T_0/2}^{t+T_0/2} \frac{di_n}{dt'} \cos \omega_0 t' \, dt' , \tag{A5.10}$$

where $T_0 = 2\pi/\omega_0$ is the period of the free oscillation. Note that the differential of the white noise source appearing in the integrands here is most definitely not slowly varying, and cannot be legitimately removed from under the integral signs. We shall return to the evaluation of $n_1(t)$ and $n_2(t)$ (or at least their power spectra) later.

Equations (A5.7) and (A5.8) are linear, simultaneous differential equations which may be solved using a standard procedure. By differentiating equation (A5.8), multiplying equation (A5.7) by $2\omega_0$ and subtracting, an equation is derived which does not contain $\psi(t)$:

$$\frac{d^3a}{dt^3} + \frac{2\omega_0}{Q_0}\frac{d^2a}{dt^2} + 4\omega_0^2\frac{da}{dt} + \frac{4\omega_0^3}{Q_0}a = \frac{\omega_0}{\pi v_0 C}\left\{\frac{dn_2(t)}{dt} - 2\omega_0 n_1(t)\right\} . \tag{A5.11}$$

On Fourier transforming both sides of this equation and solving for the transform of $a(t)$, we find that

$$A(j\omega) = \frac{\dfrac{\omega_0}{\pi v_0 C}\left\{j\omega N_2(j\omega) - 2\omega_0 N_1(j\omega)\right\}}{\left\{-j\omega^3 - \dfrac{2\omega_0}{Q_0}\omega^2 + 4j\omega\omega_0^2 + \dfrac{4\omega_0^3}{Q_0}\right\}}$$

$$\simeq -\frac{N_1(j\omega)}{2\pi v_0 C\left\{\dfrac{\omega_0}{Q_0} + j\omega\right\}} , \tag{A5.12}$$

where $N_1(j\omega)$ and $N_2(j\omega)$ are the transforms of $n_1(t)$ and $n_2(t)$, and the approximation holds when $\omega \ll \omega_0$, which is our case since $a(t)$ is a very low frequency function. From equation (A5.12), the power spectrum of $a(t)$ is

$$\overline{S_a(\omega)} = \frac{\overline{S_{n_1}(\omega)}}{4\pi^2 v_0^2 C^2 \left(\dfrac{\omega_0^2}{Q_0^2} + \omega^2 \right)}, \tag{A5.13}$$

where $\overline{S_{n_1}(\omega)}$ is the power spectrum of $n_1(t)$.

A solution for the phase fluctuation is now found from equation (A5.8) by Fourier transforming both sides and solving for the transform of $\psi(t)$ in terms of $A(j\omega)$:

$$\Psi(j\omega) = \left\{ \left(\omega^2 - \frac{2j\omega_0\omega}{Q_0} \right) A(j\omega) + \frac{\omega_0 N_2(j\omega)}{\pi v_0 C} \right\} \Big/ \left(2j\omega_0\omega \right)$$

$$\simeq -j N_2(j\omega)/(2\pi v_0 \omega C), \tag{A5.14}$$

where the approximation holds for the low frequency region of interest, where $\omega \ll \omega_0$. Thus, the power spectrum of $\psi(t)$ is

$$\overline{S_\psi(\omega)} = \frac{\overline{S_{n_2}(\omega)}}{4\pi^2 v_0^2 C^2 \omega^2}, \tag{A5.15}$$

where $\overline{S_{n_2}(\omega)}$ is the power spectrum of $n_2(t)$.

The cross-spectral density between the amplitude and phase fluctuations can be written down immediately from the Fourier transforms in equations (A5.12) and (A5.14). It is

$$\overline{S_{a\psi}(\omega)} = \frac{\overline{S_{n_1 n_2}(\omega)}}{4\pi^2 v_0^2 C^2 j\omega \left(\dfrac{\omega_0}{Q_0} + j\omega \right)}, \tag{A5.16}$$

where $\overline{S_{n_1 n_2}(\omega)}$ is the cross-spectral density between $n_1(t)$ and $n_2(t)$.

It only remains for us to specify the power and cross-spectral densities of the source fluctuations $n_1(t)$ and $n_2(t)$ in terms of the power spectrum of the noise current generator $i_n(t)$. Suppose we concentrate on $n_1(t)$ defined in equation (A5.9), then on integrating by parts we find that it can be expressed as

$$n_1(t) = - \left\{ i_n(t + T_0/2) - i_n(t - T_0/2) \right\} \sin \omega_0 t - \int_{t-T_0/2}^{t+T_0/2} i_n(t') \cos \omega_0 t' \, dt'$$

$$\tag{A5.17}$$

Thus, the covariance is

$$\overline{n_1(t)n_1(t+\tau)} = \overline{\{i_n(t+T_0/2)-i_n(t-T_0/2)\}\{i_n(t+T_0/2+\tau)-i_n(t-T_0/2+\tau)\}}$$

$$\times \sin\omega_0 t \sin\omega_0(t+\tau)$$

$$+ \omega_0 \sin\omega_0 t \int_{t-T_0/2+\tau}^{t+T_0/2+\tau} \overline{\{i_n(t+T_0/2)i_n(t')-i_n(t-T_0/2)i_n(t')\}}\cos\omega_0 t'\, dt'$$

$$+ \omega_0 \sin\omega_0(t+\tau) \int_{t-T_0/2}^{t+T_0/2} \overline{\{i_n(t+T_0/2+\tau)i_n(t')-i_n(t-T_0/2+\tau)i_n(t')\}}$$

$$\times \cos\omega_0 t'\, dt'$$

$$+ \omega_0^2 \int_{t-T_0/2}^{t+T_0/2}\int_{t-T_0/2+\tau}^{t+T_0/2+\tau} \overline{i_n(t')i_n(t'')}\cos\omega_0 t'\cos\omega_0 t''\, dt'\, dt''$$

$$\tag{A5.18}$$

and by definition the autocorrelation function of $n_1(t)$ is

$$\overline{\phi_{n_1}(\tau)} = \lim_{T\to\infty}\frac{1}{T}\int_{-T/2}^{T/2}\overline{n_1(t)n_2(t+\tau)}\, dt \ . \tag{A5.19}$$

Now, since $i_n(t)$ represents white noise, its autocorrelation function is a delta function:

$$\phi_i(\tau) = \overline{i_n(t)i_n(t+\tau)} = \kappa\delta(\tau) \ , \tag{A5.20}$$

where κ is a constant. On combining equations (A5.18) and (A5.20), all the terms on the right of equation (A5.18) are found to be negligible except that containing the double integral, which yields

$$\overline{\phi_{n_1}(\tau)} = \frac{\omega_0^2\kappa}{2}(T_0-|\tau|) \ , \quad |\tau|\le T_0$$

$$\tag{A5.21}$$

$$0 \qquad , \quad |\tau|> T_0 \ ,$$

From the Wiener–Khintchine theorem, the power spectral density of $n_1(t)$ is therefore

$$\overline{S_{n_1}(\omega)} = 2\kappa\omega_0^2\int_0^{T_0}(T_0-\tau)\cos\omega\tau\, d\tau$$

$$= 4\pi^2\kappa \ , \tag{A5.22}$$

where the cosine function in the integrand has been set equal to unity because, over the range of integration, $\omega\tau \ll 1$. Now, the power spectrum of the white noise generator, obtained from the autocorrelation function in equation (A5.20), is $\overline{S_i} = 2\kappa$ and hence from equation (A5.22)

$$\overline{S_{n_1}(\omega)} = 2\pi^2 \, \overline{S_i} \; . \tag{A5.23}$$

Similar arguments show that the power spectral density of $n_2(t)$ is the same as that of $n_1(t)$:

$$\overline{S_{n_2}(\omega)} = 2\pi^2 \, \overline{S_i} \tag{A5.24}$$

and also that the cross-spectral density between $n_1(t)$ and $n_2(t)$ is zero, from which it follows that $a(t)$ and $\psi(t)$ are independent fluctuations.

When the expressions in equations (A5.23) and (A5.24) are substituted into equations (A5.13) and (A5.15), the power spectral densities of the amplitude and phase fluctuations are found to be

$$\overline{S_a(\omega)} = \frac{\overline{S_i}}{2v_0^2 \, C^2 \left(\dfrac{\omega_0^2}{Q_0^2} + \omega^2 \right)} \tag{A5.25}$$

and

$$\overline{S_\psi(\omega)} = \frac{\overline{S_i}}{2v_0^2 \, C^2 \, \omega^2} \; , \tag{A5.26}$$

respectively. These results are identical to those obtained in Chapter 8 from the simplified argument in which either $a(t)$ or $\psi(t)$ is set equal to zero while the other is calculated. Thus, the underlying assumption employed there — that the energy in the noise generator $i_n(t)$ is equally shared in driving the amplitude and phase fluctuations — is vindicated.

APPENDIX 6 – CURRENT-VOLTAGE RELATIONS FOR A PARAMETRIC AMPLIFIER

The equivalent circuit of a parametric amplifier in Fig. 9.6 consists of three resonant circuits coupled together through a parametric diode. An alternative arrangement involving a variable inductor is the dual of the circuit shown in the figure, and hence does not require a separate analysis. In the discussion given below it is assumed that the amplifier is non-degenerate, that is, that the three resonant frequencies are well separated. The Q-factors of the three resonant branches are presumed to be sufficiently high for no overlap to occur between the three passbands. Thus at the resonance frequency of any one of the branches, the other two branches show an infinite admittance.

The charge, q_p, on the parametric diode is a function of the voltage, v_p, across its terminals. On expanding this function in a Taylor series, the charge may be expressed in the form

$$q_p(t) = a_1 v_p(t) + a_2 v_p^2(t) + a_3 v_p^3(t) + \dots , \tag{A6.1}$$

where the coefficients in the series depend on the detailed structure of the parametric diode. From the point of view of the operation of the parametric amplifier, the physical mechanism giving rise to the non-linear behaviour expressed in equation (A6.1) is not important; and even if this mechanism is quantum mechanical in nature, the amplifier itself can still be discussed in classical terms.

In a linear capacitor, all the coefficients except the first in equation (A6.1) are zero, and a_1 is the capacitance. When all the coefficients except the first and the second are zero, the charge varies quadratically with voltage:

$$q_p(t) = C_p v_p(t) + a_2 v_p^2(t) , \tag{A6.2}$$

where we have replaced a_1 by C_p, the linear capacitance of the parametric diode. This form of non-linear behaviour is particularly interesting because it gives rise to distortionless amplification (at least, to a good approximation for small signals). When cubic or higher-order terms appear in the charge–voltage relationship, distortion occurs through mixing of the harmonics of the input signal. For the purpose of our discussion, we shall assume that the charge varies quadratically with voltage, as expressed in equation (A6.2). It follows that the current flowing through the parametric diode is

$$i_p(t) = \frac{dq_p}{dt} = C_p \frac{dv_p}{dt} + 2a_2 v_p \frac{dv_p}{dt} . \tag{A6.3}$$

Now, the voltage across the non-linear capacitance can be decomposed into three components:

$$v_p(t) = v_1(t) + v_2(t) + v_3(t), \tag{A6.4}$$

where $v_1(t)$ is due to the signal, $v_3(t)$ is due to the high-frequency pump, and $v_2(t)$ arises from the non-linear action of the parametric diode, which causes mixing between the input signal and the pump. At this point, it is convenient to specify the three voltages on the right of equation (A6.4) as sinusoidal signals with frequencies equal to the resonance frequencies of the three circuit branches of the parametric amplifier. We then have

$$v_p(t) = V_1 \cos(\omega_1 t + \phi_1) + V_2 \cos(\omega_2 t + \phi_2) + V_3 \cos(\omega_3 t + \phi_3) , \quad \text{(A6.5)}$$

where V_i and ϕ_i, $i = 1, 2, 3$, are the amplitudes and phases of $v_i(t)$. The angular frequencies in equation (A6.5) are related through the expression

$$\omega_3 = \omega_1 + \omega_2 , \quad \text{(A6.6)}$$

and each can be expressed in terms of the circuit components in Fig. 9.6 as follows:

$$\omega_i = 1/\sqrt{L_i(C_p + C_i)} , \quad i = 1, 2, 3 . \quad \text{(A6.7)}$$

On combining equations (A6.3) and (A6.5), and ignoring all the terms at frequencies other than ω_1, ω_2 or ω_3, the current through the non-linear capacitor is found to be

$$i_p(t) = i_1(t) + i_2(t) + i_3(t) , \quad \text{(A6.8a)}$$

where

$$i_1(t) = -\omega_1 C_p V_1 \sin(\omega_1 t + \phi_1) - \omega_1 a_2 V_2 V_3 \sin\{\omega_1 t + (\phi_3 - \phi_2)\}$$

$$i_2(t) = -\omega_2 C_p V_2 \sin(\omega_2 t + \phi_2) - \omega_2 a_2 V_1 V_3 \sin\{\omega_2 t + (\phi_3 - \phi_1)\} \quad \text{(A6.8b)}$$

$$i_3(t) = -\omega_3 C_p V_3 \sin(\omega_3 t + \phi_3) - \omega_3 a_2 V_1 V_2 \sin\{\omega_3 t + (\phi_1 + \phi_2)\} \quad .$$

Thus, $i_i(t)$ and $v_i(t)$ are at the same frequency ω_i, $i = 1, 2, 3$.

The currents in equation (A6.8b) can be written in the alternative form

$$i_1(t) = C_p \frac{dv_1(t)}{dt} + \frac{a_2 V_2 V_3}{V_1} \left\{ \cos(\phi_3 - \phi_2 - \phi_1) \frac{dv_1(t)}{dt} - \omega_1 v_1(t) \sin(\phi_3 - \phi_2 - \phi_1) \right\}$$

$$i_2(t) = C_p \frac{dv_2(t)}{dt} + a_2 \frac{V_1 V_3}{V_2} \left\{ \cos(\phi_3 - \phi_2 - \phi_1) \frac{dv_2(t)}{dt} - \omega_2 v_2(t) \sin(\phi_3 - \phi_2 - \phi_1) \right\}$$

$$i_3(t) = C_p \frac{dv_3(t)}{dt} + a_2 \frac{V_1 V_2}{V_3} \left\{ \cos(\phi_3 - \phi_2 - \phi_1) \frac{dv_3(t)}{dt} + \omega_3 v_3(t) \sin(\phi_3 - \phi_2 - \phi_1) \right\} .$$

$$\text{(A6.9)}$$

Now, the admittance seen by each of the three resonant circuits due to the non-linear capacitance is

$$Y_i = I_i(j\omega_i)/V_i(j\omega_i) ,\qquad\qquad (A6.10)$$

where $I_i(j\omega_i)$ and $V_i(j\omega_i)$ are the Fourier transforms of $i_i(t)$ and $v_i(t)$, respectively, at the frequencies ω_i. On Fourier transforming each of the currents in equation (A6.9) and dividing by the transform of the corresponding voltage, we find that

$$Y_1 = j\omega_1 C_p + j\omega_1 a_2 \frac{V_2 V_3}{V_1} \exp j(\phi_3 - \phi_2 - \phi_1)$$

$$Y_2 = j\omega_2 C_p + j\omega_2 a_2 \frac{V_1 V_3}{V_2} \exp j(\phi_3 - \phi_2 - \phi_1) \qquad (A6.11)$$

$$Y_3 = j\omega_3 C_p + j\omega_3 a_2 \frac{V_1 V_2}{V_3} \exp -j(\phi_3 - \phi_2 - \phi_1) .$$

The first terms on the right of equations (A6.11) are the admittances of the capacitance C_p, and the second terms represent the contribution from the quadratic component in the charge–voltage relation.

The current–voltage relations for the three resonant branches of the parametric amplifier follow directly from equations (A6.11). They are

$$I_s(j\omega) = \left\{ G_T + j\omega_1 a_2 \frac{V_2 V_3}{V_1} \exp j(\phi_3 - \phi_2 - \phi_1) \right\} V_1 (j\omega_1)$$

$$0 = \left\{ G_2 + j\omega_2 a_2 \frac{V_1 V_3}{V_2} \exp j(\phi_3 - \phi_2 - \phi_1) \right\} V_2 (j\omega_2) \qquad (A6.12)$$

$$I_{pp}(j\omega_3) = \left\{ G_3 + j\omega_3 a_2 \frac{V_1 V_2}{V_3} \exp -j(\phi_3 - \phi_2 - \phi_1) \right\} V_3 (j\omega_3) ,$$

where $I_s(j\omega_1)$ and $I_{pp}(j\omega_3)$ are the Fourier transforms of the signal and pump generators, respectively, at the frequencies ω_1 and ω_3, and

$$G_T = (G_1 + G_S + G_L) . \qquad (A6.13)$$

By eliminating $V_2 \exp j\phi_2$ and $V_3 \exp j\phi_3$ from the first of equations (A6.12), the admittance of the signal circuit, Y_s, is found to contain a negative conductance, due to the action of the parametric diode:

$$Y_s = (G_T - G) , \qquad (A6.14)$$

where

$$G = \frac{\omega_1 \omega_2 a_2^2}{G_2 G_3^2} \frac{|I_{pp}|^2}{\left[1 + \dfrac{\omega_2 \omega_3}{G_2 G_3} a_2^2 V_1^2\right]^2} . \tag{A6.15}$$

In this expression, $|I_{pp}|$ is the amplitude of the pump current generator. The term in the denominator containing V_1 gives rise to distortion in the parametric amplifier, though for small signals satisfying the condition

$$\frac{\omega_2 \omega_3 a_2^2}{G_2 G_3} V_1^2 \ll 1 \tag{A6.16}$$

the effect is negligible. This inequality also indicates that for distortionless amplification the non-linearity should be small and the idling and pump conductances should be large.

The analysis given above may be extended to include the case where the signal frequency, ω, is not necessarily equal to the resonance frequency, ω_1, of the signal circuit. Then, the current–voltage relations in equations (A6.12) must be generalized by replacing ω_1 with ω, ω_2 with $(\omega_3 - \omega)$, and by including the susceptances of the resonant branches specified by equations

$$B_1 = G_1 Q_1 \left(\frac{\omega}{\omega_1} - \frac{\omega_1}{\omega}\right)$$
$$B_2 = G_2 Q_2 \left(\frac{\omega_3 - \omega}{\omega_2} - \frac{\omega_2}{\omega_3 - \omega}\right) \tag{A6.17}$$

where Q_1 and Q_2 are the quality factors of the signal and idler circuits. The susceptance of the pump circuit is zero, irrespective of the signal frequency, since the frequency of the pump is presumed to be held fixed at the resonance frequency ω_3.

The admittance of the signal circuit is now

$$Y_S = (G_T - G) + j\left(B_1 - \frac{B_2 G}{G_2}\right) , \tag{A6.18}$$

where the negative conductance is

$$G = \frac{\omega(\omega_3 - \omega) a_2^2 |I_{pp}|^2}{G_2 G_3^2 \left\{\left(1 + \dfrac{V_1^2}{V_0^2}\right)^2 + (B_2/G_2)^2\right\}} \tag{A6.19}$$

and

$$v_0^2 = \frac{G_2 G_3}{a_2^2 \omega_3(\omega_3 - \omega)} . \tag{A6.20}$$

APPENDIX 7 – MINIMUM NOISE IN AN AMPLIFIER

All linear amplifiers show some noise, as established in § 11.3 from an argument based on the uncertainty principle. An extension of the argument, discussed below for the case of a high signal/noise ratio at the output, leads to an expression for the minimum output-noise power spectral density involving the gain of the amplifier and the quantum of energy hf_0, where f_0 is the frequency of the signal.

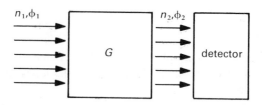

Fig. A7.1 – Amplifier with power gain G, followed by an 'ideal' detector.

Figure A7.1 shows the complete amplifier system consisting of an amplifier with gain $G > 1$, followed by a detector. The latter is 'ideal' in that the uncertainties it introduces into measurements of the number of photons and the phase are at the limit set by the uncertainty principle:

$$\Delta n_d \, \Delta \phi_d = 1/2 \ . \tag{A7.1}$$

In order to achieve the best detection performance, the uncertainties in the measurements of the number of photons and the phase of the signal at the input of the amplifier must be such that

$$\Delta n_1 \Delta \phi_1 = 1/2 \ . \tag{A7.2}$$

Letting Δn_a and $\Delta \phi_a$ represent the r.m.s. fluctuations in the number of photons and the phase at the output of the amplifier, the first step in the argument is to determine the condition on the product $\Delta n_a \, \Delta \phi_a$ which matches the detector to the amplifier.

As the uncertainties, or noise, introduced by the amplifier and the detector are independent, the uncertainty in the measurement of the output signal can be expressed as

$$\Delta n_2^2 = \Delta n_a^2 + \Delta n_d^2 = G^2 \, \Delta n_1^2 \tag{A7.3a}$$

and

$$\Delta \phi_2^2 = \Delta \phi_a^2 + \Delta \phi_d^2 = \Delta \phi_1^2 \ . \tag{A7.3b}$$

By forming the product of these two equations, it follows that

$$Q \equiv \Delta n_i^2 \, \Delta\phi_i^2 = \frac{1}{G^2} (\Delta n_a^2 + \Delta n_d^2)(\Delta\phi_a^2 + \Delta\phi_d^2)$$

$$= \frac{1}{G^2} \left(\frac{\Delta n_a^2}{w} + \frac{1}{2} \right) \left(w\Delta\phi_a^2 + \frac{1}{2} \right), \qquad (A7.4)$$

where

$$w = \Delta n_d / \Delta\phi_d . \qquad (A7.5)$$

In arriving at the result in equation (A7.4) the condition in equation (A7.1) has been employed. Note that although the product $\Delta n_d \, \Delta\phi_d$ is fixed, the ratio in equation (A7.5) is not. The matching condition mentioned above is achieved when w takes the value for which Q is a minimum. (It is clear that Q as a function of w goes through a minimum from the form of equation (A7.4), and the value at the minimum is known to be $Q = 1/4$ from equation (A7.2).) On differentiating equation (A7.4) with respect to w and equating the result to zero, the minimum is found to occur when

$$w \equiv \frac{\Delta n_d}{\Delta\phi_d} = \frac{\Delta n_a}{\Delta\phi_a} . \qquad (A7.6)$$

Thus, the detector and amplifier are matched when the relative uncertainties in the two devices are equal. When this result is substituted back into equation (A7.4) with Q set equal to $1/4$, it is easily shown that

$$\Delta n_a \, \Delta\phi_a = \frac{(G-1)}{2} , \qquad (A7.7)$$

which is the optimal condition that was being sought.

Equation (A7.7) is the basis of the derivation of the minimum power in the output noise of the amplier. As the noise is thermal in origin, it is assumed to have a bilateral power spectral density, S_0, which extends uniformly over positive and negative frequencies. Thus, the total noise power in a frequency interval df is the sum of the negative and positive frequency components, equal to $2S_0 df$. The ultimate aim of the following analysis is to determine the minimum value of the noise power, on the assumption that the noise level is very much less than that of the harmonic output signal.

Let the output field, $e(t)$, consist of a strong harmonic component with amplitude a_0 and angular frequency ω_0, and a thermal noise component, $n(t)$, due to the amplifier. Thus,

$$e(t) = a_0 \cos \omega_0 t + n(t) . \qquad (A7.8)$$

It is well known and easily proved† that this type of expression, involving a sine wave plus noise, can be written in the alternative form

$$e(t) = r(t) \cos \{\omega_0 t + \phi(t)\} \quad , \tag{A7.9}$$

where $\phi(t)$ is the phase fluctuation and $r(t)$ is the randomly varying envelope function of the field. Equation (A7.9) follows from writing

$$n(t) = n_c(t) \cos \omega_0 t - n_s(t) \sin \omega_0 t \quad , \tag{A7.10a}$$

where the two new noise functions on the right are defined as

$$n_c(t) = n(t) \cos \omega_0 t \tag{A7.10b}$$

$$n_s(t) = -n(t) \sin \omega_0 t \quad . \tag{A7.10c}$$

The phase and envelope of $e(t)$ are related to $n_c(t)$ and $n_s(t)$ through the equations

$$r(t) \cos \phi(t) = a_0 + n_c(t) \tag{A7.11a}$$

$$r(t) \sin \phi(t) = n_s(t) \quad , \tag{A7.11b}$$

from which it follows that

$$\phi(t) = \tan^{-1} \left\{ \frac{n_s(t)}{a_0 + n_c(t)} \right\} \simeq \frac{n_s(t)}{a_0} \tag{A7.12a}$$

$$r(t) = \{ [a_0 + n_c(t)]^2 + n_s^2(t) \}^{\frac{1}{2}} \simeq a_0 + n_c(t) \quad , \tag{A7.12b}$$

where the approximations are valid when the signal/noise ratio is large.

From equation (A7.12a), the power spectral density of the phase fluctuations is

$$\overline{S_\phi(\omega)} = \frac{\overline{S_{n_s}(\omega)}}{a_0^2} \quad , \tag{A7.13}$$

where $\overline{S_{n_s}(\omega)}$ is the power spectral density of $n_s(t)$. Now it follows directly from equation (A7.10c) that

$$\overline{S_{n_s}(\omega)} = 1/4 \{ \overline{S_n(\omega - \omega_0)} + \overline{S_n(\omega + \omega_0)} \}$$

$$= S_0/2 \quad , \tag{A7.14}$$

where $\overline{S_n(\ )}$ is the power spectral density of the amplifier noise, which has already been characterized as independent of frequency with a level equal to

† See for example S. O. Rice (1948), Statistical properties of a sine wave plus random noise, *Bell Syst. Tech. J.*, **27**, 109–157.

S_0. When the two results given above are combined, they give the mean-square phase fluctuations in a frequency interval df equal to

$$\Delta\phi_a^2 = \frac{S_0 df}{2a_0^2} . \tag{A7.15}$$

This expresses the phase fluctuations in terms of the power in the noise. It now remains for us to derive a similar expression for the number fluctuation Δn_a^2.

This is achieved by examining the *fluctuations in the power* of the envelope function $r(t)$. If the power fluctuation is defined as

$$\eta(t) = [r^2(t) - \overline{r^2(t)}] , \tag{A7.16}$$

then we are interested in the autocorrelation function

$$\overline{\phi_\eta(\tau)} = \overline{\eta(t)\eta(t+\tau)} . \tag{A7.17}$$

On substituting for $r(t)$ from equation (A7.12b), this is found to be

$$\overline{\phi_\eta(\tau)} = \overline{n_c^2(t) n_c^2(t+\tau)} - \overline{n_c^2(t)}^2 + 4a_0^2 \overline{n_c(t)n_c(t+\tau)} + \dots , \tag{A7.18}$$

where the terms that are not shown are all identically zero. As the signal level is assumed to be very much greater than that of the noise, the first two terms on the right of equation (A7.18) are negligible compared with the third, and hence

$$\overline{\phi_\eta(\tau)} \simeq 4a_0^2 \overline{n_c(t) n_c(t+\tau)}$$

$$= 4a_0^2 \overline{\phi_{n_c}(\tau)} , \tag{A7.19}$$

where $\overline{\phi_{n_c}(\tau)}$ is the autocorrelation function of $n_c(t)$. From the Wiener–Khintchine theorem, the spectral density of $\eta(t)$ is therefore

$$\overline{S_\eta(\omega)} = 4a_0^2 \overline{S_{n_c}(\omega)} , \tag{A7.20}$$

where $\overline{S_{n_c}(\omega)} \equiv S_0/2$ is the power spectral density of $n_c(t)$. Thus,

$$\overline{S_\eta(\omega)} = 2a_0^2 S_0 . \tag{A7.21}$$

Equation (A7.21) gives the spectral density of the power fluctuations of the field. Now, the noise energy in the field is made up of two components, a 'classical' part due to the interaction between waves in the field, and a 'quantum' contribution associated with the particle-like character of the radiation. At the low-temperature limit, the classical contribution is zero, but the quantum noise level remains finite and hence sets the minimum level of the output noise of the amplifier. At the frequency of the signal, f_0, the energy of a field with n_a photons is

$$E = n_a h f_0 , \tag{A7.22}$$

and hence the quantum noise power (i.e. the r.m.s. power in the energy fluctuations about the mean) in a frequency interval df is

$$\Delta P = \Delta n_a h f_0 \, df \ . \tag{A7.23}$$

Since this represents the smallest possible energy fluctuation in the field, ΔP^2 can be set equal to $\overline{S_\eta(\omega)} \, df$ in order to determine the minimum output noise level:

$$\Delta n_a^2 \, h^2 \, f_0^2 \, df^2 = 2a_0^2 \, S_0 \, df \ . \tag{A7.24}$$

By combining this result with that of equation (A7.15) and employing the condition in equation (A7.7), we arrive finally at the following expression for the minimum noise power at the output of the amplifier:

$$\begin{aligned}
\Delta P_{\min} &= 2S_0 \, df \\
&= (G-1) h f_0 \, df
\end{aligned} \tag{A7.25}$$

which is the result we set out to prove.

Index